Electric Power Engineering

Electric Power Engineering

Edited by **Tim Kurian**

CLANRYE INTERNATIONAL

New Jersey

Published by Clanrye International,
55 Van Reypen Street,
Jersey City, NJ 07306, USA
www.clanryeinternational.com

Electric Power Engineering
Edited by Tim Kurian

International Standard Book Number: 978-1-63240-559-3 (Hardback)

The publisher's policy is to use permanent paper from mills that operate a sustainable forestry policy. Furthermore, the publisher ensures that the text paper and cover boards used have met acceptable environmental accreditation standards.

Trademark Notice: Registered trademark of products or corporate names are used only for explanation and identification without intent to infringe.

Printed in the United States of America.

Contents

Preface

Electric power is the product of current and voltage and it is generally produced by electric generators and electric batteries. Electric power engineering deals with the production, distribution, transmission, and utilization of electric power. This branch of engineering draws its base from electrical engineering. It includes electric motors, transformers, power grid and power electronics. This book elucidates the concepts and innovative models around prospective developments with respect to electric power engineering. Most of the topics introduced in this book cover new techniques and applications of this discipline. It presents researches performed by experts from across the globe to provide an in-depth understanding of the latest advances within this field. It is a resource guide for engineers, researchers and students.

The information contained in this book is the result of intensive hard work done by researchers in this field. All due efforts have been made to make this book serve as a complete guiding source for students and researchers. The topics in this book have been comprehensively explained to help readers understand the growing trends in the field.

I would like to thank the entire group of writers who made sincere efforts in this book and my family who supported me in my efforts of working on this book. I take this opportunity to thank all those who have been a guiding force throughout my life.

Editor

Analysis of Hybrid Rechargeable Energy Storage Systems in Series Plug-In Hybrid Electric Vehicles Based on Simulations

Karel Fleurbaey, Noshin Omar, Mohamed El Baghdadi, Jean-Marc Timmermans, Joeri Van Mierlo

Department of Electric Engineering and Energy Technology, Vrije Universiteit Brussel, Brussels, Belgium
Email: kfleurba@vub.ac.be

Abstract

In this paper, an extended analysis of the performance of different hybrid Rechargeable Energy Storage Systems (RESS) for use in Plug-in Hybrid Electric Vehicle (PHEV) with a series drivetrain topology is analyzed, based on simulations with three different driving cycles. The investigated hybrid energy storage topologies are an energy optimized lithium-ion battery (HE) in combination with an Electrical Double-Layer Capacitor (EDLC) system, in combination with a power optimized lithium-ion battery (HP) system or in combination with a Lithium-ion Capacitor (LiCap) system, that act as a Peak Power System. From the simulation results it was observed that hybridization of the HE lithium-ion based energy storage system resulted from the three topologies in an increased overall energy efficiency of the RESS, in an extended all electric range of the PHEV and in a reduced average current through the HE battery. The lowest consumption during the three driving cycles was obtained for the HE-LiCap topology, where fuel savings of respectively 6.0%, 10.3% and 6.8% compared with the battery stand-alone system were achieved. The largest extension of the range was achieved for the HE-HP configuration (17% based on FTP-75 driving cycle). HP batteries however have a large internal resistance in comparison to EDLC and LiCap systems, which resulted in a reduced overall energy efficiency of the hybrid RESS. Additionally, it was observed that the HP and LiCap systems both offer significant benefits for the integration of a peak power system in the drivetrain of a Plug-in Hybrid Electric Vehicle due to their low volume and weight in comparison to that of the EDLC system.

Keywords

Plug-In Hybrid Electric Vehicle, Hybrid Energy Storage System, High Energy Battery, High Power Battery, Electrical Double-Layer Capacitor, Lithium-Ion Capacitor

1. Introduction

The generation of energy in the society today is unsustainable from an environmental and economic point of view. The global economy is subjected to an increased price of fossil fuels and the public has shown a growing concern about environmental aspects. In the transport sector in particular, which accounts for 19% of the global energy use and 23% of the energy related CO_2 emissions [1], internal combustion engines (ICE) based on fossil fuels dominate the market. Recent research in the automobile industry however has spurred the development of cleaner and more efficient technologies. Its main result is the increased use of (partially) electrically propelled vehicles such as Battery Electric Vehicles (BEVs), Hybrid Electric Vehicles (HEVs) and Plug-in Hybrid Electric Vehicles (PHEVs) [1]-[5]. PHEVs combine a Rechargeable Energy Storage System (RESS) with an ICE such that they can drive a certain distance purely electrical without being subjected to the drawback of a limited range. The future development of the various types of electric vehicles will however be highly dependent on the performance of the RESS [3] [5]-[7].

Currently, lithium-ion batteries are the preferred type of RESS for BEV and PHEV applications due to their high energy and power densities. They encompass a variety of technologies based on the materials that are used as anode, cathode and electrolyte. According to [4] [7]-[11], Lithium nickel cobalt aluminum oxide (NCA), Lithium iron phosphate (LFP), Lithium nickel manganese cobalt oxide (NMC) and Lithium manganese spinel oxide (LMO) show the highest potential for large scale application in BEVs and PHEVs (see **Table 1**). However, Burke and Omar *et al.* concluded respectively in [6] and [10] that none of the current batteries can yet fulfill the performance criteria proposed by the United States Advanced Battery Consortium (USABC). Critical aspects hereby are the cost, the lifetime and the energy content (thus the range of the electric vehicle). Moreover, battery design inherently incorporates trade-offs between these characteristics: for example, optimizing a battery for high power density results in the use of thin electrodes with a large surface area in order to make fast intercalation possible, while energy optimized batteries need thick electrodes in order to store the ions [5]-[11].

A number of hybrid RESS topologies have been proposed in the last decade in order to improve the performance of the RESS [12]-[19]. The most known dual source RESS combines an energy optimized (HE) battery with an Electrical Double-Layer Capacitor (EDLC) system. EDLC cells are characterized by a very high power density but have a low energy density. Their non-faradaic mechanism results in a low internal resistance, a high efficiency (95% - 98%) and a lifetime up to one million cycles [20] [21]. These characteristics make them an interesting candidate for use as Peak Power System (PPS) in a hybrid RESS. It is reported in [13]-[15] that the addition of an EDLC system to a Valve Regulated Lead-Acid (VRLA) battery results in an extension of the lifetime of the HE battery and enhances the power capabilities and energy efficiency of the RESS.

However, due to the superior energy density of lithium-ion batteries, most of the research for BEV and PHEV applications has focused on the combination of an energy optimized lithium-ion battery (HE) with an EDLC system. In [15]-[17], the authors found that the addition of an EDLC system in parallel to a lithium-ion battery results in an improvement in power capability of the hybrid RESS, in energy efficiency of the RESS and in cycle life of the HE battery. The low energy density and cell voltage of EDLCs however oblige the use of large amounts of cells connected in series, thus resulting in a large volume, weight and cost of the module.

The performance of the dual source RESS containing a HE battery and a power optimized (HP) lithium-ion battery is currently the subject of the European FP7 SuperLIB project [18] [19]. Its advantage over the HE-EDLC topology is the higher cell voltage and higher energy content of the HP battery, what can result in a smaller and lighter PPS module. However, the cycle life of HP batteries is significantly lower than that of EDLC cells.

The recent commercialization of Lithium-ion Capacitors (LiCaps) also introduces the need for its investigation as PPS in a hybrid RESS. LiCap cells contain a porous activated carbon positive electrode, as conventional EDLCs, and a negative carbon electrode that contains pre-doped lithium ions, as lithium ion batteries (see **Figure 1**) [22]-[26]. The result is a hybrid storage device with characteristics in between lithium-ion batteries and EDLCs. Its main advantages with respect to EDLCs are its higher energy density (10 - 15 Wh/kg versus 5 - 7 Wh/kg) and higher operating voltage (3.8 V versus 2.7 V), meaning that it is possible to reduce the amount of cells connected in series for certain high power applications [27]-[30].

In this paper, a model of a PHEV with series hybrid drivetrain topology is proposed in Matlab/Simulink environment. This model is used to analyze and compare the performance of different RESS architectures: the HE (non hybrid) battery stand-alone system, HE-EDLC, HE-HP and HE-LiCap. The focus of the RESS perfor-

Table 1. Comparison of different lithium-ion technologies [11].

	Chemistry: Neg/Pos Electrode	Cell Voltage (V) Max/Nom	Capacity (mAh/g) Neg/Pos Electrode	Energy Density (Wh/kg)	Cycle Life (Deep)	Thermal Stability
NMC	Graphite/NiCoMnO$_2$	4.2/3.6	360/180	100 - 170	2000 - 3000	Fairly stable
LMO	Graphite/Mn Spinel	4.0/3.6	360/110	100 - 120	1000	Fairly stable
NCA	Graphite/NiCoAlO$_2$	4.2/3.6	360/180	100 - 150	2000 - 3000	Least stable
LFP	Graphite/Iron Phosphate	3.65/3.25	360/160	90 - 115	>3000	Stable
LTO	Lithium Titanate/Graphite	2.8/2.4	180/110	60 - 75	>5000	Most stable

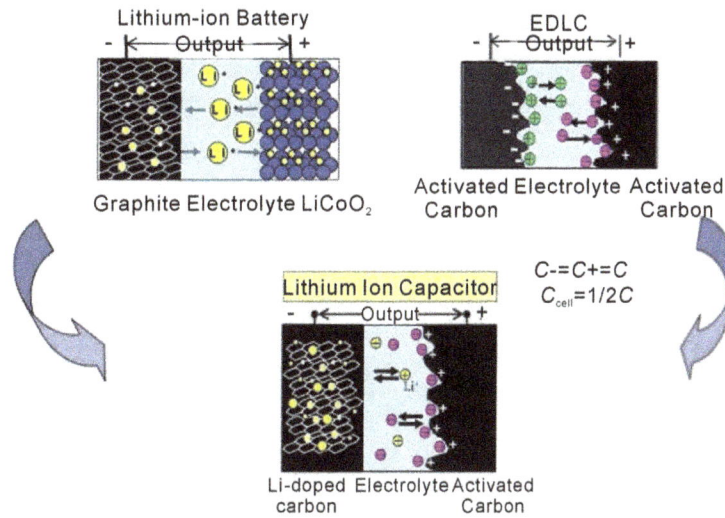

Figure 1. Mechanism of LiCaps [30].

mance analysis lies on the energy content, power capacity, energy efficiency, cycle life, volume and weight of the RESS. The case study and the proposed model of the Series PHEV are highlighted respectively in Section 2 and Section 3, while special attention to the modeling of the RESS is given in Section 4. The control strategy is explained in Section 5 and Section 6 contains the results and discussion of the simulations. The conclusions of the research are provided in Section 7.

2. Case Study

The performance of the different RESS is analyzed for a PHEV with a series drivetrain topology shown in **Figure 2(a)**. The PHEV drives in all electric Charge Depletion (CD) mode as long as the State of Charge (SoC) of the battery is high enough, after which the ICE starts and the vehicle runs in Charge Sustaining (CS) mode. The advantage of the series hybrid configuration is the decoupling of the ICE from the wheels, meaning that it can run constantly in its most efficient operating area. The extra energy conversion of the mechanical power from the ICE to electrical power at the DC-bus however reduces the drivetrain efficiency [2].

The drivetrain of a Series PHEV with a hybrid RESS is shown in **Figure 2(b).**While driving in CD mode, the HE battery will act as the main energy source while the PPS provides/absorbs power during accelerations/ decelerations. The DC/DC converter in between the DC-bus and the PPS is introduced in order to allow controlling the power flows inside the hybrid RESS by regulating the power/current of the PPS with respect to the DC-bus voltage [16].

The considered vehicle in this study is the Chevrolet Volt. The specifications of this vehicle are shown in **Table 2**, where m is the mass of the vehicle (kg), C_d the drag coefficient (-), A_f the frontal surface (m^2), C_r the rolling resistance coefficient (-), δ_J the rotational inertia coefficient (%), η_t the efficiency of the transmission (%), P_{aux} the mechanical auxiliary power (W) and P_{HVAC} the power of the heating, ventilation and air conditioning (W).

Table 2. Specification Chevrolet Volt [31].

Parameter	Chevrolet Volt
Published Data	
m	1715 kg
C_d	0.287
A_f	2.16 m^2
Assumed Data	
C_r	0.010
δ_J	1.08
η_t	95%
P_{aux}	700 W
P_{HVAC}	1857 W

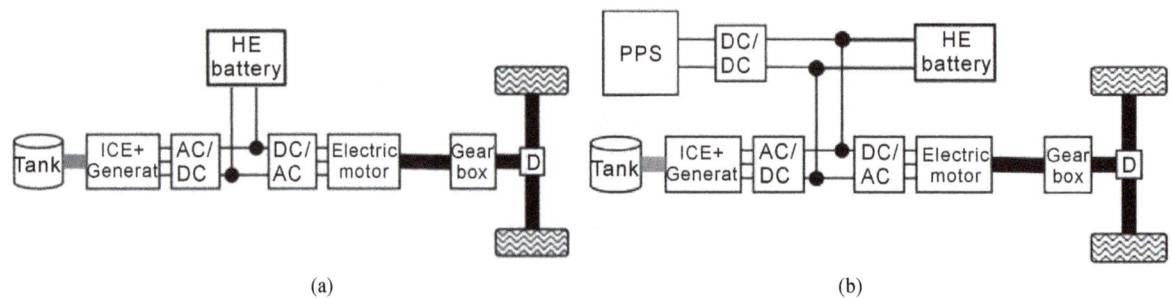

Figure 2. Drivetrain Series PHEV. a) Drivetrain Series PHEV [32]; b) Drivetrain Series PHEV with hybrid RESS.

3. Series PHEV Model

Regarding the modeling of vehicles, two different approaches can be distinguished in literature according to the direction of calculation of the power: the backwards or effect-cause method calculates the power in the direction opposite to the physical direction of the traction power. It starts by following the velocity profile of a specific driving cycle and computes the power that has to be delivered by the energy sources backwards through each component of the drivetrain. The main advantage of the backwards approach is its simplicity and fast computation time. The forward or cause-effect method computes the power according to the physical direction of the power flow, starting from the energy source through the drivetrain to the wheels. It uses a controller (cf. driver) that tries to follow a driving cycle and is preferred for real-time simulations as it allows obtaining best effort performances [32]-[35].

For this analysis, an efficient combined backward-forward model is designed to simulate the above-mentioned Series PHEV (see **Figure 3**). The main part of the program is written according to the backwards approach – starting from the driving cycle (green) through the drivetrain (blue) to the energy sources (yellow) and controlled by the control strategy (red) - for reasons of simplicity and computation time [34]. The longitudinal dynamics are used to compute the wheel velocity and torque. Constant efficiencies for the transmission and converters and efficiency maps for the electrical motor, generator and ICE are used to characterize the components of the drivetrain. A forward model (purple) is introduced in order to take the maximum performance of the components into account. The forward subsystem is only activated at those moments that the performance of one of the components in the drivetrain is restricted, such that its effect on the computation time is limited.

An example of the influence of the forward part of the simulation is presented in **Figure 4**. Around timestamp 1110s, the battery stand-alone system cannot provide adequate power to reach the velocity of the driving cycle. The maximum achievable velocity of the PHEV is calculated. Starting around 1120s, the real achieved velocity of the vehicle is higher than that of the driving cycle in order to catch up the non-driven distance during the period

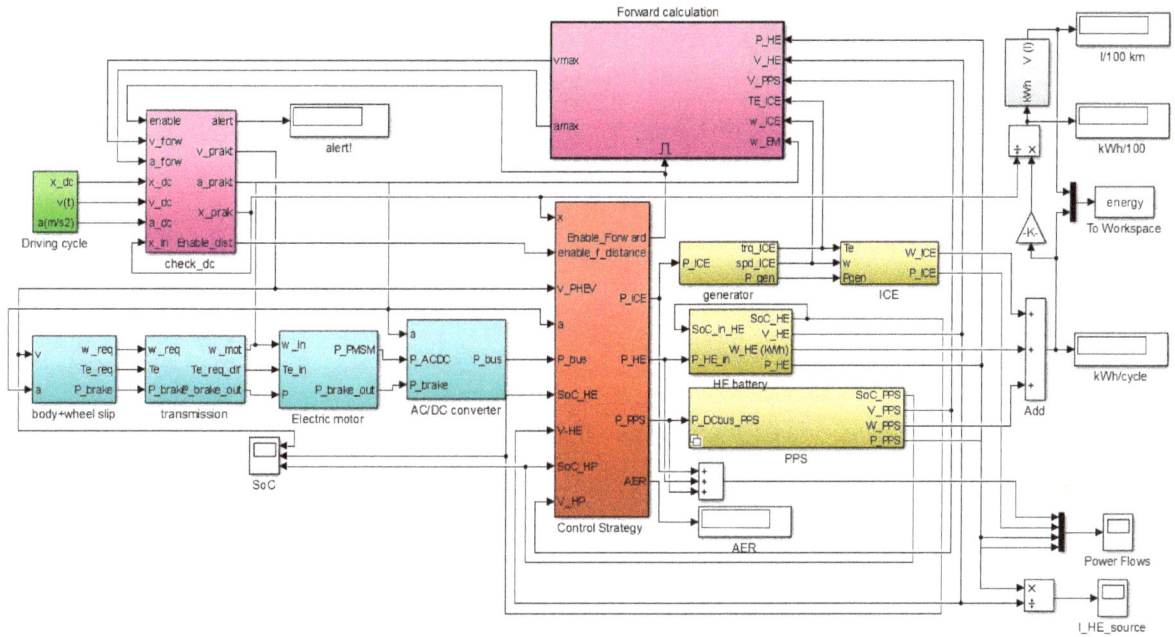

Figure 3. Combined backward/forward model of Series PHEV with hybrid RESS.

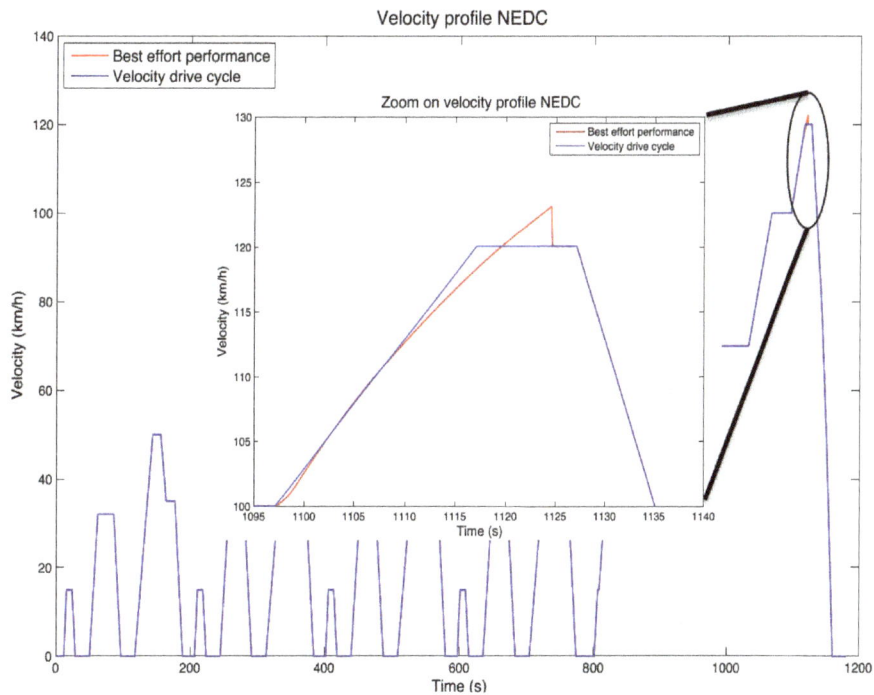

Figure 4. Influence forward approach.

that the driving cycle could not be followed.

4. RESS Models

4.1. HE and HP Battery

To model the batteries, the Rint-model with varying internal resistance in function of state of charge (SoC) and current demand is chosen (see **Figure 5(a)**). The main parameters of the HE and HP battery system are hig-

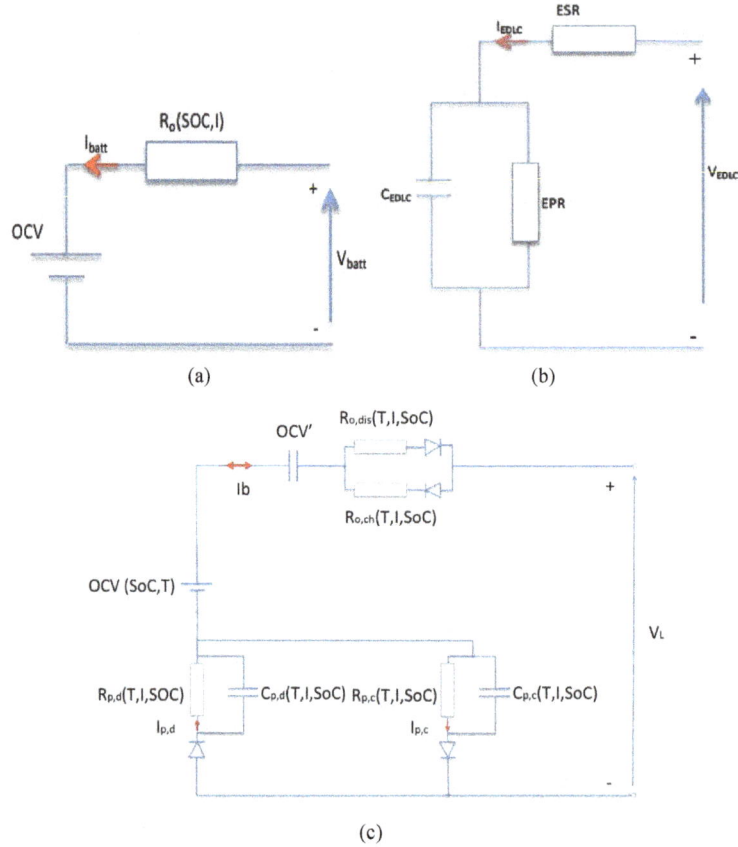

Figure 5. RESS models. a) Rint battery model; b) EDLC model; c) LiCap model [30].

hlighted in **Table 3**. Both the high energy (HE) and high power (HP) battery are lithium iron phosphate (LFP) batteries. The energy optimized LFP battery has a capacity of 45 Ah and an energy density of 138 Wh/kg. The power capability of the HE battery is limited by the maximum charge and discharge current rates of respectively 2C and 3C. The power optimized cell is characterized by a much lower reference capacity (7 Ah) but higher charge and discharge current rate up to 20C. The HE and HP battery systems contain both one string of 100 cells connected in series, resulting in a system voltage level of 330V and an energy content of 14.8 kWh for the HE battery system and 2.3 kWh for the HP battery system, of which respectively 10.4 kWh and 1.8 kWh are usable due to the limitation of maximum Depth of Discharge (DoD).

4.2. EDLC Model

The equivalent circuit presented in **Figure 5(b)** describes the behavior of the EDLC cells and the parameters of the EDLC system are presented in **Table 4**. The Equivalent Series Resistance (ESR) denotes the ohmic resistance of the EDLC system, while the Equivalent Parallel Resistance (EPR) takes the static losses due to self-discharge into account.

The dimensioning of the EDLC system is based on the strategy proposed in [16] and shown in **Figure 6**. It is based on the Maxwell 125V heavy transport module [36]. The EDLC system is sized such that the kinetic energy of a full brake from 120 km/h to 0 km/h can be stored in the PPS, taking into account the losses in the drivetrain:

$$E_{tot} = \frac{m * V_{max}^2}{2} * \eta_t * \eta_{EM} * \eta_{DC/AC} * \eta_{DC/DC} \tag{1}$$

The EDLC cells are controlled such that their voltage can vary from their maximum voltage $U_{EDLC,max}$ to $U_{EDLC,max/2}$, meaning that the EDLC cells can release or store 75% of their total energy content during operation:

Table 3. HE and HP battery parameters.

Parameter	HE Battery	HP Battery
m_{batt} (kg)	171.8	44.4
V_{batt} (dm^3)	109	26
$U_{n,batt}$ (V)	330	330
$C_{batt,ref}$ (Ah)	45	7
Max C-Rate (dis/ch)	3/2	20/20

Table 4. EDLC module characteristics [36].

Parameter	EDLC
m (kg)	121
V (dm^3)	1.02
C (F)	31.5
U_n (V)	250
$U_{min} - U_{max}$ (V)	136 - 272
ESR (mΩ)	36

$$E_{EDLC} = \frac{C_{EDLC} * \left(U^2_{EDLC,max} - U^2_{EDLC,min}\right)}{2} \tag{2}$$

The required total capacitance of the EDLC system can be calculated from Equation (2). Ideally, the cell capacitances are all identical such that

$$N_{cell} = \frac{U_{max}}{U_{n,cell}} \tag{3}$$

and

$$C_{cell} = \frac{N_{series}}{N_{par}} * C_{EDLC} \tag{4}$$

If the cell cannot provide the required current, the methodology should be performed again with an extra stack in parallel. Following the explained methodology resulted in the connection of two 125V modules in series.

4.3. LiCap Model

The model used to describe the behavior of the LiCap system is presented in **Figure 5(c)** and was proposed by Omar *et al.* in [30]. They presented a modified Freedom CAR first order model specifically designed for the 3300F prismatic LiCap cells of JSR Micro [37]. The hysteresis has been taken into account by separating the charge and discharge ohmic resistances and polarization circuits.

In order to allow comparing the performance of the HE-LiCap and HE-EDLC configurations in the Series PHEV, the LiCap system is sized such that its energy content is the same as that of the EDLC system:

$$n_{LiCap} = \frac{2 * E_{EDLC,sys}}{C_{n,LiCap} * U^2_{n,LiCap}} \tag{5}$$

Equation (5) resulted in the use of 59 LiCap cells in series. To obtain the mass and volume of the LiCap module, a packing factor of respectively 0.63 and 0.37 are used based on data released for one cell and for a module of 15 cells. The parameters of the LiCap system are shown in **Table 5**.

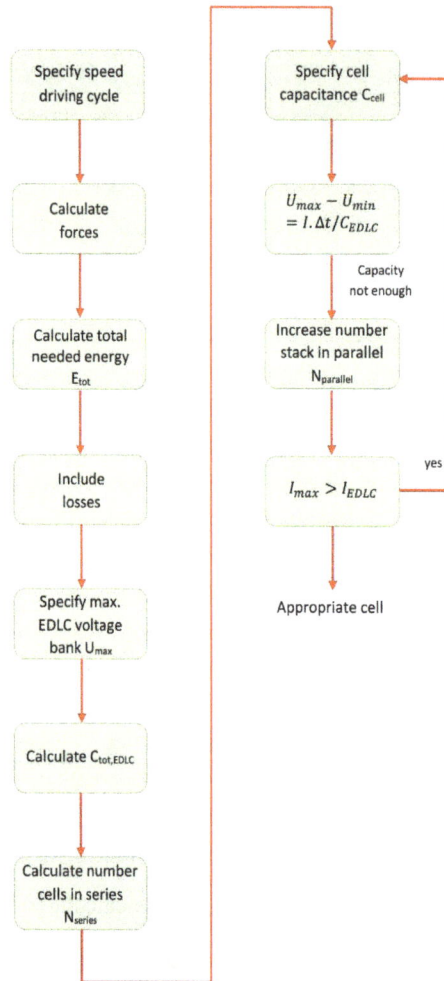

Figure 6. Dimensioning strategy EDLC system [16].

Table 5. LiCap parameters [37].

Parameter	LiCap
m (kg)	121
V (dm^3)	1.02
C (F)	31.5
U_n (V)	250
$U_{min} - U_{max}$ (V)	136 - 272
$R_{0,licap}$ (mΩ)	36

5.2. Hybrid RESS Control

5.2.1. Hybrid RESS Control

The idea behind hybridization of the RESS is to reduce the stress on the HE battery by adding an energy storage system that can provide the power peaks during acceleration. This distinction between peak power provided by the PPS and the moving average power supplied by the HE battery is made by the low pass filter (LPF) (see **Figure 7**). The LPF blocks the high frequency component of P_{dc}, resulting in the moving average power, which

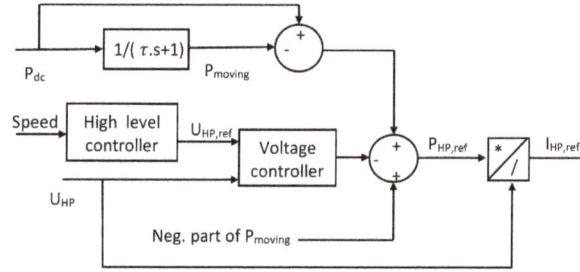

Figure 7. Control strategy hybrid RESS [14].

is provided by the HE battery:

$$P_{moving} = \frac{1}{\tau * s + 1} * P_{dc} \tag{6}$$

where τ denotes the time constant (s) of the LPF. The peak power is defined as the difference between the required and the moving average power:

$$P_{peak} = P_{dc} - P_{moving} \tag{7}$$

The high level controller regulates the SoC of the PPS in function of the speed of the vehicle in order to ensure efficient operation of the PPS. The SoC of the PPS must for example be low at high velocities of the PHEV in order to allow absorbing all the regenerative braking energy during a possible brake event, while it must be high at low velocities in order to provide power during the next acceleration. A proportional controller compares the reference SoC with the actual SoC of the PPS and produces a power difference that needs to be supplied/absorbed by the PPS. The SoC controller is slightly adjusted for the HE-HP hybrid architecture as the energy content of the HP battery is significantly higher than that of the EDLC and LiCap system (2.3 kWh versus respectively 292 and 294 Wh). It was opted to not relate the SoC of the HP with the velocity of the PHEV for this reason, but instead partially deplete the HP battery during CD mode in order to extend the AER of the vehicle. This practically means that the controller only becomes active if the SoC of the HP dropped below 30% or rose above 90%. This limitation is implemented in order to ensure safe operation of the battery and enlarge its cycle life.

5.2.2. Optimization Time Constant

The main design parameter of the control strategy is the time constant of the LPF. Implementing a large time constant results in a reduced HE battery stress but also requires a larger PPS. For the HE-EDLC case a time constant of 11s has been chosen. It allows full usage of the EDLC system within its voltage limits as can be seen in **Figure 9** and is consistent with the time constant of 8s used by Cheng *et al.* in [14] for a similar configuration. The same time constant of 11s is also used for the HE-LiCap configuration as the LiCap system is dimensioned such that it contains the same amount of energy as the EDLC system.

The optimal time constant of the HE-HP configuration is obtained by analyzing the AER in function of the time constant. It is shown in **Figure 8** that the time constant of 46s resulted in the maximum AER of 47.9 km for the simulated PHEV during the NEDC.

6. Results and Discussion

Simulations with three different driving cycles have been performed: the New European Driving Cycle (NEDC), the Federal Test Procedure Driving Cycle (FTP-75) and the Highway Fuel Economy Driving Schedule (HWFET). The FTP-75 driving cycle represents typical urban driving behavior with a lot of accelerations up to a low velocity, while the HWFET describes highway driving conditions. The NEDC combines both driving conditions by implementing four repetitions of the ECE-15 Urban Driving Cycles with an Extra-Urban driving cycle. The simulation results are presented in **Table 6** and **Figure 9** illustrates the SoC level of the PPS and the distribution of the power flows over the different energy sources in the considered hybrid RESS during the NEDC.

It can be observed in **Figure 9(a)** and **Figure 9(b)** that the behavior of the HE-EDLC and HE-LiCap RESS are similar due to the use of the same time constant. The power provided by the HE battery is the moving average of the required power, while respectively the EDLC system and the LiCap system provide the peak power. The effect of a larger time constant is illustrated in **Figure 9(c)** that represents the HE-HP RESS configuration. It is clearly visible that the power provided by the HE battery changes smoother over time, while the HP must provide the peak power during longer periods.

It is observed in **Table 6** that the consumption of the PHEV is reduced due to hybridization of the RESS. The best result was obtained for the HE-LiCap system, where fuel savings of 6.0%, 10.3% and 6.8% were achieved for respectively the NEDC, the FPT-75 and the HWFET driving cycle. This result can be explained by the high energy efficiency of the HE-LiCap configuration and the low weight of the LiCap system. It can remarked that the energy efficiency of the HE-EDLC system is higher than that of the HE-LiCap system, but the final consumption of the PHEV is larger in the first case due to the higher mass of the EDLC system.

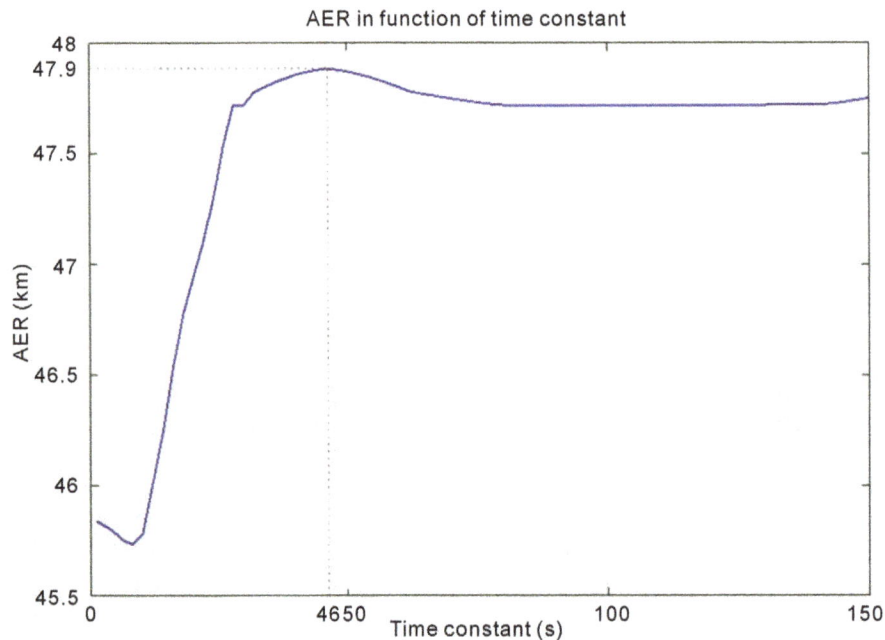

Figure 8. Optimization time constant HE-HP configuration.

Table 6. Simulation results.

	W_{dc} (kWh)	W_{dc} (l/100 km)	AER (km)	Extension AER (%)	η_{RESS} (%)	$I_{HE,RMS}$ (A)	$V_{DC,avg}$ (V)	ΔV_{DC} (V)
NEDC								
HE	2.63	2.70	42.5	/	86.5	38.9	322.3	55.1
HE-EDLC	2.50	2.57	45.2	6	92.7	32.6	322.4	44.7
HE-HP	2.52	2.59	47.9	13	90.3	26.7	322.6	24.6
HE-LiCap	2.47	2.54	45.4	7	92.1	32.6	322.4	44.3
FTP-75								
HE	4.17	2.64	43.8	/	84.2	41.3	322.9	49.3
HE-EDLC	3.82	2.41	48.6	11	90.3	30.0	322.8	40.8
HE-HP	3.91	2.47	51.3	17	86.5	22.4	322.7	16.4
HE-LiCap	3.74	2.36	49	12	90.4	29.4	322.8	47.6
HWFET								
HE	2.77	1.88	61.1	/	88.7	53.9	318.8	56.9
HE-EDLC	2.61	1.77	66.2	8	91.3	43.1	318.7	31.7
HE-HP	2.65	1.80	70.0	15	88.3	35.4	318.7	17.0
HE-LiCap	2.58	1.75	66.5	9	90.7	42.9	318.7	32.2

(a)

(b)

(c)

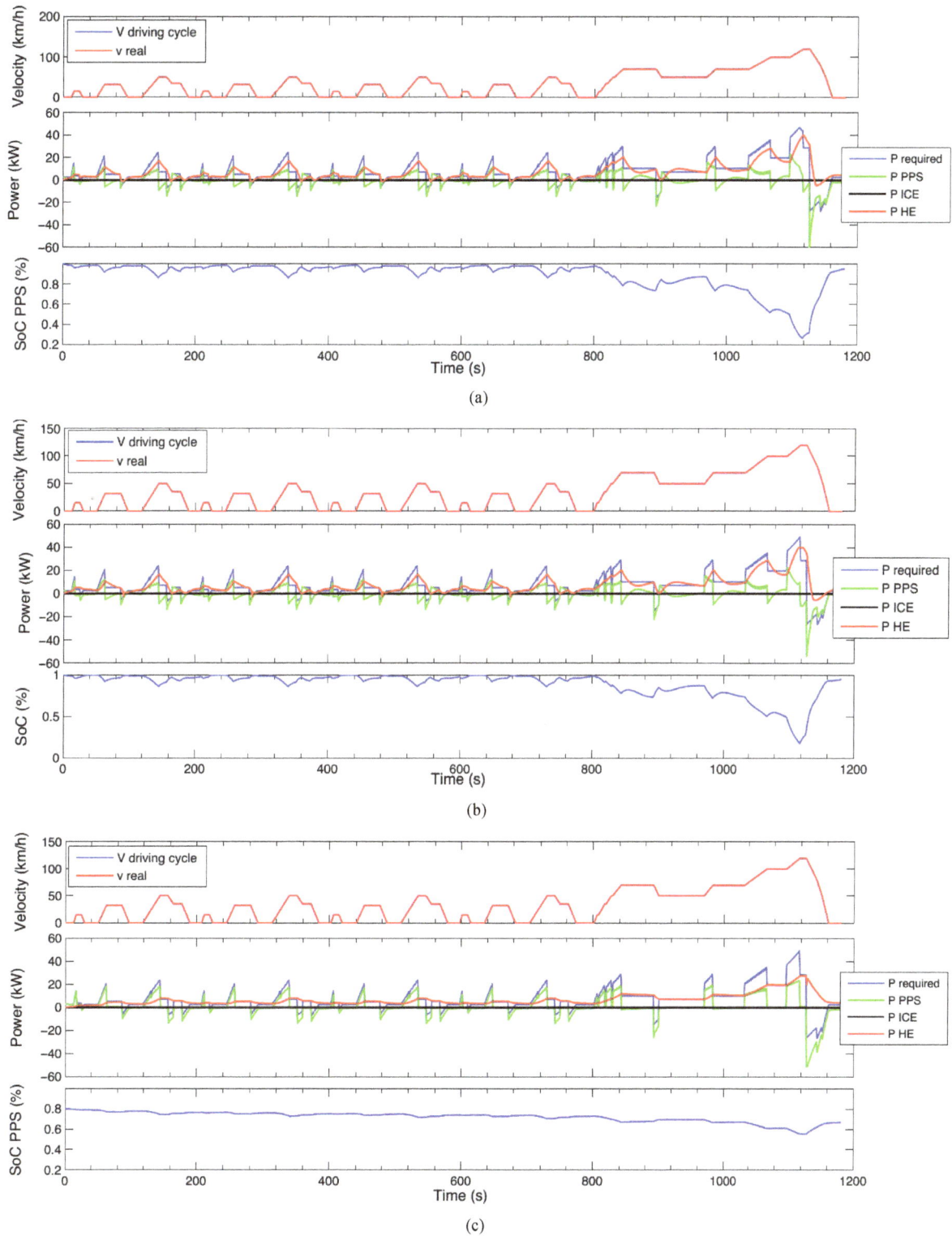

Figure 9. Power flows in the investigated hybrid RESS. a) Hybrid HE-EDLC configuration; b) Hybrid HE-LiCap configuration; c) Hybrid HE-HP configuration.

6.1. Energy Content

The AER of an electrically propelled vehicle is mainly linked to the energy content of its RESS. As mentioned in Section 4.1, the implemented HE battery system contains 14.8 kWh of which 10.4 kWh is usable. The obtained AER of the simulated series PHEV with this battery stand-alone system was 42.5 km for the NEDC. The increase in AER observed for the PHEVs with a hybrid RESS is mainly due to the increase in energy efficiency of the RESS (see Section 6.3). The 6.3% extension in AER obtained in this research for the case study of the NEDC is in accordance with an extension of the AER of 7% found in [16] for the same driving cycle. The significant increase in AER obtained for the HE-HP configuration is due to the (partial) depletion of the HP battery during CD driving mode (see Section 5.2.1). For example, during repeatedly driving of the NEDC, it was observed that the SoC of the HP battery dropped from 80% down to 45% at the moment that the ICE was started. The HP battery thus provided 0,8kWh of additional energy during CD mode. The largest extension of the AER was 17% obtained for the FTP-75 driving cycle, supporting the premise that peak shaving is more advantageous during urban driving.

6.2. Power Capability

The power capabilities of the proposed HE battery stand-alone system is limited by a 3C maximum discharge current rate, which results in a maximum discharge power of 44 kW:

$$P_{HE,max} = U_{HE,n} * I_{HE,max} = U_{HE,n} * 3C_{HE,ref} \tag{8}$$

It was already presented in **Figure 4** that the simulated Series PHEV could not always follow the velocity profile of the NEDC. The reasons are the limitation on the maximum available power as mentioned above and the implemented restriction on the maximum current slope of the HE battery ($1/10$ I_{max}). Simulations with the FTP-75 and HWFET showed the same performance restriction, which were mostly retrieved for the FTP-75 due to its frequent accelerations. The introduction of a PPS improved largely the power performance of the PHEV. The maximum available power of the different PPS can be estimated by:

$$P_{EDLC,max} = \frac{U^2_{n,EDLC}}{4 * ESR} = 434 \text{ kW} \tag{9}$$

$$P_{HP,max} = U_{HP,max} * 20C_{HP,ref} = 46 \text{ kW} \tag{10}$$

$$P_{EDLC,max} = \frac{U^2_{n,LiCap}}{4 * R_{o,LiCap}} = 232 \text{ kW} \tag{11}$$

The restrictions found for the battery stand-alone system were not observed anymore after the implementation of the PPS.

6.3. Energy Efficiency

The energy efficiency of an energy storage system is normally defined as the ratio of the energy provided by the source during discharge over the energy required to charge the source up to the same SoC:

$$\eta_e = \frac{\int_{t_{begin}}^{t_{end}} U_{RESS,dis} * I_{RESS,dis} * dt}{\int_{t_{begin}}^{t_{end}} U_{RESS,ch} * I_{RESS,ch} * dt} \tag{12}$$

In order to allow a fair comparison between the battery stand-alone system and the hybrid RESS, that also contains a PPS and a DC/DC converter in between the PPS and DC-bus, Equation (13) has been used in this research to denote the energy efficiency of the RESS:

$$\eta_e = \frac{\int_{t_{start}}^{t_{end}} P_{tot} * dt - \int_{t_{start}}^{t_{end}} P_{DC} * dt}{\int_{t_{start}}^{t_{end}} P_{tot} * dt} \tag{13}$$

where P_{tot} represents the total power delivered by the different energy sources (kW) and P_{DC} denotes the power demand at the DC-bus (kW). **Figure 10** shows the decrease of the energy efficiency for the HE battery stand-alone system in function of the applied current. The course of the plotted curve can be explained by the

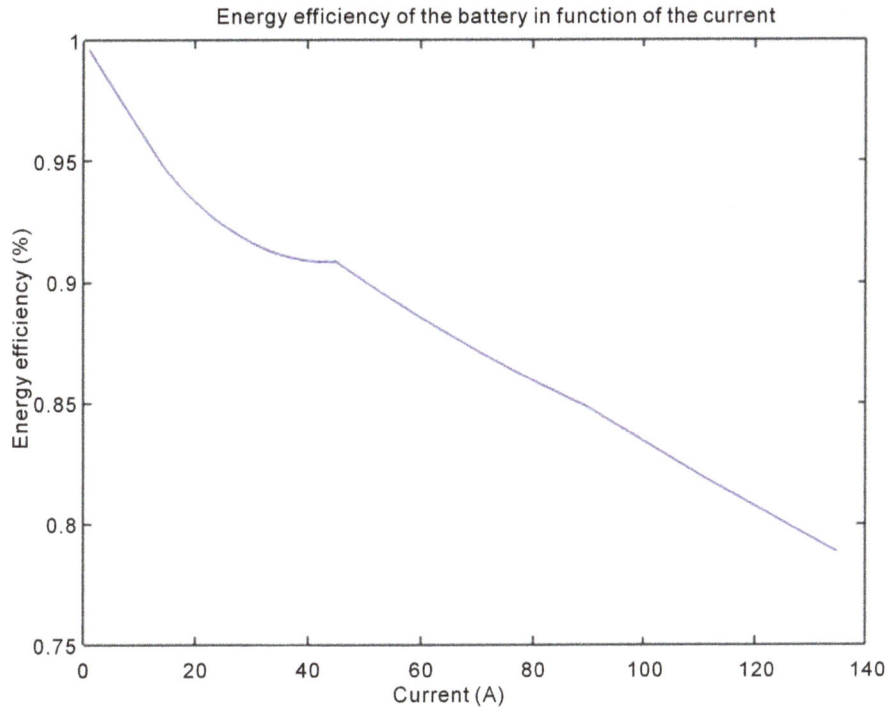

Figure 10. Energy efficiency of HE battery in function of current.

quadratic increase of the losses in function of the current.

$$P_{dis} = R_{o,HE} * I^2 \tag{14}$$

Hybridization of the RESS has thus the advantage that the HE battery, which is characterized by a relatively large internal resistance in comparison to the PPS, is subjected to less peak powers (see the RMS current of the HE battery in **Table 6**). The consequence is a significant reduction of the losses in the HE battery and the appearance of losses in the DC/DC converter and the PPS.

It was observed that the obtained energy efficiencies of the hybrid RESS for the different driving cycles were higher than those of the battery stand-alone system for all three driving cycles, except in case of the HE-HP configuration during the HWFET cycle. The HE-EDLC topology reached the highest energy efficiency of the studied configurations due to their low ESR. For the NEDC, FTP-75 and HWFET, an improvement of respectively 7%, 7% and 3% over the battery stand-alone system was observed. The slightly higher internal resistance of the LiCap compared to EDLCs system caused a small reduction in energy efficiency for the HE-LiCap configuration, while the larger internal resistance of the HP battery system resulted in a further decrease of the energy efficiency for the HE-HP topology. It was observed that the energy efficiency of the HE-HP dual source RESS even dropped below that of the battery stand-alone system in the case the HWFET cycle that is characterized by a relatively high and constant velocity. The beneficial effect of peak power shaving with the HP battery is thus limited and the added weight and extra losses in the DC/DC converter and the HP battery resulted in a higher consumption.

6.4. Cycle Life

One of the main benefits of hybridization of the RESS is the reduction of the HE battery stress, which results in an extension of the lifetime. An accurate theoretical analysis of the cycle life of the HE battery system however is very complex. The ageing of a battery system is affected by a number of parameters, such as temperature, DoD, charge and discharge current, etc. Some of these parameters cause temporary performance losses, while others can reduce the capacity of the battery permanently or even damage the cell. In order to make a statement on the magnitude of the life cycle extension, dedicated experiments should be performed. The simulation results showed the decrease in RMS current of the HE battery in the studied hybrid RESS topologies. For city driving behavior simulated with the FTP-75, the addition of the EDLC, HP and LiCap system resulted respectively in a reduction of the HE battery RMS current with 27%, 46% and 29%. The large time constant used in the HE-HP configuration resulted in a stronger filtering of the power peaks and thus further reduction of the RMS current.

Du Pasquier *et al.* compared in [38] the lifetime of an EDLC cell, a lithium ion battery cell and a lithium ion capacitor cell with LTO as negative electrode. The devices were subjected to cycles of a 3C charge rate and a constant power discharge of 675W/kg until the stored energy reached 80% of its initial value. It is observed that the cycle life of the components using an activated carbon cathode is at least one or two orders of magnitude larger than that of the lithium ion battery. However, it has to be remarked that this study already dates from 2003 and that the LiCap cells used in this research contain a different negative electrode, making it impossible to draw conclusions based on the cited research.

The datasheet of the 125V Maxwell module [36] mentions that one million cycles can be reached before the capacity dropped below 80% of its initial capacity when charging and discharging the module at a current of 100A. The data released for the 3300F LiCap cells in [37] show that the capacity did not decrease below 90% of the initial capacity after 200000 cycles when cycling at a charge and discharge current of 200A. No specific data regarding cycle life of the used HP battery were available, but different studies confirmed that the cycle life of LFP batteries do net yet reach the target of the USABC of 5000 cycles until 100% DoD [11] [38]. The cycle life of the battery can be enlarged by reducing the DoD. It was for example observed in [39] that the studied LFP battery reached 2600 cycles at 100% DoD and 34957 cycles at 20% DoD.

It can thus be concluded from the reasoning above that the cycle life of HP lithium ion battery cells is significantly lower than that of EDLC and LiCap cells. However, an accurate comparison of the lifetime of the three considered PPSs for this usage in this specific application was not possible due to the gap in literature regarding the ageing of advanced energy storage components.

6.5. Volume and Weight

The volume and weight of the RESS are considered as key issues in the design of PHEVs and are shown **in Figure 11** for the different hybrid topologies. The disadvantage of the HE-EDLC topology from the point of view of weight (EDLC system adds 120 kg) and volume (1.02 dm^3) is clearly observable. The low nominal voltage of EDLC cells resulted in the requirement of a large amount of cells in series in order to reach the required energy content. Here, it should be noted that the nominal voltages of the EDLC system is 250V in comparison to 330V for the HP battery system and 195V for the LiCap system. Moreover, the assembly of the cylindrical EDLC cells is not as efficient as the assembly of pouch cells (HP battery) or prismatic cells (LiCap).

It can thus be concluded from **Figure 11** that the weight and volume of the HP battery system and of the Li-Cap system offer significant advantages over the EDLC system from the integration point of view.

6.6. DC-Bus Voltage

Hybridization of the RESS also benefits the stability of the DC-bus voltage as demonstrated in **Figure 12** and **Table 6**. It was observed that the average DC-bus voltage of the different systems was not affected significantly by hybridization of the energy storage system, while the variations of the DC-bus voltage were drastically reduced due to the implementation of a PPS. The DC-bus voltage for the HE battery stand-alone system varied with 55.1V, 49.3V and 56.9V for the NEDC, FTP-75 and HWFET cycle respectively. This difference was reduced to 24,6V, 16,4V and 17,0V for the HE-HP configuration during the same driving cycles. The same effect was noticed in smaller magnitudes for the HE-EDLC and HE-LiCap topologies due to smaller time constant of the LPF. The lower voltage drop over the HE battery results in a higher energy efficiency of the HE battery, a reduction of the heat development due to filtering of the peak currents and in a higher efficiency of the complete drivetrain.

7. Conclusion

In this paper, a comparative analysis of the performance of a HE battery stand-alone system with that of a HE-EDLC system, a HE-HP system and a HE-LiCap hybrid RESS is presented for a PHEV with series hybrid drivetrain topology. Therefore, a combined backwards-forward simulation model in Matlab Simulink was developed in order to assess the performance of the considered RESS at different driving behaviors.

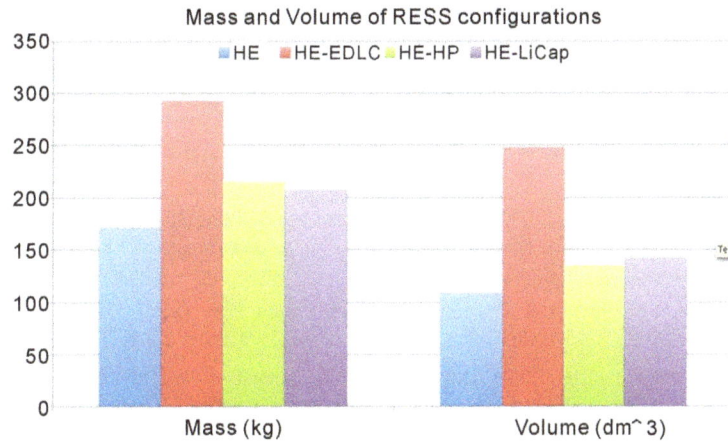

Figure 11. Volume and weight comparison of different RESS topologies.

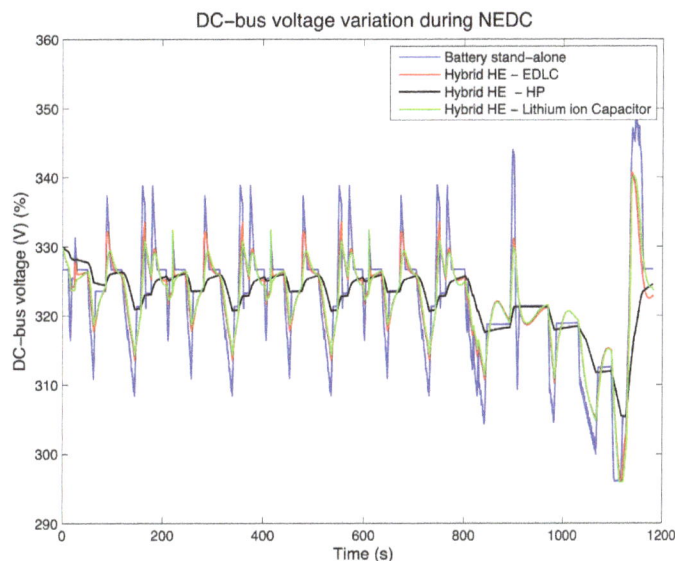

Figure 12. DC-bus voltage for different RESSS topologies.

It was found that the HE-HP configuration showed the highest potential with regard to AER extension. For the FTP-75 driving cycle, an increase of the AER with 17% was obtained. Moreover, the possibility of using a large time constant in the HE-HP configuration also resulted in a significant reduction of the HE battery RMS current and in a more stabilized the DC-bus voltage. The HE-EDLC topology proved to be the most beneficial topology from energy efficiency point of view with 92.7%, 90.3% and 91.3% for the NEDC, FTP-75 and HWFET respectively. The obtained energy efficiencies of the HE-LiCap system were 92.1%, 90.3% and 90.7%, but a significantly lower weight of the LiCap system resulted in an overall lower consumption of the series PHEV with the HE-LiCap system. Moreover, the LiCap system showed its advantage for usage in this application also from the point of view of the volume of the LiCap system, which can be considered as a key parameter in transport application. This advantage was also retrieved for the HE-HP hybrid configuration. Finally, a gap in literature with regard to cycle life analysis of different advanced energy storage systems was identified. Based on the datasheets of the different components, it was observed that the cycle life of the HP lithium ion battery is significantly lower than that of the EDLC and LiCap system, but performing an accurate lifetime analysis of the different components for use in this specific application was not possible.

It can thus be concluded that hybridization of the RESS resulted in an improved performance in terms of power capability, energy efficiency, all electric range and of battery stress. The different hybrid RESS all have their specific advantages, but the HE-LiCap topology resulted in the lowest consumption during the different simulated drive cycles due to their high energy efficiency and low weight.

References

[1]   International Energy Agency (2009) Transport, Energy and CO_2: Moving toward Sustainability. OECD Publishing, Paris.

[2]   Maggetto, G. and Van Mierlo, J. (2000) Electric and Electric Hybrid Vehicle Technology : A 2000/2010 Perspective. *The Challenge for Cities in the 21st Century: Transport, Energy and Sustainable Development*, DG TREN, Programme Thermie, Bilbao/Guggenheim.

[3]   Van Mierlo, J., Timmermans, J.-M., Matheys, J. and Van den Bossche, P. (2006) Milieuvriendelijke Voertuigen. *Vlaams Wetenschappelijk Economisch Congress*, Brussels, 19-20 October 2006.

[4]   Axsen J., Burke A. and Kurani K. (2008) Batteries for Plug-In Hybrid Electric Vehicles (PHEVs): Goals and the State of Technology Circa 2008. Institute of Transportation Studies, University of California, Davis, Research Report UCD-ITS-RR-08-14.

[5]   International Energy Agency (2011) Technology Roadmap: Electric and Plug-In Hybrid Electric Vehicles. OECD Publishing, Paris.

[6]   Axsen J., Kurani K.S. and Burke A. (2010) Are Batteries Ready for Plug-In Hybrid Buyers? *Transport Policy*, **17**, 173-182. http://dx.doi.org/10.1016/j.tranpol.2010.01.004

[7]   The Boston Consulting Group (2011) Batteries for Electric Cars: Challenges, Opportunities and the Outlook to 2020. www.bcg.com

[8]   Omar, N., Verbrugge, B., Mulder, G., Van den Bossche, P., Van Mierlo, J., Daowd, M., *et al.* (2010) Evaluation of Performance Characteristics of Various Lithium-Ion Batteries for Use in BEV Application. *IEEE Vehicle Power and Propulsion Conference*, Lille, 1-3 September 2010, 1-6.

[9]   Omar, N., Daowd, M., Van Den Bossche, P., Hegazy, O., Smekens, J., Coosemans, Th. and Van Mierlo, J. (2012) Rechargeable Energy Storage Systems for Plug-In Hybrid Electric Vehicles—Assessment of Electrical Characteristics. *Energies*, **5**, 2952-2988. http://dx.doi.org/10.3390/en5082952

[10] Omar, N. and Van Den Bossche, P. (2011) Assessment of Performance of Lithium Iron Phosphate Oxide, Nickel Manganese Cobalt Oxide and Nickel Cobalt Aluminum Oxide Based cells for Using in Plug-In Battery Electric Vehicle Applications. *IEEE Vehicle Power and Propulsion Conference (VPPC)*, Chicago, 6-9 September 2011, 1-7.

[11] Burke, A. and Miller, M. (2009) Performance Characteristics of Lithium-Ion Batteries of Various Chemistries for Plug-In Hybrid Vehicles. *EVS24*, Stavanger, 13-16 May 2009, 1-13.

[12] Pay, S. and Baghzouz, Y. (2003) Effectiveness of Battery-Supercapacitor Combination in Electric Vehicles. *2003 IEEE Power Tech Conference*, Bologna, 23-26 June 2003.

[13] Omar, N., Van Mierlo, J., Verbrugge, B. and Van den Bossche, P. (2010) Power and Life Enhancement of Battery-Electrical Double Layer Capacitor for Hybrid Electric and Charge-Depleting Plug-In Vehicle Applications. *Electrochimica Acta*, **55**, 7524-7531. http://dx.doi.org/10.1016/j.electacta.2010.03.039

[14] Cheng, Y., Van Mierlo, J., Van den Bossche, P. and Lataire, P. (2006) Super Capacitor Based Energy Storage as Peak Power Unit in the Applications of Hybrid Electric Vehicles. *3rd IET International Conference on Power Electronics, Machines and Drives*, Dublin, 4-6 April 2006, 404-408. http://dx.doi.org/10.1049/cp:20060140

[15] Omar, N., Al Sakka, M., Daowd, M., Coosemans, T., Van Mierlo, J. and Van den Bossche, P. (2010) Assessment of Behavior of Active EDLC-Battery System in Heavy Hybrid Charge Depleting Vehicles. *ESSCAP'10: Fourth European Symposium on Super Capacitors and Applications*, Bordeaux, 21-22 October 2010.

[16] Omar, N., Daowd, M., Hegazy, O., Van den Bossche, P., Coosemans, T. and Van Mierlo, J. (2012) Electrical Double-Layer Capacitors in Hybrid Topologies—Assessment and Evaluation of Their Performance. *Energies*, **5**, 4533-4568. http://dx.doi.org/10.3390/en5114533

[17] Wu, Z., Zhang, J., Jiang, L., Wu, H. and Yin, C. (2012) The Energy Efficiency Evaluation of Hybrid Energy Storage System Based on Ultra-Capacitor and $LiFePO_4$ Battery. *WSEAS Transactions on Systems*, **11**, 95-105

[18] Omar, N., Coosemans, T., Martin, J., Sauvant-Moynot, V., Salminen, J., Kortschak, B., *et al.* (2012) SuperLib Project: Advanced Dual-Cell Battery Concept for Battery Electric Vehicles. *EVS26*, Los Angeles, 6-9 May 2012, 1-6.

[19] Omar, N., Fleurbaey, K., Kurtulus, C., Van den Bossche, P. and Van Mierlo, J. (2013) SuperLIB Project—Analysis of the Performances of the Hybrid Lithium HE-HP Architecture For Plug-In Hybrid Electric Vehicles. *EVS27*, Barcelona, 17-20 October 2013.

[20] Burke, A. (2007) R & D Considerations for the Performance and Application of Electrochemical Capacitors. *Electro-*

chimica Acta, **53**, 1083-1091. http://dx.doi.org/10.1016/j.electacta.2007.01.011

[21] Kötz, R. and Carlen, M. (2000) Principles and Applications of Electrochemical Capacitors. *Electrochimica Acta,* **45**, 2483-2498. http://dx.doi.org/10.1016/S0013-4686(00)00354-6

[22] Amatucci, G.G., Badway, F., Pasquier, A.D. and Zheng, T. (2001) An Asymmetric Hybrid Nonaqueous Energy Storage Cell. *Journal of the Electrochemical Society,* **148**, A930-A939. http://dx.doi.org/10.1149/1.1383553

[23] Wang, Y., Luo, J., Wang, C. and Xia, Y. (2006) Hybrid Aqueous Energy Storage Cells Using Activated Carbon and Lithium-Ion Intercalated Compounds. *Journal of the Electrochemical Society,* **153**, A1425-A1431. http://dx.doi.org/10.1149/1.2203772

[24] Wang, Y., Yu, L. and Xia, Y. (2006) Electrochemical Capacitance Performance of Hybrid Supercapacitors Based on $Ni(OH)_2$Carbon Nanotube Composites and Activated Carbon. *Journal of the Electrochemical Society,* **153**, A743-A748. http://dx.doi.org/10.1149/1.2171833

[25] Karthikeyan, K., Aravindan, V., Lee, S.B., Jang, I.C., Lim, H.H., Park, G.J., *et al.* (2010) A Novel Asymmetric Hybrid Supercapacitor Based on Li_2FeSiO_4 and Activated Carbon Electrodes. *Journal of Alloys and Compounds,* **504**, 224-227. http://dx.doi.org/10.1016/j.jallcom.2010.05.097

[26] Pasquier, A.D., Plitz, I., Gural, J., Menocal, S. and Amatucci, G. (2003) Characteristics and Performance of 500 F Asymmetric Hybrid Advanced Supercapacitor Prototypes. *Journal of Power Sources,* **113**, 62-71. http://dx.doi.org/10.1016/S0378-7753(02)00491-3

[27] Gualous, H., Alcicek, G., Diab, Y., Hammar, A., Venet, P. and Adams, K. (2008) Lithium Ion Capacitor Characterization and Modelling. *ESCAP'08: 3rd European Symposium on Supercapacitors and Applications,* Rome, 6-7 November 2008.

[28] Omar, N., Al Sakka, M., Smekens, J. and Van Mierlo, J. (2013) Electric and Thermal Characterization of Advanced Hybrid Li-Ion Capacitor Rechargeable Energy Storage System. *IEEE 4th International Conference on Power Engineering, Energy and Electrical Drives,* Istanbul, 13-17 May 2013, 1574-1580.

[29] Omar, N., Daowd, M., Hegazy, O., Sakka, M., Coosemans, T., Van den Bossche, P. and Van Mierlo, J. (2012) Assessment of Lithium-Ion Capacitor for Using in Battery Electric Vehicle and Hybrid Electric Vehicle Applications. *Electrochimica Acta,* **86**, 305-315. http://dx.doi.org/10.1016/j.electacta.2012.03.026

[30] Omar, N., Ronsmans, J., Firouz, Y., Monem, M.A., Samba, A., Gualous, H., *et al.* (2013) Lithium-Ion Capacitor—Advanced Technology for Rechargeable Energy Storage Systems. *EVS27,* Barcelona, 17-20 October 2013.

[31] General Motors (2013) Chevrolet Volt Specifications.

[32] Barrero, R., Coosemans, T. and Van Mierlo, J. (2009) Hybrid Buses: Defining the Power Flow Management Strategy and Energy Storage System Needs. *EVS24,* Stavanger, 13-16 May 2009.

[33] Van Mierlo, J. and Maggetto, G. (2004) Innovative Iteration Algorithm for a Vehicle Simulation Program. *IEEE Transactions on Vehicular Technology,* **53**, 401-412. http://dx.doi.org/10.1109/TVT.2004.823534

[34] Wipke, K.B., Cuddy, M.R. and Burch, S.D. (1999) ADVISOR 2.1: A User-Friendly Advanced Powertrain Simulation Using a Combined Backward/Forward Approach. *IEEE Transaction on Vehicular Technology,* **48**, 1751-1761. http://dx.doi.org/10.1109/25.806767

[35] Gao, D.W., Mi, C. and Emadi, A. (2007) Modeling and Simulation of Electric and Hybrid Vehicles. *Proceedings of the IEEE,* **95**, 729-745. http://dx.doi.org/10.1109/JPROC.2006.890127

[36] Maxwell Technologies (2014) 125V Heavy Transportation Modules: Product Specifications.

[37] JSR Micro NV (2014) JSR Micro Prismatic Lithium Ion Capacitor. http://www.jsrmicro.be/en/lic_prismatic_cell

[38] Pasquier, A.D., Plitz, I., Gural, J., Badway, F. and Amatucci, G. (2004) Power-Ion Battery: Bridging the Gap between Li-Ion and Supercapacitor Chemistries. *Journal of Power Sources,* **136**, 160-170. http://dx.doi.org/10.1016/j.jpowsour.2004.05.023

[39] Omar, N., Monem, M.A., Firouz, Y., Salminen, J., Smekens, J., Hegazy, O., *et al.* (2014) Lithium Iron Phosphate Based Battery—Assessment of the Aging Parameters and Development of Cycle Life Model. *Applied Energy,* **113**, 1575-1585. http://dx.doi.org/10.1016/j.apenergy.2013.09.003.

2

Gravito-Electric Power Generation

Roger Ellman

The-Origin Foundation, Inc., Santa Rosa, CA, USA
Email: RogerEllman@The-Origin.org

Abstract

It is now possible to deflect gravitational action away from an object so that the object is partially levitated. That effect makes it possible to extract energy from the gravitational field, which makes the generation of *gravito-electric* power technologically feasible. Such plants would be similar to hydro-electric plants and would have their advantages of not needing fuel and not polluting the environment. However, gravito-electric plants could be much smaller than hydro-electric plants; their location would not be restricted to suitable water elevations, and the plants and their produced energy would be much less expensive. Gravito-electric power can be placed into operation now. It can replace all existing nuclear and fossil fuel plants, and would essentially solve the problem of global warming to the extent it is caused by fossil fuel use. The physics development is comprehensively presented. That is followed by the engineering design.

Keywords

Gravitation, Power Generation, Global Warming

1. Summary Development

Light normally travels in a straight direction. But, when some effect slows a portion of the light wave front the direction of the light is deflected. In **Figure 1**, the shaded area propagates the arriving light at a slower velocity, v', than the original velocity, v, [its index of refraction, n', is greater] so that the direction of the wave front is deflected from its original direction.

A slowing of part of its wave front is the mechanism of all bending or deflecting of light. In an optical lens, as in **Figure 2**, light propagates more slowly in the lens material than outside the lens. The amount of slowing in different parts of the lens depends on the thickness of the lens at each part. In the figure the light passing through the center of the lens is slowed more than that passing near the edges of the lens. The result is the curving of the light wave front.

"Gravitational lensing", **Figure 3**, is an astronomically observed effect in which light from a cosmic object too far distant to be directly observed from Earth becomes observable because a large cosmic mass [the "lens"], located between the Earth observers and that distant object, deflects the light from the distant object as if focusing it, somewhat concentrating its light toward Earth enough for it to be observed from Earth. The light rays are so bent because the lensing object slows more the portion of the wave front that is nearer to it than it slows the farther away portion of the wave front.

The same effect occurs on a much smaller scale in the diffraction of light at the two edges of a slit cut in a flat

The slower speed,
v', in region #2
bends the wave
front back as
shown.

Region #1
[v]
[n]

v' < v
n' > n
∅' < ∅

Region #2
[v']
[n']

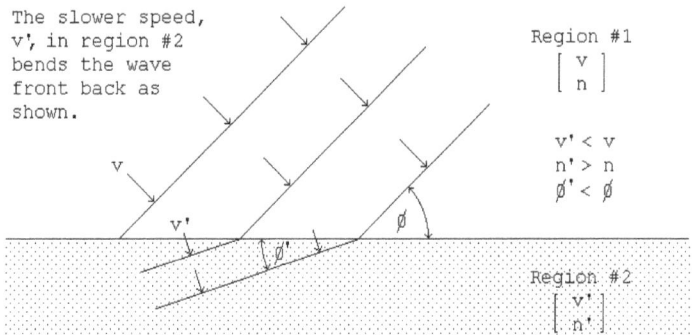

Figure 1. Deflection of Light's Direction by Slowing of Part of Its Wave Front.

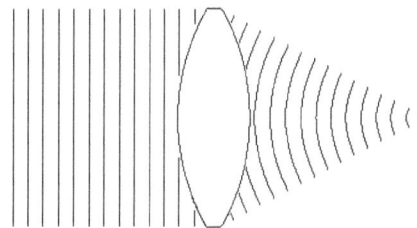

Figure 2. The Bending of Light's Wave Front by an Optical Lens.

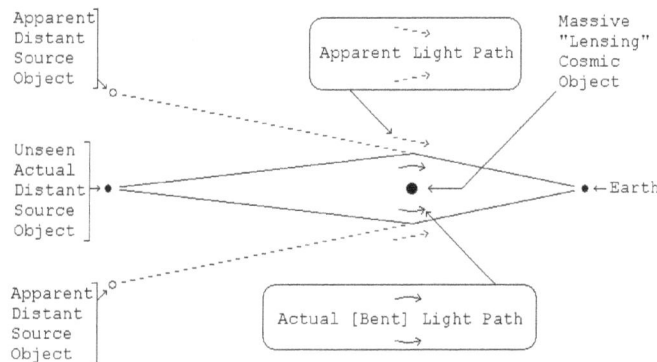

Apparent
Distant
Source
Object

Apparent Light Path

Massive
"Lensing"
Cosmic
Object

Unseen
Actual
Distant
Source
Object

•←Earth

Apparent
Distant
Source
Object

Actual [Bent] Light Path

Figure 3. Gravitational Lensing Bending of Light Rays.

thin piece of opaque material as in **Figure 4**. The bending is greater near the edges of the slit because the slowing is greater there. The effect of the denser material in which the slit is cut slows the portion of the wave front that is nearer to it more than the portion of the wave front in the middle of the slit.

In both of these cases, gravitational lensing and slit diffraction, the direction of the wave front is changed because part of the wave front is slowed relative to the rest of it. In the case of gravitational lensing the part of the wave front nearer to the "massive lensing cosmic object" is slowed more. In the case of diffraction at a slit the part of the wave front nearer to the solid, opaque material in which the slit is cut is slowed more.

But, neither of the cases, gravitational lensing and slit diffraction, involves the wave front passing from traveling through one substance to another as in the original illustration of **Figure 1**. The wave front in the gravitational lensing case is traveling only through cosmic space. The wave front in the slit diffraction case is traveling only through air. There is no substance change to produce the slowing. What is it that slows part of the wave front thus producing the deflection?

In the case of gravitational lensing the answer is that the effect is caused by gravitation. There is no other physical effect available. But how does gravitation produce slowing of part of the incoming wave front so as to deflect it? Gravitation, at least as it is generally known and experienced, causes acceleration, not slowing.

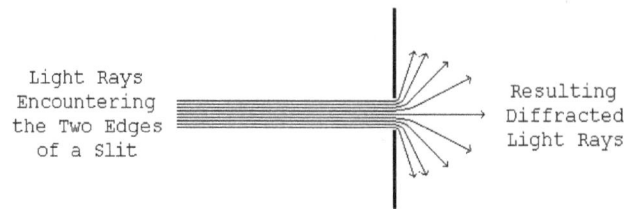

Figure 4. Diffraction at a Slit Causing Bending of Light Rays.

2. Electro-Magnetic Field (Light) and Gravitational Field (Gravity)

2.1. Light

Given two particles [e.g. electrons or protons] that have electric charges, the particles being separated and with the usual electric [Coulomb] force between them, if one of the charged particles is moved the change can produce no effect on the other charge until a time equal to the distance between them divided by the speed of light, c, has elapsed.

For that time delay to happen there must be something flowing from one charge to the other at speed c [a fundamental constant of the universe] and each charge must be the source of such a flow.

That electric effect is radially outward from each charge, therefore every charge must be propagating such a flow radially outward in all directions from itself, which flow must be the "electric field".

When such a charge moves with varying velocity it propagates a pattern called electromagnetic field outward into space. Light is that pattern, that field traveling in space. Since light's source is a charged particle that, whether the particle is moving or not, is continuously emitting its radially outward flow that carries the affect of its charge, then light's electromagnetic field is a pattern of variations in that flow due to the charge's varying velocity.

2.2. Gravity

Given two masses, *i.e.* particles that have mass [e.g. electrons or protons], being separated and with the usual gravitational force [attraction] between them, if one of the masses is moved the change can produce no effect on the other mass until a time equal to the distance between them divided by the speed of light, c, has elapsed.

For that time delay to happen there must be something flowing from one mass to the other at speed c and each particle, each mass must be the source of such a flow.

That gravitational effect is radially outward from each mass, therefore every mass must be propagating such a flow radially outward in all directions from itself, which flow must be the "gravitational field".

2.3. That Flow

We therefore find that the fundamental particles of atoms, of matter, which have both electric charge and gravitational mass, must have something flowing outward continuously from them and:

- Either the particles have two simultaneous, separate outward flows, one for the effects of electric charge and another for gravitation, or
- They have one common universal outward flow that acts to produce all of the effects: electric and electro-magnetic field [light] and gravitational field [gravity].

There is clearly no contest between the alternatives. It would be absurd for there to be two separate, but simultaneous, independent outward flows, for the two different purposes. And, the single outward *Flow* from particles, carrying both the electric and electromagnetic field and the gravitational field, means that gravitational field can have an affect on light, on electro-magnetic field because they both are the same medium – the universal outward *Flow*.

Gravitational lensing is experimentally observed gravitational field affecting light [1].

3. Gravitational Slowing/Deflection of Light

Because that universal outward flow originates at each particle and flows radially outward in all directions its density or concentration decreases inversely as the square of distance from the source of the flow. At a large distance from the source the wave front of a very small portion of the total spherical outward flow is effectively flat—a "plane flow".

As presented in detail in *The Origin and Its Meaning* [1], two such universal flows encountering each other [flowing through each other] interfere with each other, that is each slows the flow of the other. The effect is proportional to the density or concentration of each flow.

Picturing Flow #1 of **Figure 5**, as that from a "lensing" gravitational mass and Flow #2 as that of the light from a distant object, then the figure depicts how the flow of the "lens" slows part of the wave front of the flow of the propagating light. The slowing is greater for rays of light that pass close to the lens and is less for those farther out. Thus the wave front of the light is deflected or bent as in the actually observed "gravitational lensing".

In "gravitational lensing" gravitation produces deflection of the flow that carries light. That deflected flow is the same flow that also simultaneously carries gravitation. Thus the gravitational flow from one mass can produce deflection of the gravitational flow from another mass.

Therefore, a properly configured material structure can deflect gravitation away from its natural action, reducing the natural gravitation effect on objects that the gravitation would otherwise encounter and attract.

That same effect, on a vastly reduced scale, produces the deflection, the bending of the light direction that is seen in slit diffraction. In the diffraction effect the role of the "massive lensing cosmic object" is performed by the individual atoms making up the opaque thin material in which the slit is cut. That effect shows that the gravitational lensing process, involving immense cosmic masses, can be implemented on Earth on a much smaller scale practical for human use.

4. The Energy Aspect and the Source of the Flow

But, changing the "natural gravitation effect" means changing the gravitational potential energy of objects in the changed gravitational field. If the energy is changed where does the difference come from or go to?

The potential energy for an object of mass, m, at a height, h, in a gravitational field is truly <u>potential</u>. It is the kinetic energy that the mass <u>would acquire</u> from being accelerated in the gravitational field <u>if it were to fall</u>. The greater the mass, m, the greater the kinetic energy, $\frac{1}{2} \cdot m \cdot v2$. The greater the distance, h, through which the mass would fall the greater the time of the acceleration, the greater the velocity, v, achieved, the greater the kinetic energy, $\frac{1}{2} \cdot m \cdot v2$.

While at rest at height h [as on a shelf] the total mass of the object is the same as its rest mass. The object has no actual "potential energy". It is merely in a situation where it could acquire energy, acquire it by falling in the gravitational field. Falling, the mass of the object increases as its velocity increases, reflecting its gradually acquired kinetic energy.

Since, <u>until it falls, the object does not have the energy that it will acquire when it falls</u> in the gravitational field <u>the energy that it acquires must come from the gravitational field</u>.

The energy of gravitational field is in its flow radially outward from all gravitational masses. The flow is a flow of the potential for energy, realized at any encounter with another gravitational mass

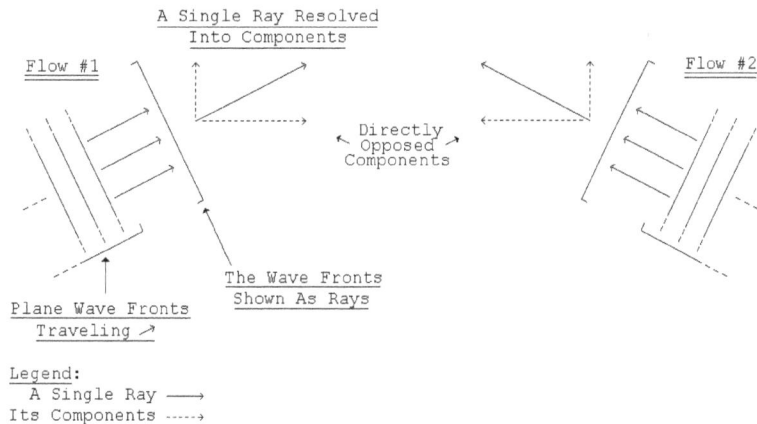

Figure 5. The Encounter of Two Flows.

- That flow creates potential energy, <u>creates the situation where kinetic energy could be acquired,</u> at any gravitational mass that it encounters.
- It does so continuously, replenishing and replenished by the on going continuing outward flow.
- It does so continuously, regardless of the number or amount of masses encountered and regardless of their distance from the source of the flow.
- At each encountered mass the amount of the flow varies with the magnitude of its source mass and varies inversely as the square of the distance from it.

But, for there to be a continuous flow outward from each mass particle, each must be a supply, a reservoir, of that which is flowing. The original supply of the flow of gravitational potential energy, came into existence at the "Big Bang" the beginning of the universe.

If that immense reservoir of energy could be tapped by tapping some of its appearance in its outward flow, which is the gravitational field, it could supply all of civilization's energy needs cheaply, cleanly, and permanently without [for practical human/Earth purposes] being used up.

Since the original "Big Bang" the outward flow has been very gradually depleting the original supply. That process, an original quantity gradually depleted by flow away of some of the original quantity is an exponential decay process and the rate of the decay is governed by its time constant. In the case of the overall universal decay, appearing among other places in the outward flow from every gravitating mass, the time constant is about τ = 3.57532·1017 *seconds* ($\approx$ 11.3373·109 *years*). [1].

5. Tapping the Energy of the Gravitational Field

The general vertically upward outward flow of gravitational energy can be tapped by deflecting part of a local region's gravitational flow away from its normal vertical direction. That produces above that local region a region of lesser gravitation than its surroundings of normal gravitation. That can be configured to produce an imbalance in a rotary device above it powering its rotation analogously to a water wheel. That rotational energy, connected to an electric generator, can generate electrical energy, *i.e.* useful electric power. **Figure 6**, below, [**Figure 4** but now rotated 90˚] illustrates such deflection using a single slit.

Multiple such slits parallel to each other would spread the deflection left and right in the figure. Additional multiple such slits at right angles to the first ones would spread the deflection over a significant area.

6. Gravitation Deflector Design

The edges of the slit in **Figure 6** are actually rows of atoms. A cubic crystal, such as of Silicon, consists of such rows of atoms, multiple rows and rows at right angles, all equally spaced, **Figure 7**—a naturally occurring configuration of the set of slits required for deflection of gravitation.

The flow from each of the cubic crystal's atoms is radially outward. Therefore its concentration falls off as the square of distance from the atom. The amount of slowing of an incoming gravitational flow and therefore the amount of its resulting deflection, depends on the relative concentration of the atoms' flow and of the overall gravitational flow.

In the case of diffraction of the flow of light at a slit the concentration of the flow from the atoms of the slit material is comparable to the concentration in the horizontal flow of the light, because the light originates from a local source, not from the Earth's immense gravitation.

But for the flow from the atoms of the slit to deflect the much more concentrated vertically upward flow of

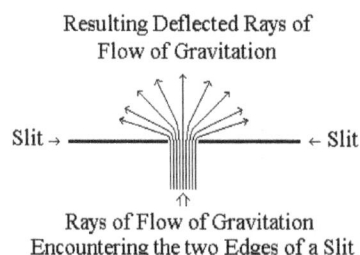

Resulting Deflected Rays of
Flow of Gravitation

Slit → ———————— ← Slit

Rays of Flow of Gravitation
Encountering the two Edges of a Slit

Figure 6. Slit Diffraction, the Basic
Element of a Gravitational Deflector.

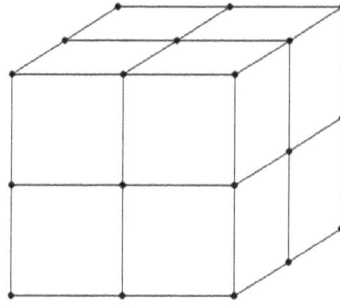

Figure 7. A Small Piece of a Cubic Crystal.

Earth's gravitation the flow from the atoms of the slit must also be much more concentrated. The only way to achieve that more concentrated flow is to create a configuration in which the flow of Earth's gravitation is forced to pass much closer to the atoms of the slit so that, per the inverse square variation in the atoms' flow, it will pass through a concentration of the slit atom's flow comparable to the concentration in the Earth's gravitational flow.

The spacing between the edges of the diffracting slit is about 5×10^{-6} *meters*. The spacing of the atoms at the corners of the "cubes" in a Silicon cubic crystal is 5.4×10^{-10} *meters*. As developed in the references, an inter-atomic spacing of less than 3×10^{-19} *meters*, much closer than the natural spacing in the Silicon cubic crystal, is required to obtain deflection of a major portion of the incoming Earth's gravitational flow.

Such a close atomic spacing cannot be obtained by directly arranging for, or finding a material that has, such a close atomic spacing. However, that close an atomic spacing can be effectively produced relative to just the vertical flow of gravitation by slightly tilting the Silicon cubic crystal's cubic structure relative to the vertical.

Figure 8, illustrates the tilting, schematically and not to scale, and shows how it increases the number of crystal atoms closely encountered by the upward gravitational flow.

Pure, monolithic, Silicon cubic crystals are grown commercially in diameters from 25.4 millimeters (1 inch) to 300 millimeters (11.8 inches) for making the "chips" used in many electronic devices. By appropriate tilting of the cubic structure each of its 5.4×10^{-10} *meters* inter-atomic spaces is effectively sub-divided into 1010 "sub-spaces" each of them 5.4×10^{-20} *meters* long and with an atom in each. A 4.5 *millimeters* shim on a 300 *millimeters* diameter Silicon cubic crystal ingot produces such an effect, producing a *tilt tangent* = 0.015 for a *tilt angle* = 0.86° that produces the objective effective sub-division of the crystals' natural inter-atomic spacing, a sub-division that acts only on vertical flow, as of gravitation.

The gravitational deflector requires a large, thick piece of Silicon cubic crystal rather than the thin wafers sawed from the "mother" crystal for "chip" making.

Per the detailed analysis in the references, The Silicon cubic crystal ingot for the deflector is to be:
- Diameter appropriate to the application,
- 0.5 *meters* or more thick,
- with the orientation of the cubic structure marked for proper placement of tilt-generating shims, and
- with the bottom face of the cylinder sawed and polished flat at a single cubic structure plane of atoms.

Mean free path [*MFP*] is the average straight line distance a moving particle travels between encounters with another particle. For atoms in solid matter the mean free path is

$$MEP = \frac{1}{[\text{Atoms Per Volume}] \cdot [\text{Atom Cross-Section}]} \tag{1}$$

For the Earth the atoms per unit volume is on the order of 5×10^{28} per cubic meter.

In the cubic crystal deflector the atomic spacing produced by the tilt is on the order of 10 - 20 meters. Therefore each atom has cross-sectional space available to it of a circle of that diameter so that for this purpose the atom's cross-section area is $[\pi/4] \times [10^{-20}]2 = 8 \times 10^{-39}$ *square meters*.

For targets as fine as those in the cubic crystal deflector, the mean free path in the Earth's outer layers is, therefore 2.5×10^{9} *meters*

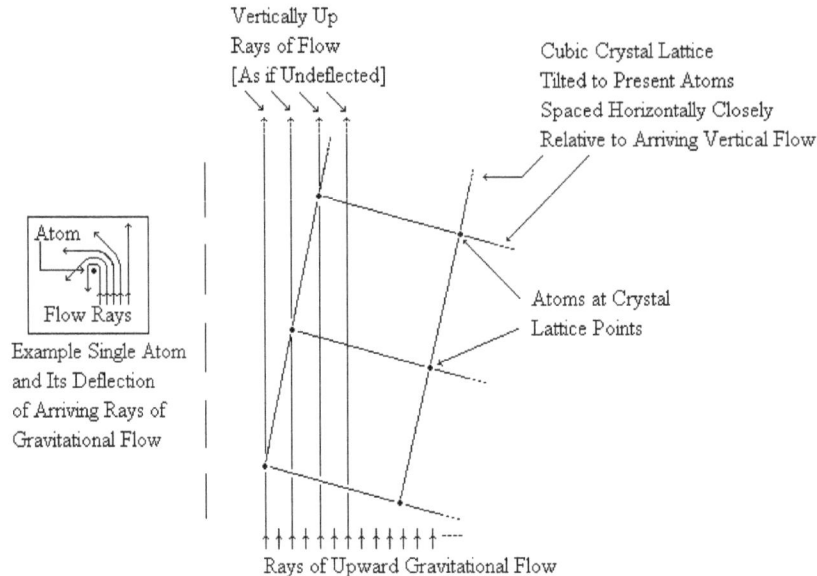

Figure 8. Cubic Crystal Lattice Tilted for Effective Gravitational Flow Deflection.

The mean free path in the 0.5 *meters* thick minutely tilted Silicon cubic crystal ingot for intercepting Earth's natural <u>vertically</u> outward gravitation is ½ the 0.5 *meters* thickness of the ingot. Therefore, the gravitational deflector is about 1010 times more effective than the natural Earth at intercepting Earth's natural gravitation.

However, that effectiveness is only for vertical rays of flow.

The Silicon crystal's mean free path for non-vertical flow—flow already once deflected within the crystal—is that of Earth, 2.5×10^9 *meters*, which takes the once-deflected flow out of the crystal.

The overall deflector consists of:

- A support having a verified perfectly horizontal upper surface for the cubic crystal deflector bottom face to rest upon;
- The Silicon cubic crystal ingot specified above; and
- Precision shims for producing the tilt of the cubic crystal ingot, the shims located at the mid-point of two adjacent sides of the horizontal plane of the cubic structure as in **Figure 9**.

For an array of ingots for a larger area than a single ingot can provide, the individual ingots can be machined to fit snugly together. That could be done by machining to a square cross-section or a hexagonal one.

The manner of the deflection is curving of the path of rays of gravitational Flow as they pass close to atoms of the deflector with the direction to which curved depending on the relative positions of the ray and an atom and the amount of the curving depending on how close the ray passes to the atom. Because of the range of those variables and their various combinations the "deflection" is essentially a "scattering" in various amounts in various directions, all scattering being away from the perfectly vertical upward which the deflector is designed to solely deflect.

The "scattering" is illustrated two-dimensionally in **Figure 10**. Three dimensionally it can be visualized as that figure viewed from the top while rotated through a full circle.

The physical example of the "scattering" is the diffraction pattern of light diffracted by a slit. **Figure 11**, presents the diffraction pattern produced by a slit that is 5.4×10^{-6} *meters* wide with incoming light of wavelength 4.13×10^{-7} *meters*. The peaks and valleys of the pattern, the interference pattern, are a phenomenon of the light imprint on the *Flow* that carries it. <u>The envelope of the pattern is the relative amounts of the underlying *Flow* carrying the light</u>.

For that reason, while the interference pattern varies according to the wavelength of the light involved, <u>the form of the envelope of that pattern is always the same</u>.

The *Flow* concentration produced by the two slit edges falls off with distance from the edge inversely as the square of distance from its atoms. The Cauchy-Lorentz Distribution is an inverse square function of its variable. Its Density Function can represent the relative *Flow* intensity pattern produced by the diffraction process by representing the envelope of the diffraction pattern. In **Figure 12** the Cauchy-Lorentz distribution is fitted to the

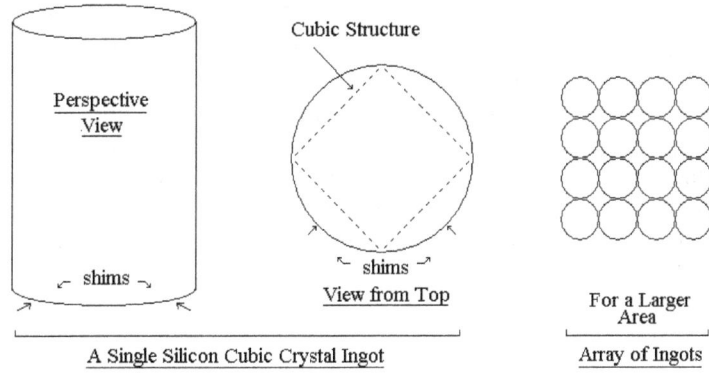

Figure 9. The Overall Deflector.

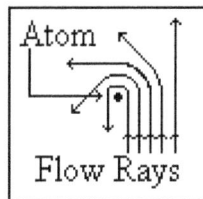

Figure 10. Single Atom Deflection of Rays of Gravitational Flow.

Diffraction Pattern
·Slit = $5.4 \cdot 10^{-6}$ Meter Wide
·Light Wavelength = $4.13 \cdot 10^{-7}$ Meter

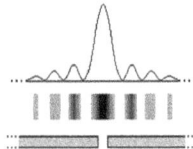

Figure 11. A Slit Light Diffraction Pattern.

The Envelope of the Relative Intensities of the
Light Diffraction Pattern Is the Actual Amount
of the *Flow* Relative Intensities.

Diffraction Pattern of   Cauchy-Lorentz Distribution
$5.4 \cdot 10^{-6}$ Meter Wide Slit   Density Function

$$f = \left[\frac{\gamma}{(d - mid)^2 + \gamma^2} \right]$$

The Two Over-Layed

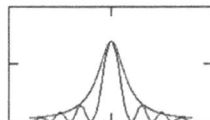

Figure 12. The Cauchy-Lorentz Distribution Diffraction Pattern Envelope.

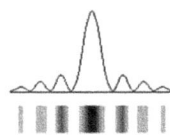

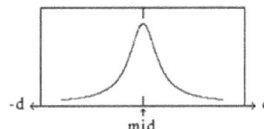

diffraction pattern by the appropriate choice of value of its distribution parameter γ [Greek *gamma*].

The deflection angle, Φ, is the angle of deflection of the rays to any particular point on the diffraction pattern and of intensity per the Cauchy-Lorentz Distribution at that point.

The interest here is not in the location of the light interference maxima and minima, but in the deflection angles the diffraction imposes on the *Flow*. However, calculation of the deflection angles to the minima provides a good indication of the amount of *Flow* deflection obtained over the overall diffraction pattern. **Table 1** below presents that data for the 5.4×10^{-6} *meters* wide slit with incoming light of wavelength 4.13×10^{-7} *meters*. [The minimums are counted outward from the center peak of the diffraction interference pattern].

$Sin(\Phi) = n \cdot$ [*light wavelength/slit width*], $n = 1, 2, ...$

Table 1 demonstrates that the deflection of the *Flow* is at least in amounts up to 90°. That deflection may well extend to angles beyond 90°. There is no way of determining that from the diffraction pattern. However, while the light of the diffraction pattern cannot be deflected beyond 90° in any case because the light cannot penetrate the material containing the slit, the *Flow* readily penetrates and permeates all of material reality.

The tilt of the cubic crystal structure divides the slit into 1010 sub regions the first and last of which are at the slit's edge and produce the maximum deflection. The tilt so arranges that ultimately all of the vertical components of the incoming vertical Flow must pass through one of those "at the edge of the slit" regions and must experience maximum deflection.

The overall average effect is equivalent to every ray's vertical component curving at least 90° because the crystal tilt causes every ray to pass extremely close to an atom at some point in the crystal, as the extreme rays in **Figure 13**, below.

There does not appear to be any way to analyze, calculate, or evaluate in advance the overall deflection that is achieved other than by actual experiment. With the overall average effect equivalent to every ray's vertical component curving 90°, *i.e.* to the horizontal, the overall total net effect of the vertical components after deflection is zero. Then the overall amount of deflection is 100% of the natural un-deflected gravitation reducing the gravitation to essentially zero.

7. Gravito-Electric Power Generation

Gravito-electric power generation is similar to hydro-electric power generation in which the energy of water

Table 1. Diffraction Minimums Deflection Angle.

Minimum #	$\Phi°$	Minimum #	$\Phi°$
1	4.39	8	37.72
2	8.80	9	43.50
3	13.26	10	49.89
4	17.81	11	57.28
5	22.48	12	66.60
6	27.36	13	83.86
7	32.37	14	$Sin(\Phi) > 1.0$

Resulting Deflected Rays of
Flow of Gravitation

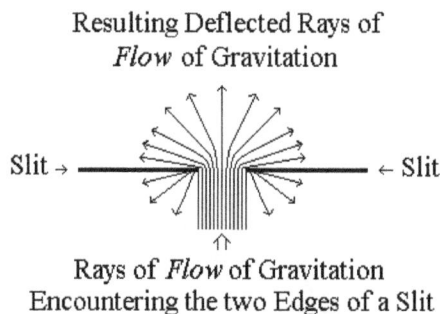

Rays of *Flow* of Gravitation
Encountering the two Edges of a Slit

Figure 13. Single Slit Gravitation Deflection.

falling in Earth's gravitational field powers water-turbines that drive electric generators.

In gravito-electric power, depicted schematically in **Figure 14** below, a gravitation deflector makes the water in the central region of the mechanism lighter than that in the outer region, which is acted on by natural gravitation. The lighter reduced gravitation water floats up on the in-flow under it of the heavier natural gravitation water. The result is continuous circulation of the water, like a continuous waterfall.

Water turbines like those used in hydro-electric plants can be placed in the gravito-electric continuous water flow to drive electric generators as in hydro-electric plants.

8. Practical Aspects—Design Engineering

While the net gravitational field is vertically upward, *i.e.* radially outward from the Earth's surface, local gravitation is radially outward from each particle of matter. As in **Figure 15** below, a mass above the Earth's surface receives rays of gravitational attraction from all over its surrounding surface and the underlying body of the Earth.

The net effect of all of the rays' horizontal components is their cancellation to zero. The net effect of all of the rays' vertical components is Earth-radially-outward gravitation.

8.1. Gravitational Ray's Horizontal and Vertical Components

The net gravitational effect on objects is due to the vertical component of all of the myriad rays of gravitational field *Flow* at a wide variety of angles to the horizontal. This "components aspect" is valid because of the "components aspect" appearance in the "Gravitational Lensing" effect on cosmic light.

The various rays of the *Flow* propagation from the individual particles of the gravitating body [for example the Earth] are from each individual particle of it to the selected point [above the gravitating body] on which their action is being evaluated. That is the point *P* in **Figure 15** directly above the "*A*" at height *h* in the figure.

The Earth's gravitational action along a ray of *Flow* takes place from the Earth's surface to deep within the Earth. The inverse square effect, that the strength of a *Flow* source is reduced as the square of the increase in the radial distance of it from the object acted upon, is exactly offset by that the number of such sources acting [per "ray" so to speak] increases as the square [non-inverse] of that same radial distance. That is, the volume, hence the number, of *Flow* sources for a ray of propagation at the object is contained in a conical volume, symmetrically around the ray with its apex at the object acted upon.

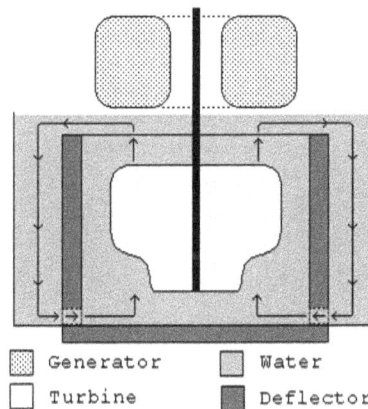

Figure 14. A Gravito-Electric Generator.

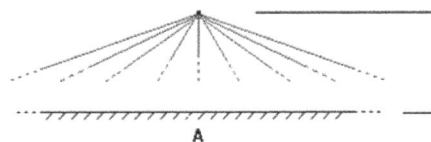

Figure 15. Rays of Gravitation from the Surroundings.

However, because the net gravitational effect is produced only by the vertical component of each ray of *Flow* propagation, the effectiveness of each ray is proportional to the Cosine of the angle between that ray and the perfectly vertical as the angle θ in **Figure 16**, below.

The actual total gravitational action includes all rays from $\theta = 0$ through to $\theta = 90°$. That range would require an infinitely large deflector to act on all such rays, that is the deflector would have to be a disk of infinite radius. For lesser values of the maximum θ addressed, the portion of the total gravitation sources included is the integral of $Cos\ \theta \cdot d\theta$ from $\theta = 0$ *to* $\theta = Chosen\ Lesser\ Value$. The integral of the *cosine* is the *sine*. Example lesser portions of the total gravitational action addressed as θ varies are presented in **Table 2**, below.

The gravitational deflector as a disk beneath the *Object* to be levitated must extend horizontally far enough to intercept and deflect the *Chosen Lesser Value* of angle θ rays of gravitational wave *Flow* that are able to act on the *Object* of the deflection as depicted in **Figure 17**, below.

For the perfectly vertically traveling rays of gravitation waves the required vertical distance that must be traveled within the cubic crystal is the previously presented 0.50 *meters* and 0 horizontal distance is traversed in so doing. But a ray at angle θ, in order to traverse the required 0.50 *meters* vertically, must traverse horizontally $0.50 \cdot Tan[\theta]$ *meters*, at the same time. For θ more than 45° that can become quite large and the deflector likewise.

Because the deflector disk must extend over a large area to deflect most of the gravitation, an alternative, and better, solution to the problem of rays of gravitation arriving over the range from $\theta = 0$ *to* $\theta = 90°$ is to wrap the deflector up the sides of the *Object* to be levitated as **Figure 18**.

In this configuration the deflector takes up little more space than the *Object* levitated. However, the non-perfectly vertical traveling rays must still travel within the cubic crystal the horizontal distance $0.50 \cdot Tan[\theta]$ *meters*. That requires that the horizontal thickness of the vertical sides of the cup-shaped deflector must be of that $0.50 \cdot Tan[\theta]$ *meters* thickness.

Because the value of $Sin\theta$ and, therefore, the fraction of the total gravitational action, increases relatively little above $\theta = 60°$ whereas the value of $Tan[\theta]$ increases quite rapidly, from 1.7 *to* ∞ above $\theta = 60°$ that $\theta = 60°$ is the appropriate value to which to design. The thickness of the "walls" of the "cup" would then be $0.50 \cdot Tan[60°]$ = 0.85 *meters*. The deflector would be only slightly larger than the *Object* levitated.

Table 2. The Effect of θ.

θ	$Sin\ \theta$ = Fraction of Total Maximum Gravitational Action
0°	0.000
30°	0.500
45°	0.707
60°	0.866

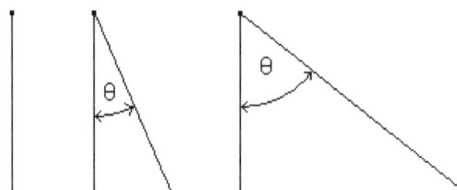

Figure 16. The Angle θ.

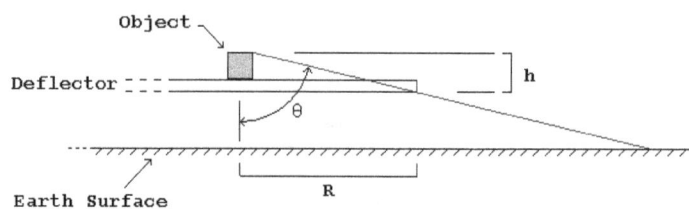

Figure 17. The Deflector as a Disk.

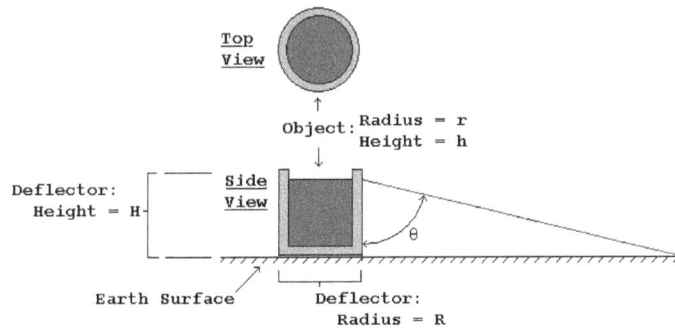

Figure 18. Cup-Shaped Deflector.

8.2. Gravitation Deflector Design Parameters

The Deflector is a cup shaped array of monolithic Silicon cubic crystals. The crystals forming the flat "base" of the "cup" must be 0.50 *meters* in height. The "sides" of the "cup" will be the same kind of 0.50 *meter* crystals stacked and aligned vertically. The thickness of the "sides" must be 0.85 *meters*.

The crystals are grown with circular cross-section and in diameters up to 0.30 *meters*; however, those cylindrical pieces must then be machined to hexagonal or square cross section for a number of them to fit together with negligible open space. The cross-section area of these crystals is $\pi \cdot d^2/4 = 0.785 \cdot d^2$ before machining.

For a circular deflector the configuration is poorly compatible with arranging the crystals in a close-fitting array unless it involves a large number of crystals each of small cross-section relative to the horizontal cross-section of the overall deflector. For that case the crystals should be machined to hexagonal cross-section. For smaller deflectors the configuration should be of rectangular cross-section and the crystals machined to square cross-section, **Table 3**.

8.2.1. Circular Cross-Section Gravitation Deflector Structure
A circular cross-section gravitation deflector structure to provide deflection for an object of height, *h*, and diameter, *d* meters would have the **Table 4** parameters.

From the above the examples of **Table 5** obtain. The 0.85 *meters* thickness of the "cup" "sides" requires 20 layers horizontally of 50.8 *millimeters* [2*inch*] crystals.

8.2.2. Square Cross-Section Gravitation Deflection Structure
A square cross-section gravitation deflector structure to provide deflection for an object of square cross-section side, *s*, and height, *h* meters would have the parameters of **Table 6**.

From the above the examples of **Table 7** obtain. The 0.85 *meter* thickness of the "cup" "sides" requires 3 layers horizontally of 300 *millimeters* [11.8 *inch*] crystals. **Figure 19** depicts a resulting deflector.

8.3. Calibrating the Individual Silicon Crystals

The individual crystals cannot be grown exactly identical to each other. In each the orientation of the long axis of the cubic crystal structure may vary minutely from each of the others.

To find the optimum tilt and orientation for a single crystal the tilt must be varied over the range of possibilities while the effect from exactly below it is observed on a balance scale. But most of the effect of gravitation on a single crystal is not from exactly below it.

The solution to that problem is to conduct the optimization atop a structure, that relying on the inverse square effect, effectively isolates the crystal from most of the gravitation from surrounding sources except that exactly below it—a high pedestal having a cross section comparable to that of the crystal, as in **Figure 20**.

To conduct that calibration on thousands of crystals should not be necessary if a method can be developed to exactly measure the long axis orientation in any given crystal. The process can then determine the optimum orientation of the crystal tilt relative to the actual long axis of a few cubic crystals being calibrated. That same crystal tilt relative to the actual long axis can then be applied to each of the other crystals by means of appropriate shims once the particular long axis orientation of each has been exactly measured.

Table 3. Crystal Cross-Sections.

Deflector	Crystal Cross-Section	Crystal Cross-Section Area	Percent of Crystal
Circular	Hexagonal	$[\sqrt{3}/3]\cdot d^2 = 0.577\cdot d^2$	73.5
Rectangular	Square	$d^2/2 = 0.500\cdot d^2$	63.7

Table 4. Round Deflector Parameters.

Part	Dimension	Size
Cup Sides	Thickness, t	0.85 m
	Inside diameter, ID	d
	Outside diameter, OD	$d + 2\cdot t = d + 1.7$
	Height	h
	Height Layers	$h \div 0.5$
	Layer Area	$\pi\cdot[OD^2 - ID^2] \div 4 == 0.785\cdot[OD^2 - ID^2]$
Base Disk	Thickness	1 crystal layer = 0.5 m
	Diameter	$d + 2\cdot t = d + 1.7$
	Area	$\pi\cdot[OD^2 - ID^2] \div 4 == 0.785\cdot[OD^2 - ID^2]$

Table 5. Example Round Deflectors.

d	h	Cup Disk Base		Cup Sides		Total Volume	Nr. of 2" Hex Crystals
		Area	Volume	Area	Volume		
1	1	5.72	5.75	4.94	4.94	10.7	13,280
10	10	131	1310	29	290	319	779,570

Table 6. Square Deflector Parameters.

Part	Dimension	Size
Cup Sides	Thickness, t	0.85 m
	Square inside, IS	s
	Square outside, OS	$s + 2\cdot t = s + 1.7$
	Height	h
	Height Layers	$h \div 0.5$
	Layer Area	$OS^2 - IS^2$
Base Square	Thickness	1 crystal layer = 0.5 m
	Side	$s + 2\cdot t = s + 1.7$
	Area	$[s + 2\cdot t]^2 = [s + 1.7\cdot t]^2$

Table 7. Example Square Deflectors.

s	h	Cup Disk Base		Cup Sides		Total Volume	Nr. of 11.8" Square Crystals
		Area	Volume	Area	Volume		
1	1	7.3	7.3	6.3	6.3	13.6	195
10	10	137	1370	36.9	369	1,739	2,486

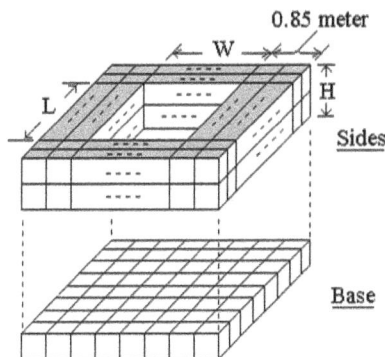

Figure 19. Highly Schematic Depiction of a Rectangular Gravitation Deflector.

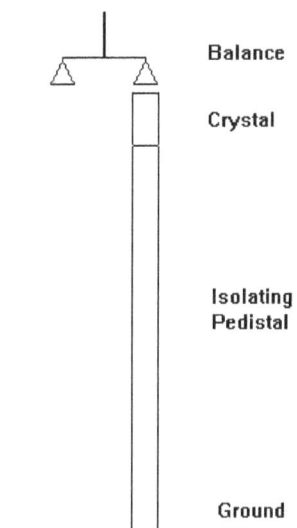

Figure 20. Crystal Calibration.

8.4. Alternative to Calibration

Monolithic silicon cubic crystals are commercially available with the ends nearly a single plane, that is within 0.2 *degrees* of the (100) plane of the cubic structure. In view of the various effects analyzed in Appendix C to the book, *Gravitics—The Physics of the Behavior and Control of Gravitation*, and their resolution in its section *The Random Distribution Solution to The Crystal Tilt* [2], that amount or moderately more of inaccuracy in the crystal tilt is of no significance except that it potentially may call for crystal thicknesses moderately greater than 0.5 *m*.

The task, then, is research and development to better optimize the design so that practical implementation can begin

See the references [1-3], below, for the detailed development of mass, field, gravitation, the *flow*, the exponential decay and the time constant.

References

[1]   Ellman, R. (2004) The Origin and Its Meaning. 2nd Edition, The-Origin Foundation, Inc., Santa Rosa.

[2]   Ellman, R. (2008) Gravitics—The Physics of the Behavior and Control of Gravitation. 2nd Edition, The-Origin Foundation, Inc., Santa Rosa.

[3]   Ellman, R. (2011) Gravito-Electric Power Generation. US Patent No. 13/199,867.

Study on Surface Electric Field Simulation of High Voltage Transmission Line Assembly

Liankai Chen[1], Wenqing Lai[1], Jun Wang[2], Guoyi Jiang[1], Yan Zhou[1], Yong Chen[2], Haibo Liu[1], Zhaoyu Qin[2], Lei Ke[2], Lei Wang[2], Yang Shen[2]

[1]East Inner Mongolia Electric Power Company Limited, Inner Mongolia, Hohhot, China
[2]Wuhan NARI Limited Liability Company of State Grid Electric Power Research Institute, Wuhan, China
Email: chenliankai@md.sgcc.com.cn

Abstract

For the studies in the field of high voltage power transmission, this paper has adopted the method of finite element node potential, and put forward two kinds of high pressure sensor-fixture modeling scheme for the sensor-fixture of the high voltage side, the simulation analysis shows that the sensor-fixture surface should be smooth, and should not appear the conclusion of edges and corners. While through establishing the four clamps assembly optimized model, and simulates the strain gages, fixtures and conductor surface field strength and electric field distribution in the model as a whole in turn, this paper Finally got the optimal size of fixture structure and assembly of each part reasonable location layout.

Keywords

High Voltage Assembly; Finite Element Method; Simulation Calculation; Electric Field Distribution

1. Introduction

Transmission line is the material basis for the development of electric power industry. The transmission grid facilities are burdened with the task of transmission and distribution of electrical energy, the reliability and operating conditions of which directly determine the stability and security of the entire system, but also determine the power quality and reliability of supply. Currently, with the difficult inspection and heavy maintenance workload, if the problems on the transmission line cannot be promptly found and solved, it will affect the country's production and residents' life, which has become an imperative social problem that ensures the safety and smooth flow of the power "lifeline" [1].

With the increasing of the grid capacity and transmission voltage rating, it's becoming increasingly important that the on-line monitoring of the performance of the high voltage power transmission equipment in operating [2].

2. Calculation Principle

2.1. Finite Element Analysis of Electrostatic Field

At present, the electromagnetic field numerical calculation method commonly used at home and abroad is the charge simulation method of finite element method, boundary element method, finite difference method, and a variety of methods of coupling or mixed application method etc. [3].

Calculation of the model in this study is composed of a variety of media, and the model is more complex, which can be calculated by finite element method Finite element method based on variation principles, combined with the Difference method and developed a numerical method.

Electrostatic field potential energy can be expressed as a function and its derivative pending integral of the integration region (or solving the field), in accordance with the differential method of discrete method, it is divided into finite sub-regions (called cells). And then use these discrete units, the electrostatic field energy be approximated by a finite number of nodes in the potential of the function [4]. Thus, the electrostatic field energy demand is reduced to the problem of extreme multi-function extremum problem, which usually boils down to a set of multivariate linear algebraic equations finite element equation. Finally, the specific characteristics of equations, using appropriate algebraic methods, and seek each node potential, to achieve a discrete solution of variation problems.

In the electrostatic field, generally solving potential of φ is a direct object. In isotropic, linear, homogeneous medium, the potential to meet the Poisson equation or the Laplace equation:

$$\nabla^2 \varphi = \frac{\rho}{\varepsilon} \tag{1}$$

Or

$$\nabla^2 \varphi = 0 \tag{2}$$

where, ρ is free charge density, ε is the dielectric constant.

Based on the principle of least action electrostatic field, in the medium a fixed charge system, the surface charge distribution of the electric field tends to always make synthesis with minimal electrostatic energy [5].

The entire field for a given potential values on the boundary of the first class of boundary value problems, energy integral expression is:

$$W_e = \iint_D \left\{ \frac{\varepsilon}{2} \left[\left(\frac{\partial \varphi}{\partial x} \right)^2 + \left(\frac{\partial \varphi}{\partial y} \right)^2 \right] - \rho \varphi \right\} dx dy dz = \min \tag{3}$$

Finite element method is a continuous field via split into a finite number of discrete units, in two-dimensional field problems commonly used to split triangular and quadrilateral elements, within each cell of the potential function approximated by polynomial interpolation.

After the split, the secondary functional can be expressed as the sum of all the cells energy function:

$$F(\varphi) = \sum_{e=1}^{q} F_e(\varphi) \tag{4}$$

In the formula, $F(\varphi)$ represents the element ε of the corresponding triangular energy integral, which is a unit of energy integration [6]:

$$F_e(\varphi) \approx F_e(\varphi^e) = \iint_D \frac{\varepsilon}{2} \left[\left(\frac{\partial \varphi^e}{\partial x} \right)^2 + \left(\frac{\partial \varphi^e}{\partial y} \right)^2 \right] dx dy = \min \tag{5}$$

While:

$$\varphi^e = \sum_{i=1}^{q} N_i^e \varphi_i^e \tag{6}$$

Then:

$$\iint_D \frac{\varepsilon}{2} \left(\frac{\partial \varphi^e}{\partial x} \right)^2 dx dy = \iint_D \frac{\varepsilon}{2} \left(\sum_{i=1}^{q} \frac{\partial N_i^e}{\partial x} \varphi_i^e \right)^2 dx dy \tag{7}$$

If $\dfrac{\partial N_i^\varepsilon}{\partial x}$ is independent of x, then:

$$\iint_D \frac{\varepsilon}{2}\left(\frac{\partial \varphi^e}{\partial x}\right)^2 \mathrm{d}x\mathrm{d}y = \varphi^{e\mathrm{T}} \boldsymbol{K}_x^e \varphi^e \tag{8}$$

Of which:

$$\varphi^e = \begin{bmatrix} \varphi_1^e \\ \vdots \\ \varphi_q^e \end{bmatrix} \tag{9}$$

$$\boldsymbol{K}_x = \frac{\varepsilon S^e}{2}\begin{bmatrix} \dfrac{\partial N_1}{\partial x}\dfrac{\partial N_1}{\partial x} & \dfrac{\partial N_1}{\partial x}\dfrac{\partial N_2}{\partial x} & \cdots & \dfrac{\partial N_1}{\partial x}\dfrac{\partial N_q}{\partial x} \\ \dfrac{\partial N_2}{\partial x}\dfrac{\partial N_1}{\partial x} & \dfrac{\partial N_2}{\partial x}\dfrac{\partial N_2}{\partial x} & \cdots & \dfrac{\partial N_2}{\partial x}\dfrac{\partial N_q}{\partial x} \\ \vdots & \vdots & \ddots & \vdots \\ \dfrac{\partial N_q}{\partial x}\dfrac{\partial N_1}{\partial x} & \dfrac{\partial N_q}{\partial x}\dfrac{\partial N_2}{\partial x} & \cdots & \dfrac{\partial N_q}{\partial x}\dfrac{\partial N_q}{\partial x} \end{bmatrix} \tag{10}$$

S^e is the split region enclosed area. Similarly, if $\dfrac{\partial N_i^\varepsilon}{\partial y}$ has nothing to do with y, then:

$$\left(\frac{\partial \varphi^e}{\partial y}\right)^2 = (\sum_{i=1}^q \frac{\partial N_i^e}{\partial y}\varphi_i^e)^2 = \varphi^{e\mathrm{T}} \boldsymbol{K}_y^e \varphi^e \tag{11}$$

$$\boldsymbol{K}_y = \frac{\varepsilon S^e}{2}\begin{bmatrix} \dfrac{\partial N_1}{\partial y}\dfrac{\partial N_1}{\partial y} & \dfrac{\partial N_1}{\partial y}\dfrac{\partial N_2}{\partial y} & \cdots & \dfrac{\partial N_1}{\partial y}\dfrac{\partial N_q}{\partial y} \\ \dfrac{\partial N_2}{\partial y}\dfrac{\partial N_1}{\partial y} & \dfrac{\partial N_2}{\partial y}\dfrac{\partial N_2}{\partial y} & \cdots & \dfrac{\partial N_2}{\partial y}\dfrac{\partial N_q}{\partial y} \\ \vdots & \vdots & \ddots & \vdots \\ \dfrac{\partial N_q}{\partial y}\dfrac{\partial N_1}{\partial y} & \dfrac{\partial N_q}{\partial y}\dfrac{\partial N_2}{\partial y} & \cdots & \dfrac{\partial N_q}{\partial y}\dfrac{\partial N_q}{\partial y} \end{bmatrix} \tag{12}$$

Therefore:

$$\mathrm{F}_e(\varphi^e) = \varphi^{e\mathrm{T}}(\boldsymbol{K}_x^e + \boldsymbol{K}_y^e)\varphi^e \tag{13}$$

All sub-regions combined simultaneous equations, we obtain:

$$\mathrm{F}(\varphi) \approx \varphi^{\mathrm{T}} \boldsymbol{K} \varphi \tag{14}$$

where φ and K is the set of simultaneous equations after all the variables and coefficient matrix. So the problem is discretized into functional multivariate quadratic function extremum problem:

$$\varphi^{\mathrm{T}} \boldsymbol{K} \varphi = \min \tag{15}$$

By the function extreme value theory, we have:

$$\frac{\partial \mathrm{F}}{\partial \varphi_i} = 0 \tag{16}$$

The formula is:

$$\boldsymbol{K}\varphi = 0 \tag{17}$$

The above equation is the finite element equations, which can solve the equation by using the boundary conditions; and you can get the desired approximate solution.

In summary, the finite element method calculation steps are as follows:

1) Clear field range, the field split into a finite element (two-dimensional field can be trilateral or quadrilateral

base unit, three-dimensional field available tetrahedral or hexahedral basic unit);

2) Electric energy calculation unit elements of the coefficient matrix;
3) Calculate the total electric energy coefficient matrix elements;
4) Finite element equations are listed;
5) Solving the finite element equations, find the node potential;
6) Potential can be calculated based on each node's remaining amount of the electric field.

2.2. ANSYS Software in the Application of the Electrostatic Field Calculation

ANSYS is the financial structure, fluid, electric, magnetic, acoustic analysis in one large general-purpose finite element analysis software. Its electromagnetic field analysis includes several modules: low frequency, high frequency, high frequency TV size (MOM), cable bundle electromagnetic compatibility (EMC) and signal integrity (SI), PCB's and other EMC and SI [7]. It works with most CAD software interface, data sharing and exchange, such as Pro/Engineer, AutoCAD, Solid works, etc.

ANSYS analysis process consists of three main steps: pre-treatment (create or read into the finite element model, defining material properties, mesh); loading and solving (applied load and set constraints solving); post-treatment (see Analysis a result, the check result is correct) [8].

Among them, the pre-treatment of the mechanical structure of the system is mainly made-up of nodes and elements into the mix into a finite element model of the various parts of its defined element type, material properties and real constants, then meshing, finite element model to obtain the unknown nodes and elements. Loading and Solution mainly defined properly, the correct load system components in order to understand the structure after the reaction by the external load, load on the finite element model, then select the appropriate direct solver and iterative solver for solving computing nodes [9]. The subsequent processing is mainly analysis node inspection and unit results.

3. Calculation Models and Parameters Established

3.1. Research Contents

In order to effectively avoid the transmission line conductor aeolian vibration caused by the wire off shares accidents, it requires real-time monitoring of wire in the breeze under the vibration information, so need to install the wire strain measuring device measuring wire strain. Strain gauge the middle part of the metal, composite material with metal wire ends, fixture and fixed with strain gage. Strain gauge on the wire installation diagram as shown in **Figure 1**. Different structure of the metal fixture assembly surface field strength will produce different effects, so the need for fixture structure modeling of different options to select the optimal metal clamp structure to reduce vibration and point discharge breeze on high-voltage transmission wires resulting damage.

n this study, a high-pressure side sensor fixture design two programs, In order to verify the merits of two kind of fixture, respectively, finite element analysis of the phenomenon of charge concentration of two kinds of fixture scheme. The following is the detailed modeling program.

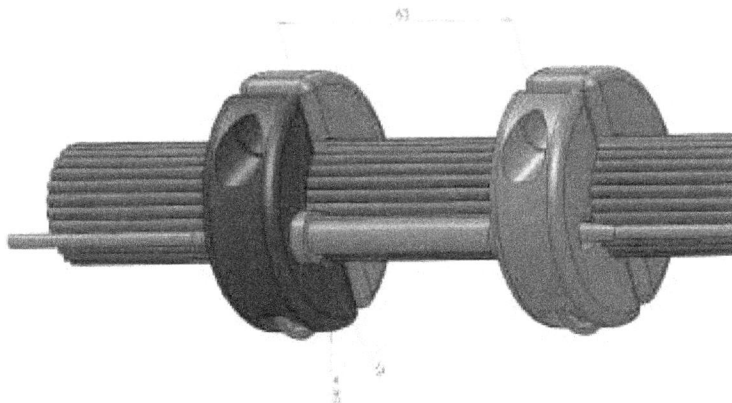

Figure 1. Strain gauge assembly simulation diagram.

3.2. Sensor Fixture Modeling

The surface of the clamp tip scheme I will have the electric field concentration, and easily lead to burn out the wire tip discharge, while the scheme II uses the methods to avoid the tip, the fixture body with circular arc design, the design of bolt connection parts of sinking, completely hidden in the fixture screw bolt, Compact structure design of high voltage side as shown in **Figure 2**.

   Since two briquetting symmetrical structure and coupling docking, we are concerned about the charge distribution is mainly concentrated in the outer surface of the arc. To save modeling time is taken in the right briquetting COMSOL modeling and analysis, shown in **Figure 3**. Similarly, removal of the air domain compact structure only concerned with compact contour of the air within the inner boundary of the electric field distribution. The simulation parameters are shown in **Table 1**.

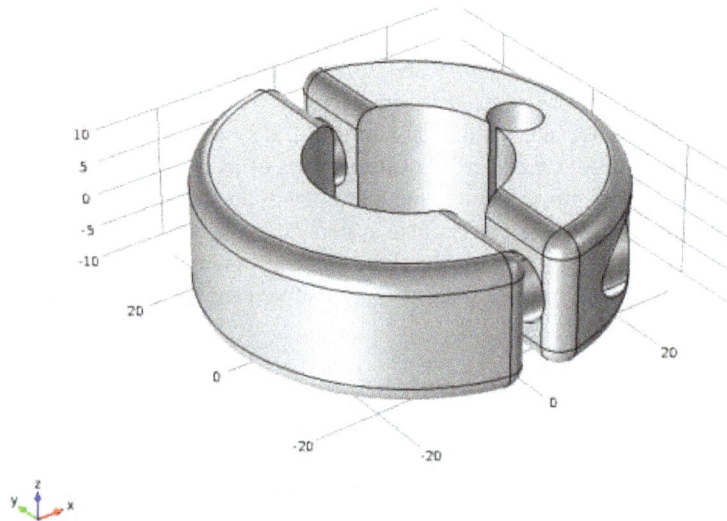

Figure 2. Compact structure design of high voltage.

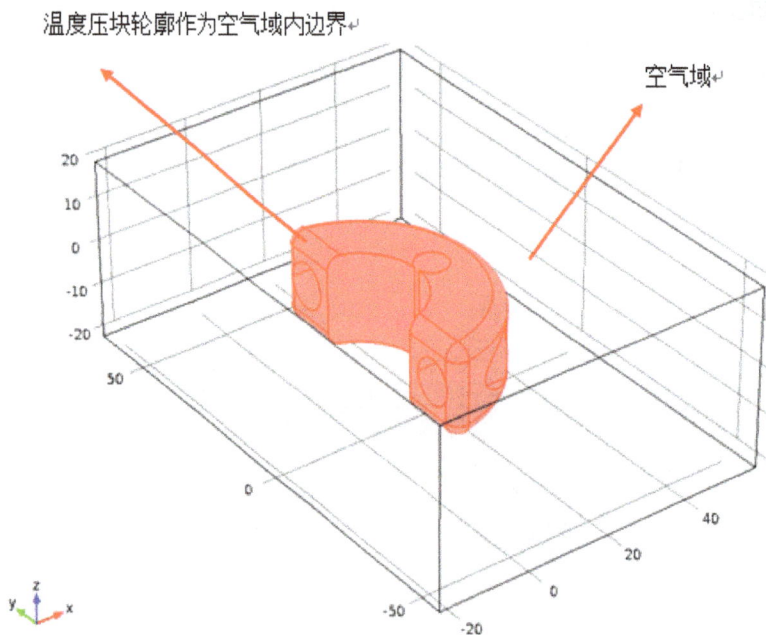

Figure 3. The compacts parcels by air domain.

Table 1. Scheme of B simulation parameter list.

simulation parameter	Parameter values
Geometric parameters	*With length* 80 mm, *width* 120 mm, 60 mm *high air field surrounded fixture. Air domain and the pressing block center symmetry.*
Material parameters	*air*
Meshing	*Using user-controlled mesh, the largest unit length* 2 mm, *the smallest unit of length* 1 mm, *cell growth rate of* 1.3. *Finally mesh degrees of freedom* 1,854,100.
Boundary Conditions	*Fixture inner contour of the air domain boundary, surface potential* 220 kV, *air domain boundaries are deemed to be grounded.*

4. Results and Analysis

4.1. Analysis of Sensor Modeling Scheme I of Fixture Results

A sensor jig modeling scheme potential distribution of the simulation results shown in **Figure 4**, we can clearly see the outline edge clamps outwardly from the 220 kV to zero potential (ground) of the uniform field in the ambient air.

The electric field distribution simulation results shows that the electric field concentration at the tip of the metal clamp surface corner. The surface potential of 220 kV case, the tip surface of the field about 5.32×10^7 v/m, and in the circular surface is uniform electric field distribution. So, in order to avoid the discharge temperature and humidity in the extreme case, fixture surface should be smooth.

4.2. Analysis of Sensor Modeling Scheme II of Fixture Results

The results of simulation of air field potential distribution scheme of II as shown in **Figure 5**, potential along the air domain boundary from 220 kV to 0 steady decline, and the scheme of simulation results similar to I.

The electric field distribution simulation results as shown in **Figure 8**, the electric field intensity around the right arc is about $1.81e^7$ - $1.87e^7$ V/m. It is much lower than the electric field of a rectangular boundary scheme of fixture design focus on the value, and the uniform electric field distribution, no surge spikes.

Comparison of the simulation results of plan I and II, the electric field concentration at the tip of the metal clamp surface corner. The surface potential of 220 kV case, the tip surface of the electric field is about 5.32×10^7 v / m, and in the circular surface is uniform electric field distribution. Peak electric arc chamfer is about $2.18e^7$ V/m - $2.36e^7$ V/m, and the rectangular clamp tip electric field peak compared to $5.32e^7$ V/m is reduced by about 50%, and the boundary of uniform electric field distribution. Arc pressing block is not easy to generate point discharge, optimization design.

In order to avoid the discharge temperature and humidity in the extreme case, fixture surface should be smooth [10]. The optimization scheme is to use arc chamfering, and as far as possible to ensure that the surface finish. Give extended application, low end fixture, especially from the wire clamp close, in order to avoid the tip electric caused by induction, the edges and corners also need passivation [11].

4.3. Results and Analysis of the Electric Field Distribution Simulation Model Assembly

This study is based 220 kV transmission line of the finite element model was established by the four kinds of mode, Model A is the original model, model B is the only changes bolt model, model C in the second model, based on the composite metal and the sensor connection portion is placed jig hole, model D is changed rounded 0.8 mm bolt hole wall model, outsourcing air cylinder modeled by a radius of 100 m.

Using ANSYS software based on a section of the model diagram of four models for the combination of data modeling. Model A is the original model; model B is changed bolt model; C model based on B model, the sensor metal and composite joints on the fixture hole; model of D is modified bolt hole wall fillet 0.8 mm model; the overall distribution of the electric field then the simulation of the four types of models, the whole field distribution results are shown in **Figures 6-9**.

From this point of view of the whole electric field distribution, models A and the electric field under the other three models are quite different, model B, model C and model D is close to. The maximum field intensity model A appeared on the ends of the bolts is 64.19 kV/cm, maximum field strength, model B, model C and model D is

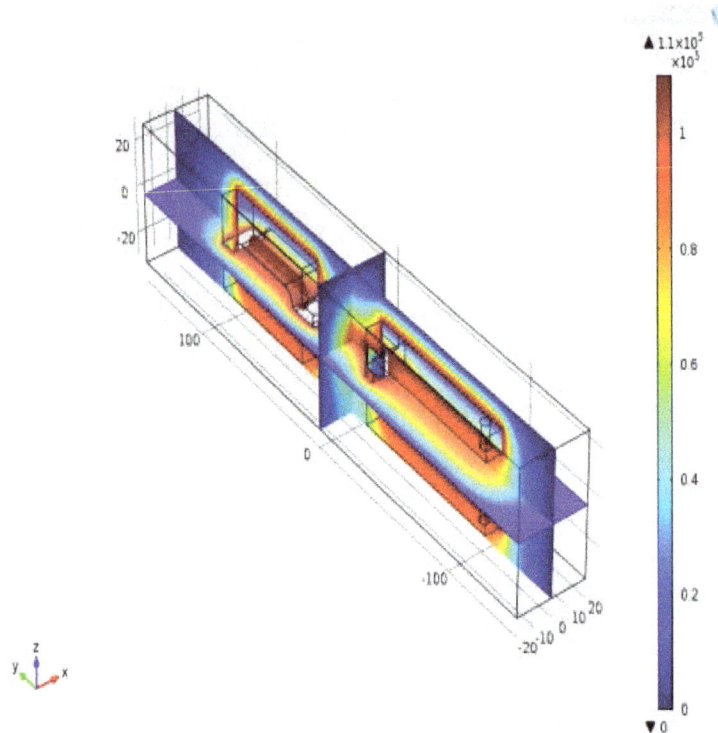

Figure 4. Distribution of the air domain potential surrounding the fixture in scheme I.

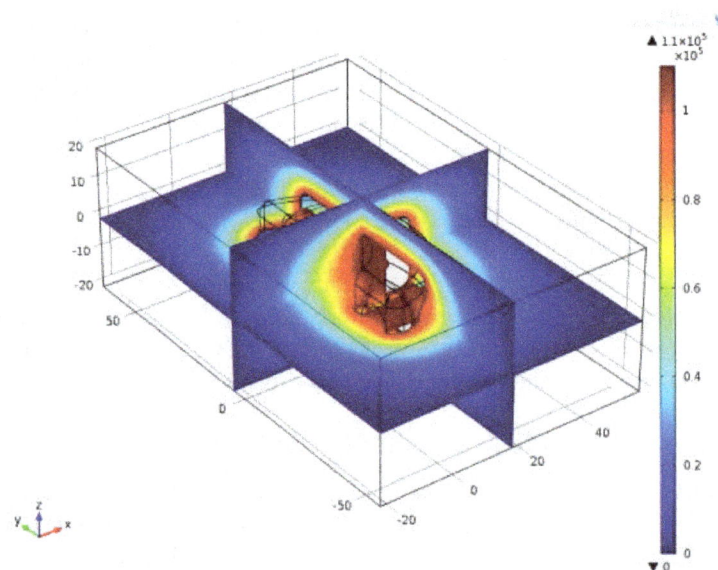

Figure 5. Distribution of the air domain potential surrounding the fixture scheme in B.

appeared in the top corner of fixture, the maximum electric field strength model for B 28.14 kV/cm, the maximum field strength model for C 28.16 kV/cm, the maximum field strength model of D is 28.26 kV/cm, the maximum electric field strength of three kinds of model are very small.

In order to compare the difference between different models of the maximum field intensity, listed in **Table 2** field in different parts of the different model maximum.

As can be seen from the **Table 2**, the potential of the composite section in both ends of strain gauge uneven

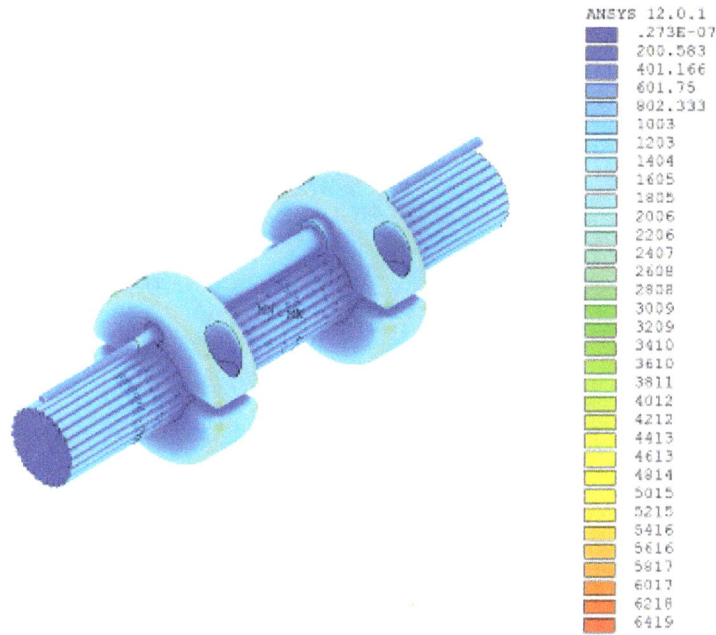

Figure 6. Model A whole field distribution.

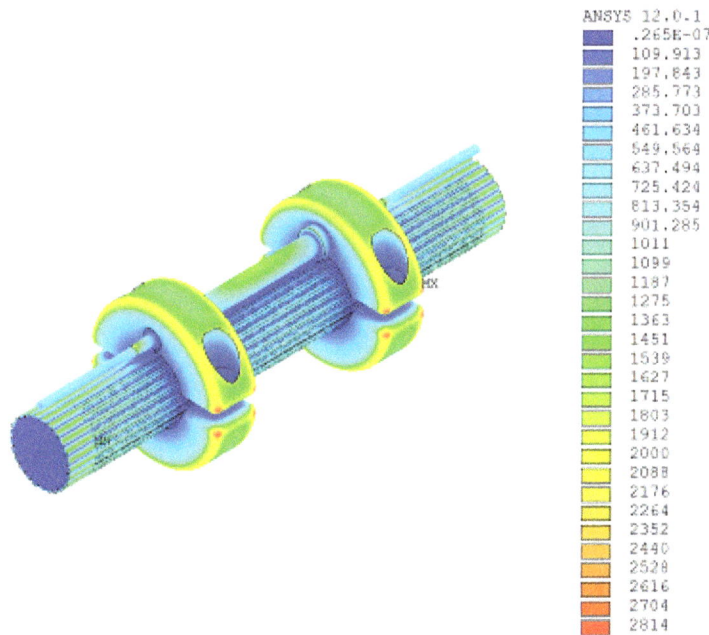

Figure 7. Model B whole field distribution.

Table 2. Different models of different parts of the maximum value.

	Model A	Model B	Model C	Model D
The maximum electric field value (kV/cm)	64.19	28.14	28.16	28.26
Conductor surface field maximum (kV/cm)	15.77	15.74	15.77	15.82
Clamp the maximum surface electric field (kV/cm)	27.64	28.14	28.06	2 8.26
The maximum strain gage surface electric field (kV/cm)	21.13	21.75	16.65	15.94

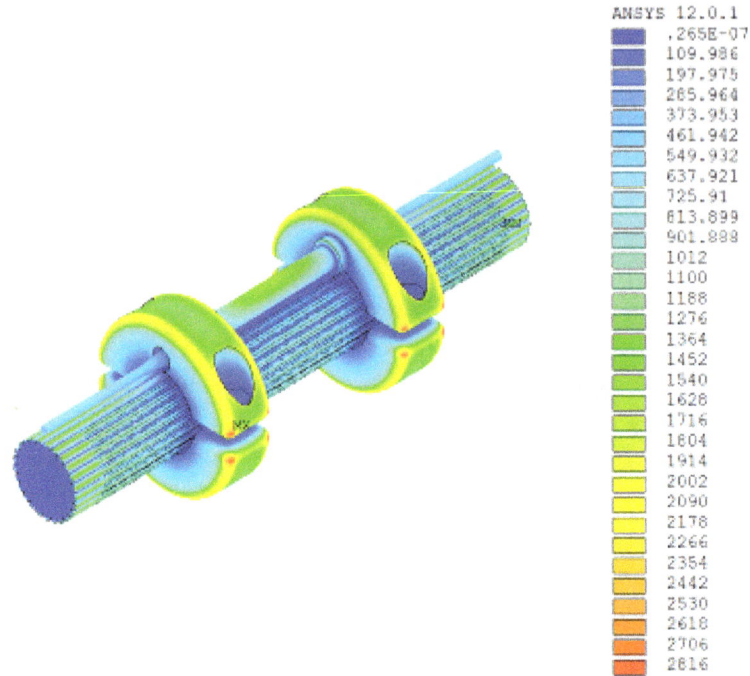

Figure 8. Model C whole field distribution.

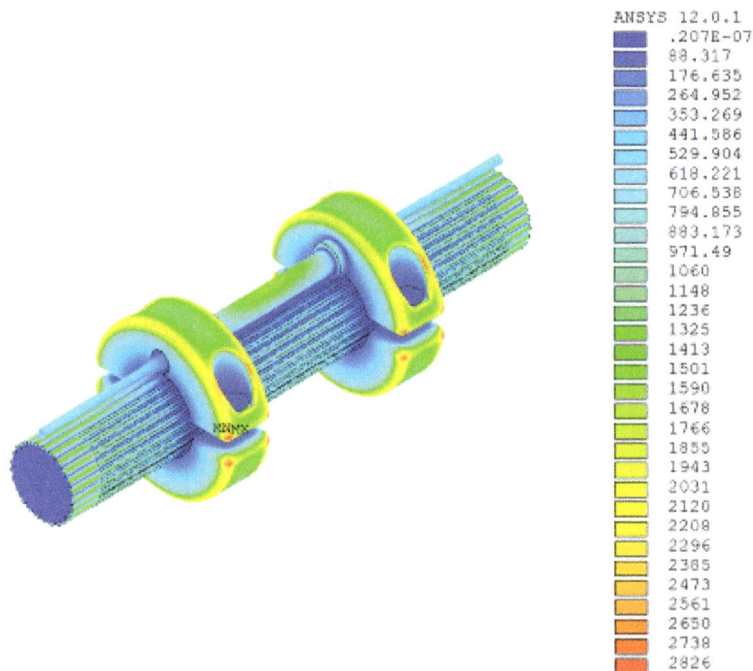

Figure 9. Model D whole field distribution.

distributed in the four models. The potential is highest at the part connect with the metal, and lowest at both ends. The overall integral maximum field value in model one was significantly higher than the other three models. At the same time, the maximum field strength value of strain gauges in model one and model two was greater than the other two models. Comprehensive comparison of the four models, proposed to shorten the clamp bolt to place the end portion inside the clamp in order to avoid excessive field strength appears. The fillet radius of clamp portion can be appropriately enlarged to reduce the surface field strength [12].

5. Summary

Through the simulation and analysis and comparison of the two modeling programs of sensor fixture, that the electric field distributed more concentrated at the corner of the tip of the metal clamp's surface, while it's more even at the arc surface. The surfaces of metal clamps should be as smooth as possible, do not appear at the angular [13], the optimal approach is to use curved chamfer, which can effectively avoid the discharge at the tips under the extreme temperature conditions. In the design of the model assembled, place the joints which connected with metal and composite material inside the clamps, to prevent the field strength is too large on the surface of the strain gauge. And the metal fixture can reduce its surface field strength by properly increasing the fillet radius, making the design of model assembly optimized.

References

[1] Wang, J.-Y., Fang, C.-E., Dai, Y.-S., Li, W. and Hou, X.-Y. (2007) Electric Field Calculation and Optimal Design of 35 kV Contactor Box. *High Voltage Engineering*, **33**, 63-65.

[2] Zhang, Y.-J., Li, X., Liu, H.-B. and Sun, H.-H. (2012) Calculation and Optimization of Electric Field for 220 kV Composite Hollow Insulators. *North China Electric Power*, **6**, 16-19.

[3] Yang, Y. (2013) Calculation and Analysis of Total Electric Field at Ground Level beneath 800 kV DC Transmission Lines With Out-of-Operational Negative Pole Line. *Power System Technology*, **37**, 1531-1535.

[4] Huang, D.C., Ruan, J.J., Chen, Y., Huo, F., Yu, S.F. and Liu, S.B. (2008) Study on the Voltage and Electric Field Distribution along Composite Insulator of 1000 kV AC Transmission Line. *Proceedings of the CSEE*, 52-57.

[5] Li, Y. and Zhu, L.-X. (2013) Main Insulation Electric Field Calculation and Analysis Software for Power Transformer. *Transformer*, **50**, 52-56.

[6] Jiang, X. and Wang, Z.Y. (2004) Calculation and Analysis for the Electric Field of Composite High Voltage Bushing. *High Voltage Engineering*, **30**, 17-21.

[7] Lin, X., Cai, Q., Xu, J.-Y. and Li, S. (2011) Large Scale of Electric Field Calculation of 1100 kV Disconnector Based on Domain Decomposition Method. *High Voltage Apparatus*, **47**, 1-6.

[8] Zhou, X.X., Lu, T.B., Cui, X., Zhen, Y.Z and Luo, Z.N. (2011) A Hybrid Method for the Simulation of Ion Flow Field of HVDC Transmission Lines Based on Finite Element Method and Finite Volume Method. *Proceedings of the CSEE*, **31**, 127-133.

[9] Huang, D.-C., Ruan, J.-J. and Liu, S.-B. (2010) Potential Distribution along UHV AC Transmission Line Composite Insulator and Electric Field Distribution on the Surface of Grading Ring. *High Voltage Engineering*, **36**, 1442-1447.

[10] Huang, D.-C., Wei, Y.-H., Zhong, L.-H., Ruan, J.-J. and Huang, F.-C. (2007) Discussion on Several Problems of Developing UHVDC Transmission in China. *Power System Technology*, **31**, 6-12.

[11] Sun, C.-H., Zong, W., Li, S.-Q., Peng, Y.-H. and Ren, W.-W. (2006) A More Accurate Calculation Method of Surface Electric Field Intensity of Bundled Conductors. *Power System Technology*, **30**, 92-96.

[12] Liu, G., Li, L., Zhao, X.J., Li, W.P., Li, B., Sun, Y.L., Ji, F. and Li, J.Z. (2012) Analysis of Nonlinear Electric Field of Oil-Paper Insulation under AC-DC Hybrid Voltage by Fixed Point Method Combined with FEM in Frequency Domain. *Proceedings of the CSEE*, **32**, 154-161.

[13] Zhen, Y.Z., Cui, X., Luo, Z.N., Lu, T.B., Lu, J.Y., Yang, Y., Liu, Y.Q., Han, H. and Ju, Y. (2011) FEM for 3D Total Electric Field Calculation near HVDC Lines. *Transactions of China Electrotechnical Society*, **26**, 153-160.

4

Impact of Parasitic Resistances on the Output Power of a Parallel Vertical Junction Silicon Solar Cell

Nfally Dieme, Moustapha Sane

Laboratory of Semiconductors and Solar Energy, Department of Physics, Faculty of Science and Technology, Cheikh Anta Diop University, Dakar, Senegal
Email: nfallydieme@yahoo.fr

Abstract

This paper describes the theoretical model for calculating IV-curve of parallel vertical silicon solar cells (SCs) based on solving diffusion-recombination equation for such SC, which was suggested that two IV curve zones (those which are close to the short current and open circuit points) can be linearized. This linearalization allows obtaining the values of shunt (R_{sh}) and series (R_s) resistances. The evolution of the electric power based on these resistances was illustrated to show the values that shunt and series resistances must have to obtain a good efficiency.

Keywords

Series Resistance, Shunt Resistance, Power, Vertical Junction

1. Introduction

The depletion of fossil energy sources and the heavy pollution of the atmosphere push the researchers to find an alternative such as solar energy obtained by using solar panels. The strong demand for solar power requires a deep research to increase the efficiency of solar panels. It is therefore urgent for us to find anything that may weaken the smooth functioning of the panels. The performance of these panels can be known and improved by the study of certain electrical parameters such as parasitic resistances. These parameters are derived from the junction non-ideality, volume carriers recombination, current leakages, and, on the other hand, the material resistivity, metallic contacts and collection grids [1] [2]. The aim of this work is two-fold: to show an analytical approach of the measurement of parasitic resistances and their impact on the performance of the solar cell.

2. Theoretical Background

The parallel vertical junction silicon solar cell is presented in **Figure 1** [3].

We assume that illumination is made with polychromatic light, and is considered to be uniform on the $z = 0$ plane. The contribution of the emitter is neglected.

When the solar cell is illuminated, there are simultaneously three major phenomena that happen: generation, diffusion and recombination.

These phenomena are described by the diffusion-recombination equation obtained with [3] [4]:

$$\frac{\partial^2 n(x)}{\partial x^2} - \frac{n(x)}{L^2} = -\frac{G(z)}{D}. \tag{1}$$

D is the diffusion constant:

$$D = \mu \cdot \frac{K}{q} \cdot T. \tag{2}$$

With q as the elementary charge, k the Boltzmann constant; T is the average temperature prevailing in the material.

$G(z)$ is the carrier generation rate at the depth z in the base and can be written as [5] [6]:

$$G(z) = \sum_{i=1}^{3} a_i e^{-biz}. \tag{3}$$

a_i and b_i are obtained from the tabulated values of AM1.5 solar illumination spectrum. $n(x)$, L, τ, and μ are respectively the density of the excess minority carriers, the diffusion length, lifetime and mobility.

The solution to the Equation (1) is:

$$n(x) = \theta_1 \sinh\left(\frac{x}{L}\right) + \theta_2 \cosh\left(\frac{x}{L}\right) + \sum \frac{a_i}{D} L^2 e^{-biz}. \tag{4}$$

Coefficients θ_1 and θ_2 are determined through the following boundary conditions [7]:
- at the Junction ($x = 0$):

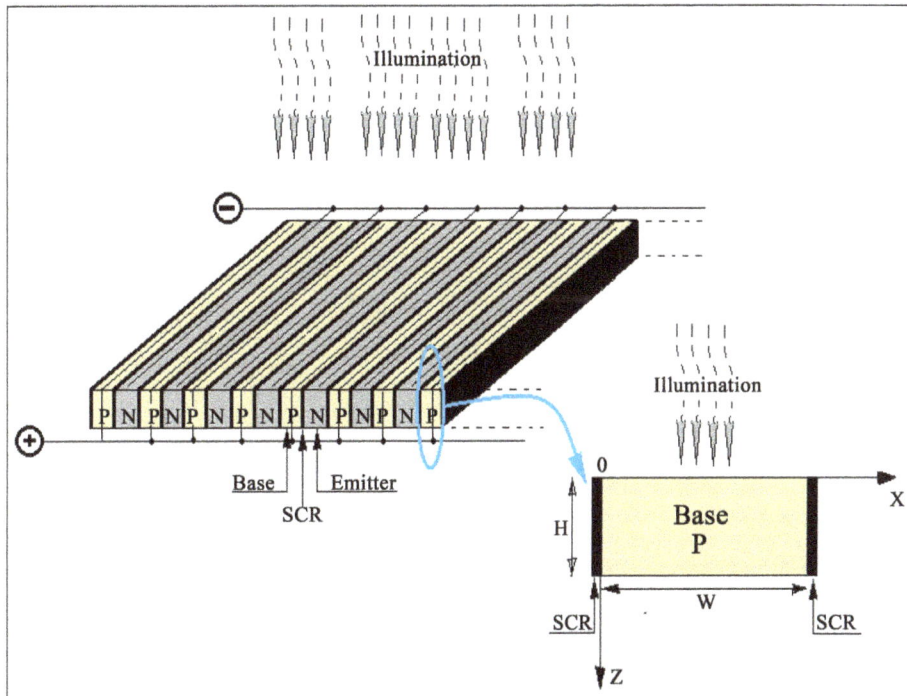

Figure 1. Vertical parallel junction solar cell ($H = 0.02$ cm; $W = 0.03$ cm).

$$\frac{\partial n(x)}{\partial x}\bigg|_{x=0} = \frac{S_f}{D} n(0). \tag{5}$$

This boundary condition introduces a parameter S_f which is called recombination velocity at the junction.
- in the middle of the base ($x = W/2$):

$$\frac{\partial n(x)}{\partial x}\bigg|_{x=\frac{W}{2}} = 0. \tag{6}$$

The photocurrent J_{ph} is obtained from the following relation [8] [9]:

$$J_{ph} = qD\frac{\partial n(x)}{\partial x}\bigg|_{x=0}. \tag{7}$$

The photo-voltage given by [8] [9]:

$$V_{ph} = \frac{k \cdot T}{q} \cdot \ln\left(N_B \cdot \frac{n(0)}{n_i^2} + 1\right) \tag{8}$$

with

$$n_i = A_n \cdot T^{\frac{3}{2}} \cdot \exp\left(\frac{Eg}{2KT}\right). \tag{9}$$

n_i refers to the intrinsic concentration of minority carriers in the base [10].

A_n is a specific constant of the material ($A_n = 3.87 \times 10^{16}$ for silicon), Eg is the energy gap and N_B is the base doping rate.

The current-voltage characteristic is illustrated by the **Figure 2** below.

This characteristic presents two very significant zones:

Area 1 is called the short-circuit operation point vicinity. The current-voltage characteristic in this area is illustrated by **Figure 3**.

It can be noticed that the current-voltage characteristic at the vicinity of the short-circuit operation point is a straight line:

$$J_{ph} = J_{sc} - \frac{1}{R_{sh}} \cdot V_{ph} \tag{10}$$

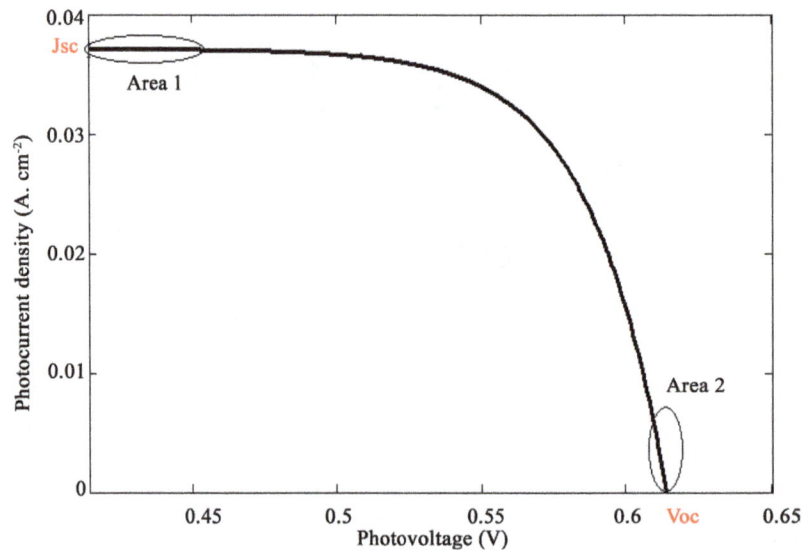

Figure 2. Current-voltage characteristic ($z = 10^{-3}$ cm).

with

J_{sc}: short-circuit current density.

R_{sh} is internal resistance, called shunt resistance.

From the Equation (10), the expression of R_{sh} is:

$$R_{sh} = \frac{V_{ph}}{J_{sc} - J_{ph}}.$$ (11)

Area 2 of **Figure 2** is called the open-circuit operation point vicinity. The corresponding characteristic is illustrated by **Figure 4** below.

The current-voltage characteristic at the vicinity of the open-circuit operation point is a straight line:

$$V_{ph} = V_{oc} - R_s \cdot J_{ph}$$ (12)

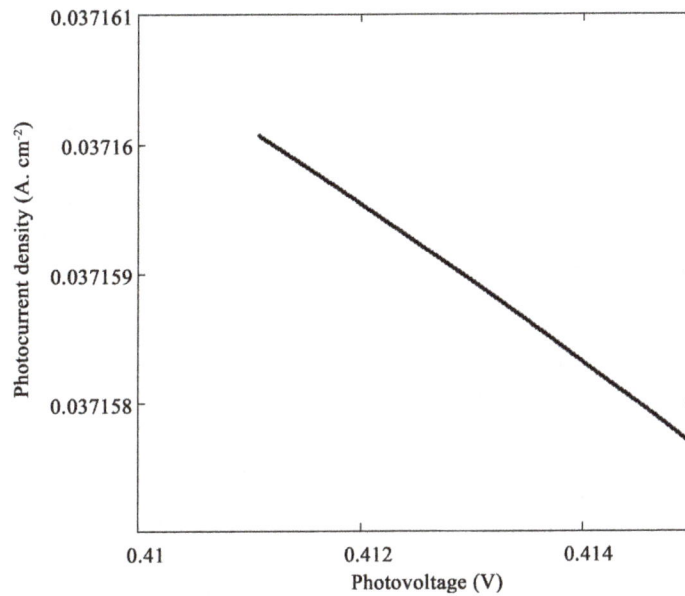

Figure 3. Current-voltage characteristic at the vicinity of the short-circuit operation point ($z = 10^{-3}$ cm).

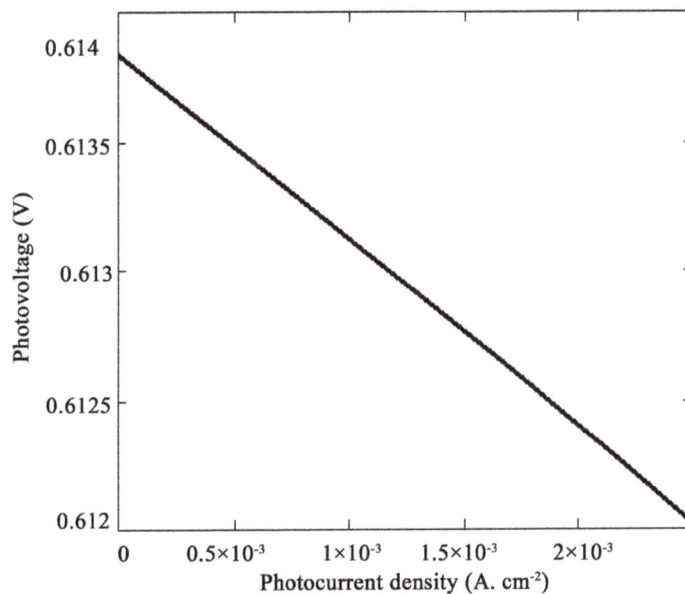

Figure 4. Current-voltage characteristic at the vicinity of the open-circuit operation point ($z = 10^{-3}$ cm).

with

R_s: is internal resistance, called series resistance.

V_{oc}: open-circuit photovoltage.

From the equation (12) the expression of the series resistance is:

$$R_s = \frac{V_{oc} - V_{ph}}{J_{ph}}.$$

(13)

The effects of these resistances are noticeable when the solar cell operates in one of the follow cases: in a short-circuit situation, R_s is negligible as compared to R_{sh}; and in an open-circuit situation R_{sh} is negligible in return [11].

3. Results and Discussion

This section is concerned with the profile of electrical power regarding parasitic resistances.

Figure 5 illustrates the profile of the output power regarding the shunt resistance.

It can be noticed that power increases along with shunt resistance (R_{sh}). The values taken by R_{sh} are high as shows it the scientific literature [12].

Indeed shunt resistance (R_{sh}) is an internal resistance that is maintained by the solar cell to be opposed to the current leakage around the junction. R_{sh} also explains the volume and surface recombination of carriers. R_{sh} is said to be high when several electrons cross the junction to be collected as photocurrent [12].

Thus volume and surface recombination and current leakage are all the lower as R_{sh} is high. Consequently the solar cell is powerful when shunt resistance is high as the curve in **Figure 5** shows.

The profile of the output power regarding the series resistance is illustrated by **Figure 6**.

It can be noticed that an increase in the series resistance causes decrease in the output power. The values taken by R_s are low. But they are in agreement with those which one finds in the scientific literature [12].

The series resistance (R_s) comes from the resistivity of the material used, from the metallic contacts and the collection grid. It also should be added that R_s emerges in a context of opposition to the diffusion of charge carriers to be collected as output-current at the junction. This same observation was also said in the scientific literature [2] where we learn that, when the output current increases the PN junction behaves as a resistance to oppose current.

High R_s decreases the flow of electrons which cross the junction. Consequently the solar cell is all the more powerful as R_s is low.

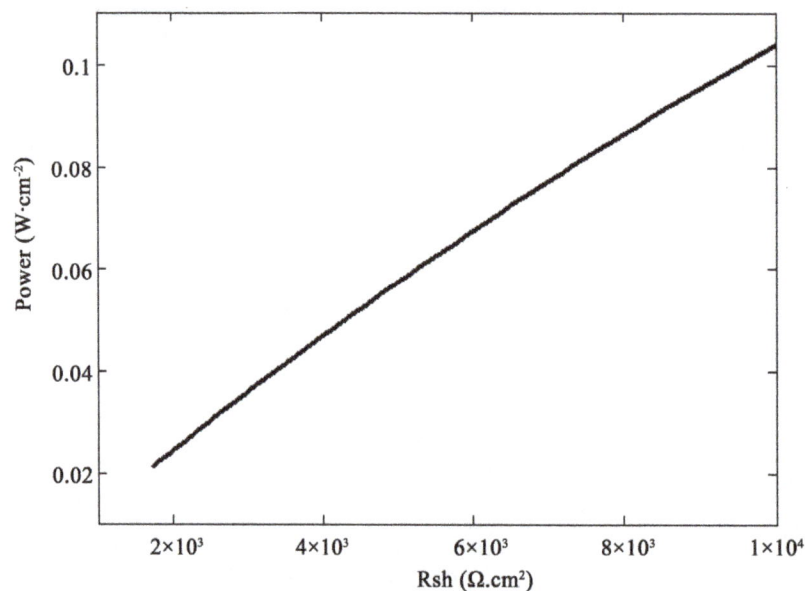

Figure 5. Electric output power versus shunt resistance ($z = 10^{-3}$ cm).

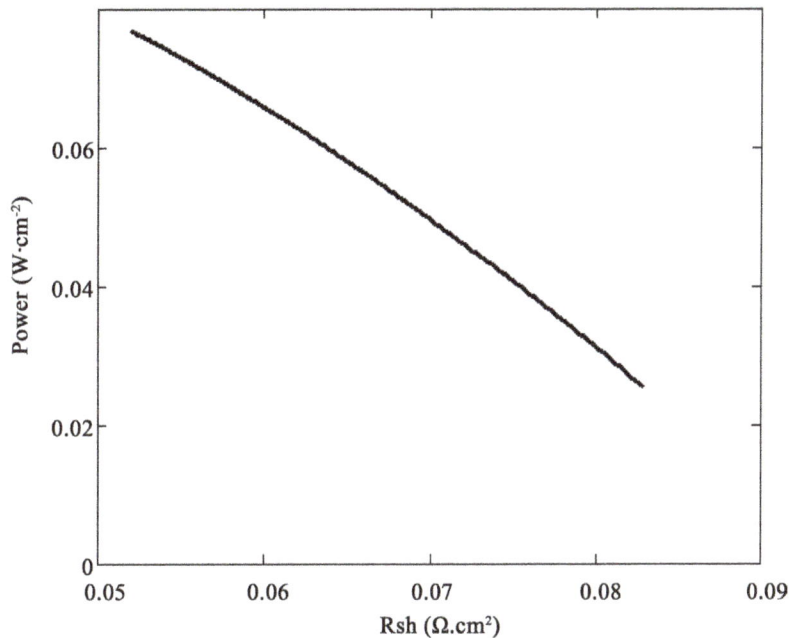

Figure 6. Electric output power versus series resistance ($z = 10^{-3}$ cm).

4. Conclusions

Through an analytical approach, the present work has allowed a measurement of the parasitic resistances (series and shunt resistances) of a vertical junction solar cell. This approach relies on linearalization of the curve through the short-circuit and open-circuit areas. It has thus shown that a solar cell is powerful when the series resistance is low and shunt resistance is high.

This work can be deepened by making a detailed study on the impact of parasitic resistances on other electrical parameters such as photocurrent, photovoltage and the diffusion capacity of solar cells.

References

[1]   Samb, M.L., Zoungrana, M., Sam, R., Dione, M.M., Deme, M.M. and Sissoko, G. (2010) Etude en modelisation à 3-d d'une photopile au silicium en régime statique placée dans un champ magnétique et sous éclairement multispectral: Détermination des paramètres électriques. *Journal des Sciences*, **10**, 23-38.

[2]   Bensalem, S. and Chegaar, M. (2013) Thermal Behavior of Parasitic Resistances of Polycrystalline Silicon Solar Cells. *Revue des Energies Renouvelables*, **16**, 171-176.

[3]   Dieme, N., Sane, M. and Barro, I.F. (2015) Photocurrent and Photovoltage under Influence of the Solar Cell Thickness. *ISJ Theoretical & Applied Science*, **7**, 1-6. http://dx.doi.org/10.15863/TAS.2015.07.27.1

[4]   Valkov, S. (1994) Electronique Analogique, Edition Castéilla, Collection A. CAPLIEZ.

[5]   Levy, F. (1995) Traité des matériaux 18: Physique et technologie des semi-conducteurs. Presses Polytechniques et Universitaires Romandes.

[6]   Dieme, N., Seibou, B., El Moujtaba, M.A.O., Gaye, I. and Sissoko, G. (2015) Thermal Behavior of a Parallel Vertical Junction Silicon Photocell in Static Regime by Study of the Series and Shunt Resistances under the Effect of Temperature. *International Journal of Innovative Science, Engineering & Technology (IJISET)*, **2**, 433-437.

[7]   Dieme, N., Zoungrana, M., Mbodji, S., Diallo, H.L., Ndiaye, M., Barro, F.I. and Sissoko, G. (2014) Influence of Temperature on the Electrical Parameters of a Vertical Parallel Junction Silicon Solar Cell under Polychromatic Illumination in Steady State. *Research Journal of Applied Sciences, Engineering and Technology*, **7**, 2559-2562.

[8]   Dieme, N. (2015) Study of the Electrons Density in the Base of the Parallel Vertical Junction Solar Cell under the Influence of the Temperature. *American Journal of Optics and Photonics*, **3**, 13-16. http://dx.doi.org/10.11648/j.ajop.20150301.13

[9]   Diallo, L.H., Dieng, B., Ly, I., Dione, M.M., Wereme, A., Ndiaye, M. and Sissoko, G. (2012) Determination of the Recombination and Electrical Parameters of a Vertical Multijunction Silicon Solar Cell. *Research Journal of Applied*

Sciences, Engineering and Technology, **4**, 2626-2631.

[10] Sze, S.M. and Ng, K.K. (2007) Physics of Semiconductor Devices. 3rd Edition, John Wiley & Sons, Hoboken.

[11] Schroder, D.K. (2006) Semiconductor Material and Device Characterization. 3rd Edition, John Wiley & Sons, Inc., Hoboken.

[12] Barro, F.I., Gaye, S., Deme, M., Diallo, H.L., Samb, M.L., Samoura, A.M., Mbodji, S. and Sissoko, G. (2008) Influence of Grain Size and Grain Boundary Recombination Velocity on the Series and Shunt Resistances of a Polycrystalline Silicon Solar Cell. *Proceedings of the 23rd European Photovoltaic Solar Energy Conference,* 612-615.

5

Analysis of Causes and Actual Events on Electric Power Infrastructure Impacted by Cyber Attack

Hongxu Yin[1], Rui Xiao[2], Fenfei Lv[1]

[1]The Power Company of Dezhou, Shandong, Dezhou, China
[2]College of Mechanical & Electrical Engineering, Jiaxing University, Jiaxing, China
Email: hongxu_yin@126.com

Abstract

With the development of electric power technology, information technology and military technology, the impact of cyber attack on electric power infrastructure has increasingly become a hot spot issue which calls both domestic and foreign attention. First, main reasons of the impact on power infrastructure caused by cyber attack are analyzed from the following two aspects: 1) The dependence of electric power infrastructure on information infrastructure makes cyber attack issues in information field likely to affect electric power field. 2) As regards to the potential threat sources, it will be considerably profitable to launch cyber attacks on electric power infrastructure. On this basis, this paper gives a classified elaboration on the characteristics and the possibilities of cyber attacks on electrical infrastructures. Finally, the recently published actual events of cyber attacks in respect of threat sources, vulnerabilities and assaulting modes are analyzed and summarized.

Keywords

Cyber Attack, Electric Power Infrastructure, Information Infrastructure, Dependence

1. Introduction

Electric power infrastructure provides public power supply services for social production and household use. Recently, with the development of computer and communication technology, information and communication systems become an essential part to support the normal operation of electric power infrastructure. According to the relevant standards set by International Electro-technical Commission (IEC), the future development of power system will be the common construction and management of electric power infrastructure and information infrastructure [1].

Cyber attack takes advantage of network vulnerability and security flaw to attack on system and resources [2]. For power system, monitoring and communication equipment of electric power infrastructure are the direct target of cyber attack. But because of the electric power infrastructure's dependency on information infrastructure, cyber attack will affect the safe and and stable operation of power system [3] [4]. In view of the importance of electric power infrastructure and cyber attack's characteristics of low cost, wide range and hidden action [5], cyber attack will become a potential threat that we may not neglect [4] [6]-[8].

Currently, many countries, mainly the U.S., have raised the impact of cyber attack on electric power infrastructure to nation-state level. The Homeland Security Department（DHS）has organized three cyber storm exercises since 2006 to 2010, and electric power infrastructure are significant imaginary target [9]. An Israeli-US strike on Iranian nuclear plants [10] becomes the focus of the international concern in 2011. According to this development trend, cyber attack on electric power infrastructure will be an important means of network information war [11], which needs to be paid high attention and intensive study.

This paper is organized focusing on the impact on power infrastructure caused by cyber attack. And the background and current situation of the problem are discussed to draw attention of national related departments, enterprises and research institutes. This paper analyses main reasons of the impact on power infrastructure caused by cyber attack, and summarizes characteristics of existing cyber attack. Furthermore, the impact of recently published actual events of cyber attacks on electric power infrastructure is analyzed.

2. Main Reasons of the Impact on Power Infrastructure Caused by Cyber Attack

2.1. The Dependence of Electric Power Infrastructure on Information Infrastructure

According to the IEC standards above, the power system can be divided into two parts, electric power infrastructure and information infrastructure. Electric power infrastructure, often called power primary system, is composed of power generation, transmission, transformation and distribution equipment, and its function is to lower the voltage of electrical energy generated step by step by transmission and transformation equipment and then transfer the electrical energy to distribution system. At last the electrical energy is supplied to customers through distribution line. Information infrastructure, usually called power secondary system, is composed of power monitoring system, power telecommunication, data network, etc. Power monitoring system are the business processing systems and smart devices based on computer and network technology, whose function is to monitor and control the production and operation process of the power grid and power plants [12].

The construction and development of information infrastructure are natural products of the development of communications and information technology, and also the objective requirements of the development of power system to a certain degree, aiming to improve the safety and economy of power system operation.

Interdependence of infrastructures is the interrelation or influence of the unions of 2 infrastructures. Generally, this dependence can be divided into 4 types [13]. The first type is physical dependence, which refers to the physical dependence existed between the unions of 2 infrastructures in the form of material flow. The second one is information dependence, which refers to the information dependence existed between the unions of 2 infrastructures in the form of information flow. The third one is geographical dependence, which means that the unions of 2 infrastructures are adjacent geographically and the incident in this site may affect both of the infrastructures. The last one is logic dependence, which means that the dependence exists between the unions of 2 infrastructures, but it can't be attributed to any of interdependence relationships discussed above.

Cyber security is the negative factors existed in information infrastructure, and severity of its influence on electric power infrastructure depends on the electric power infrastructure's dependence on information infrastructure.

Generally, the units of electricity power infrastructure can be divided into three classes, namely plants, substations, and lines, which constitute the entire electricity network. Information infrastructure, as it provides service to electric power infrastructure, exists in power plant monitoring system and substation automatic system corresponding to the units above. These systems mainly satisfy the units' own needs with a relatively small service scope, and they can realize local automation function, such as protection and operation and control functions based on station control level.

From a global perspective, the operating state of power system is adjusted based on the variation of load. So it is necessary to acquire global monitoring information and adjust generator's output power reasonably according to the variation of load to ensure safety and economic operation of the system. These functions are based on local function, and realized by supervisory control and data acquisition/energy management system (SCADA/ EMS).

The growth of load and scale expansion of power grid proposes a higher request of safety, promoting the development of wide area measurement and protection control system. The realization of wide area measurement and protection control system is also based on the collection, analysis and control of global information. Compared with SCADA/EMS, this requires more information and higher real-time character [14]-[16]. To satisfy the demand for uninterrupted power supply, it is essential to ensure the safety operation or instant recovery of information infrastructure when suffering internal fault or external attack [17]. The impact of information infrastructure on electric power infrastructure will be negative if the monitoring can't be conducted real-time or being taken advantage of.

In summary, information infrastructure serves electric power infrastructure in power system. In turn, electric power infrastructure depends on information infrastructure. The dependence's working point is substation automatic system or power plant monitoring system, but is amplified through the function of SCADA/EMS and wide area measurement and protection control system.

2.2. The Concept of Cyber Attack and Motivation Analysis

2.2.1. The Concept of Cyber Attack

According to the concept in the field of information security, cyber attack takes advantage of loophole and security vulnerabilities existing in the network to attack on system and resources.

Information infrastructure serving the power system to a great extent belongs to typical industrial control system. Cyber attack specified to those systems is generally defined as attack on the behavior of computer-based industrial control system without permission, aiming at destroying or lowering the function of industrial control system [18].

According to the physical security and network security guidelines of substation established by IEEE, cyber attack specified to substation refers to invading substation through possible network and operating or interfering with electronic equipment [19]. Equipment operated or interfed with include digital relay protection devices, fault recorder, automatics, substation-control level computer, PLC and communication interfaces [6].

2.2.2. The Benefit of Cyber Attack

Motivation of cyber attack derives from the aggressor's increasing economic or political interests. Compared with the traditional physical attack, cyber attack features with hidden action, low cost and wide range [5]. Cyber attack can implement aggressive behavior from any access network point of information infrastructure without approaching physical device. An attacker needs only relevant knowledge of the network and not special funding. Once an attack on electric power monitoring equipment implemented, some primary system equipment operate, possibly resulting in power system cascading failure and has larger influence sphere.

2.2.3. Threat Sources of Cyber Attack

Combined IEC 62351 standard [20] and reported domestic and international cyber attack, threat sources of cyber attack includes the following types.

1) Industrial espionage

Industrial espionage is becoming a major threat in higher power system marketization countries. Competition between enterprises increases motivation of illegal acquisition of information. And it is possible to interfere the operation of the competitors' equipment and improve their earnings.

2) Cyber hackers

Hackers usually take the Internet as a major gateway to attack, and gain profit by destroying cyber security. The profit may be monetary, industrial, political, social or the curiosity that whether the challenge of intrusion network will be successful [21].

3) Viruses

Similar to hackers, viruses and worms are typical attack taking advantage of the Internet. However, some viruses and worms may spread to Internet-isolated system by embedding into software or removable storage device. These viruses include middle attacks viruses, spyware gaining power system data and other trojan horse.

4) Larceny

Larceny has the most immediate purpose, namely attacker taking something (such as equipment, data or knowledge) away without permission. Generally speaking, the main motivation is to obtain economic benefits. Under the smart grid environment, the interaction of grid operators and user is emphasized. To tamper energy metering data through cyber attack is likely to develop into a new way of stealing electric power [22].

5) Terrorism

Terrorism is a threat which has minimum probability of occurrence. But it may bring serious consequences due to the purpose of maximizing the physical, financial, social and political damage.

6) Military action

Due to its specificity, electric power infrastructure is always taken as a priority target for military action. For example, US Army used graphite bombs to destroy the power system in Gulf War and the Kosovo War [23]. In recent years, with the mode and destructiveness of cyber attack increasing, physical attack gradually fades out of sight. And electric power infrastructure has become a potential target of cyber attacks, drawing military's attention [11].

3. Cyber Attacks against Electric Power Infrastructure

3.1. Classified by the Location of Attack

Depending on different location of cyber attack, attack can be classified as local attack, remote attack and pseudo remote attack.

Local attack occurs in the LAN, and remote attack occurs outside the location of the network the target belongs to. Pseudo remote attack is an attack that internal personnel covering up the identity of their attacker gain necessary information about the target from local and attack from the external, causing external invasion phenomenon [24].

Local attack needs to be physically close to the target. But local electric power monitoring system (e.g. Substation Automation System) usually adopt perfect physical isolation means, so it is difficult for attacker to reach the target, let alone attack. Remote attack can invade computers in control center and other critical equipment through any network connected to the power of information and communication network, utilizing the system vulnerabilities and unsound security and confidentiality mechanisms. In comparison, remote attack is most likely to occur [25]. Pseudo remote attack needs spies in power company, so it is less likely to occur.

3.2. Classified by Attack Mechanism

According to the mechanism, attack can be classified as denial service attack, replay attack, middle attack and reprograming the device [26].

1) Denial service attack (Dos)

Denial service attack occurs when excessive communication resources are occupied by an attacker so that the resource is temporarily unavailable when user needs to access, effecting the availability of information. In attacks against electric power infrastructure, attacker keeps sending forged packets in the communication channel, making the normal communication between the control center and RTU unavailable. The control center can't receive the power terminal information transmitted by RTU, meanwhile the control information issued by the control center can't be delivered. If the grid is in a state of emergency at the moment, the consequences would be unimaginable.

2) Replay attack

For replay attack, it is necessary to monitor network information flow first, and identify information representing key actions. The information is sent back into the network at specific times to re-simulate the previous occurrence. In electric power infrastructure, attacker can identify and intercept breaker tripping control instruction by network monitoring and replaying this instruction when necessary, resulting in breaker malfunction.

3) Middle attack

Attacker' action is imposed between two communicating nodes, and it deceives the sender that it is a true recipients or the recipients that it is the real sender. In this way, the attacker can tamper, delete or insert arbitrary information between two communicating nodes. In electric power infrastructure, attacker can be a middleman between RTU and control center, which intercept emergency fault information sent by RTU and replace with normal or alarm information, so that system does not take action in case of failure. Besides, it intercept control instruction from control center and delete, modify or insert the instruction. After such an attack, the confidentiality and integrity of information is completely lost. If the attacker is well aware of the power system operating state, then this attack would be devastating.

4) Reprograming the device

If RTU, IED and other equipment are reprogrammed, the attacker can easily implant malicious Trojans, disrupting the normal operation of equipment. This kind of attack is not common at present. The reasons mainly lie in two aspects: a) Remote programming of equipment is not allowed in the design of system generally, and local invasion programming needs approaching the equipment. But if the equipment is available, direct physical damage will be easier than reprogrammed and the destruction is more severe; b) As the equipment is numerous and geographically distributed, reprogramming attack is not suitable to launch a massive attack, and the requirement of programming capabilities of attacker is higher.

With the development of technology, the likelihood of such an attack still exists. For example, IEC 61850 standard proposes that IED equipment can be remotely configured [27]. Under the background of digital substation widely used and open of remote configuration interface in the future, such an attack would be more practical.

4. Case Study of Cyber Attack against Electric Power Infrastructure

4.1. Cyber Attack Instance

In recent years, cyber attacks on the Internet have been very serious, but reports about cyber attacks against electric power infrastructure are relatively few. Overall, the power monitoring system suffering from cyber attacks are on the rise. Some studies suggest that the current reported cyber attacks are likely just the tip of the iceberg. Due to fear of responsibility and corporate image damage, as well as commercial competition and other issues, most companies are reluctant to report such incidents [28]. Some examples of cyber attacks against electric power infrastructure are summarized as follows.

1) In the 8 - 14 blackout in USA and Canada in 2003, worm hindered the recovery from power blackout in Ontario in Canada to normal power supply [29].

2) In March 2007, U.S. Department of Defense and Department of Homeland Security (DHS) conducted a cyber attack experiment, causing generator self-destruction. The experiment was undertaken by the Idaho National Laboratory (INL) of energy, and it simulated a cyber attack against the copy of Aurora plant's control system. The attack invaded SCADA system and changed the generator's operation trajectory. Then the generator was out of control and galloping, final smoking and damaged. DHS believes that this type of cyber attack, if taking large-scale coordinated control, can damage electric power infrastructure for months. And DHS is reluctant to disclose specific details of the simulation action [30] [31].

3) On October 16, 2007, cyber security expert Ira-Winkler published a paper on the Internet Evolution site entitled "How to take down the power grid". This paper points out that it is not difficult to conduct cyber attack against power information control system, and he entered into American power control system as early as ten years ago. Ira-Winkler and his team have been hired by a power company, conducting test and evaluation of the vulnerability of computer systems for power grid. They can damage the browser, intrude into the control network of the power plant and monitor the power production and distribution, thus simulating the effects on the normal operation of the electric power infrastructure. Ira-Winkler noted that they can not only enter into SCADA system, but also download the file record CIO and CEO via cyber attack [4].

4) U.S. Central Intelligence Agency (CAI) pointed out that the blackout, which occurred in 2005 and 2007 in Brazil, was due to hacker's cyber attack against the power control system [32].

5) In an on-line special report of U.S. Public Broadcasting Company (PBS), an interviewee signatured hacker claimed that he could make the grid collapsed by clicking on some buttons. It is important that hackers, vandals and terrorist attackers think it is possible, not the truth of this sentence. And this report reflects that they have targeting the power grid to attack [33].

6) In April 2009, Wall Street Journal quoted the saying of an unnamed national security official. He said that cyber spies in some countries had came into contact with U.S. power grid and installed malicious software tool used to shut down certain services. They also expressed the concern that a malicious hacker may attack during a crisis or war in the future [33], although this intrusion hadn't disrupt the normal operation of the power grid.

7) The Bushehr nuclear power plant in Iran was attacked by the computer worm Stuxnet on October 26, 2010. This worm invaded into the internal of the nuclear plant via USB device and destroyed the centrifuge in nuclear facilities. Meanwhile, the worm used fake data to deceive the operator through replay mode. Thus the virus invasion was undetectable to the operators. The virus is specific to Simatic WinCC SCADA system produced by Siemens company, which is used as industrial control systems in many critical infrastructures in China. Due to

the powerful function of Stuxnet virus, the team behind this virus must have superb professional technical staff and strong financial backing. The United States and Israel admitted that they developed the virus jointly soon [34].

A brief analysis of the actual events of cyber attacks mentioned above in respect of threat sources, vulnerabilities and assaulting modes is available according to the reports, though the relevant information is limited.

4.2. Threat Sources Analysis

From the perspective of threat source, the instances above can be analyzed as follows. Threat of instance 1 is derived from the worm. Instance 2 and 6 are associated with military action. Instance 3, 4 and 5 are typical hacker attack. Instance 7 is relatively complex. Due to its powerful destructive capability, stuxnet is not simple computer viruses and closely related to the national military behavior. Instances show that among the threat sources of cyber attacks, hacker, viruses and military action are the most typical. For in-depth analysis, there are many inherent relationships between these three threat sources. And these relations can't be clearly grasped from the current public information.

4.3. Assaulting Mode Analysis

After the text edit has been completed, the paper is ready for the template. Duplicate the template file by using the Save As command, and use the naming convention prescribed by your journal for the name of your paper. In this newly created file, highlight all of the contents and import your prepared text file. You are now ready to style your paper.

From the perspective of assaulting mode, instance 2, 5 and 7 are more likely replay attack. And instance 2 and 5 are possibly middle attack. Instance 1 and 6 are typically denial service attacks. The assaulting mode of instance 3 and 4 are not clear, and there exist multifold possibility.

According to the occurrence mechanism of the attacks, combined with instance analysis, qualitative comparisons are done according to the implementation difficulty and the severity of consequences (shown in **Figure 1**).

1) Denial service attack is easy to conduct as its various means and no need for the attacker having a deep understanding of the power system. Judging from the consequences, the influence of denial service attack is relatively small because it can't directly operate key equipments despite its huge influence on monitoring process.

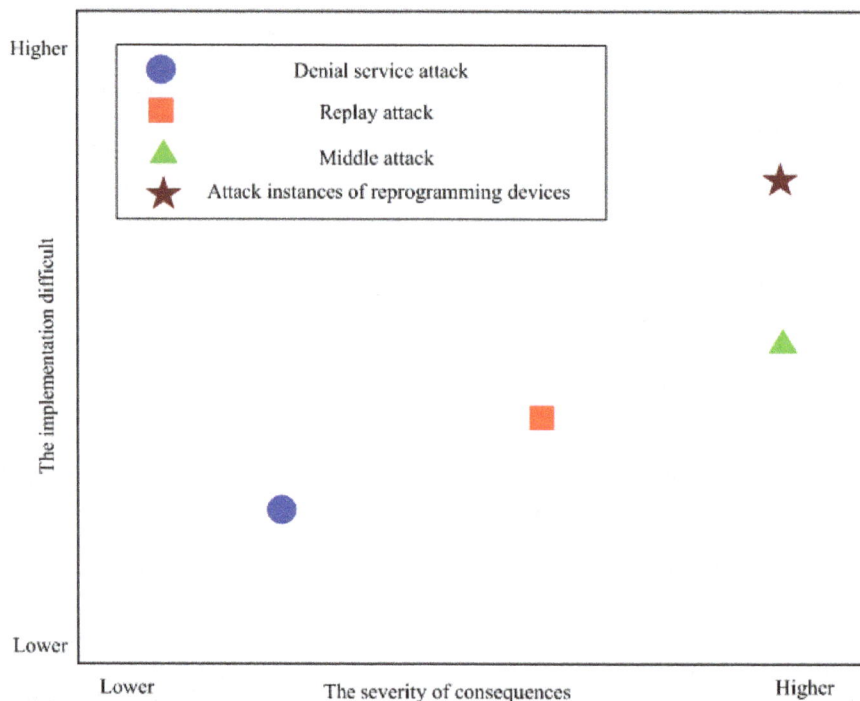

Figure 1. The comparison of impact and difficulty among the four cyber attack ways.

2) Replay attack and middle attack have much in common. And their essence is to achieve the purpose of maliciously operating critical equipment through attacker deceiving the recipient of information by certain means. But the difference between the two is that the replay attack has simpler means, and the goals able to achieve is less.

3) Attack instances of reprogramming devices haven't been found by far. But the occurrence of this kind of attack is possible in the future. Once succeed, the impact will be huge.

5. Conclusion

In summary, main reasons of the impact on power infrastructure caused by cyber attack are analyzed in respect of consequences and threats in this paper. The dependence of electric power infrastructure on information infrastructure determines the seriousness of the consequences of cyber attack. The main threats are diverse. The seriousness of the consequences improves the profit of cyber attacks, enhancing the motivation of attacks. Combined with the main assaulting mode of cyber attacks and currently reported attacks events, the impact of cyber attack on electric power infrastructure will gradually increase from the number and extent.

References

[1]   Cleveland, F. (2006) IEC TC57 Security Standards for the power system's Information Infrastructure—Beyond Simple Encryption. *IEEE PES TD* 2005/2006, 21-24 May 2006, 1079-1087.

[2]   Lin, C., Wang, Y. and Li, Q.L. (2005) Stochastic Modeling and Evaluation for Network Security. *Chinese Journal of Computers*, **28**, 1943-1956.

[3]   Niu, P.C., Kang, J.T., Li, A.W. and Li, L. (2010) New Operation Form of Power Network Started by Smart Grid. *Power System Protection and Control*, **38**, 240-244.

[4]   Li, W.W., You, W.X. and Wang, X.P. (2011) Survey of Cyber Security Research in Power System. *Power System Protection and Control*, **39**, 140-147.

[5]   Watts, D. (2003) Security & Vulnerability in Electric Power Systems. *IEEE 35th North American Power Symposium Conference*, Rolla Missouri USA, 20-21-19 October 2003, 559-566.

[6]   Lewis, J.A. (2011) The Electrical Grid as a Target for Cyber Attack.
http://csis.org/files/publication/100322_ElectricalGridAsATargetforCyberAttack.pdf

[7]   Kelic, A., Warren, D.E. and Philips, L.R. (2011) Cyber and Physical Infrastructure Interdependence.
http://prod.sandia.gov/techlib/access-control.cgi/2008/086192.pdf

[8]   Anderson, R.S. (2011) Cyber Security and Resilient Systems.
http://www.inl.gov/technicalpublications/Documents/4311316.pdf

[9]   Department of Homeland Security Office of Cyber Security and Communications National Cyber Security Division (2011) National Cyber Exercise: Cyber Storm. http://cryptome.org/cyberstorm.pdf

[10]   China News Network (2011) Iranian Officials Have Accused the USA and Israel Manufacture of Computer Viruses Destroy Iranian Nuclear Facilities. http://www.chinanews.com/gj/2011/04-17/2977981.shtml

[11]   Headquarters Department of the Army Washington, DC (2011) Infrastructure Risk Management (Army).
http://www.apd.army.mil/pdffiles/r525_26.pdf

[12]   State Electricity Regulatory Commission (2004) Power Monitoring System Security Requirements. Beijing.

[13]   Rinaldi, S.M., Peerenboom, J.P. and Kelly, T.K. (2001) Identifying, Understanding, and Analyzing Critical Infrastructure Interdependencies. *IEEE Control Systems*, **21**, 11-25. http://dx.doi.org/10.1109/37.969131

[14]   Liu, J.F., Chen, C.P. and Luo, J. (2004) Design and Application of Information's Security and Protection in Power Supervision and Control Automatic System. *Relay*, **32**, 33-35.

[15]   Wu, G.W. (2007) Information Disposal and Network Security Analysis in Digital Substation. *Relay*, **35**, 18-22.

[16]   Du, G.H. and Wang, Z.F. (2010) Design and Research on Power Network Dispatching Integration of Smart Grid. *Power System Protection and Control*, **38**, 127-131.

[17]   Stamp, J. and Mcintyre, A. (2009) Reliability Impacts from Cyber Attack on Electric Power Systems. *The* 2009 *Power Systems Conference and Exposition*, Seattle, 15-18 March 2009, 1-8.

[18]   Tatum, M. (2011) What Is a Cyber attack? http://www.wisegeek.com/what-is-a-cyberattack.htm

[19]   Substations Committee of the IEEE Power Engineering Society (2008) IEEE Guide for Electric Power Substation Physical and Electronic Security, IEEE Std 1402-2000(R2008). USA.

[20] IEC TS 62351-1 (2007) Power Systems Management and Associated Information Exchange—Data and Communications Security Part I: Communication Network and System Security—Introduction to Security Issues. Geneva.

[21] Li, J. (2011) Academician Disclose the Cyber Attack Is Very Serious in China.
http://society.people.com.cn/GB/1062/9343749.html

[22] Mu, L.H., Zhu, G.F. and Zhu, J.R. (2010) Design of Intelligent Terminal Based on Smart Grid. *Power System Protection and Control*, **38**, 53-56.

[23] Han, Y.J., Zhao, N.Q. and Liu, Y.C. (2005) Destructive Mechanism of Blackout Bomb and Its Defending Measures *Ordnance Material Science and Engineering*, **28**, 57-60.

[24] Zhang, Y.Q. (2011) Network Attack and Defense Technology. Tsinghua University Press, Beijing, 59-65.

[25] Baigent, D., Adamiak, M. and Mackiewicz, R. (2011) IEC61850 Communication Networks and Systems in Substations: An Overview for Users. http://www.sisconet.com

[26] Negrete-Pincetic, M., Yoshida, F. and Gross, G. (2009) Towards Quantifying the Impacts of Cyber Attacks in the Competitive Electricity Market Environment. *Bucharest Power Tech Conference*, Romania, 28 June-2 July 2009, 105-111.

[27] Liu, N., Zhang, J.H. and Duan, B. (2009) Design of Security Mechanism for Substation Remote Configuration Based on XML Security. *Electric Power Automation Equipment*, **29**, 113-117.

[28] Taylor, C., Krings, A. and Foss, J.A. (2011) Risk Analysis and Probabilistic Survivability Assessment (RAPSA): An Assessment Approach for Power Substation Hardening.
http://citeseerx.ist.psu.edu/viewdoc/download?doi=10.1.1.80.8323&rep=rep1&type=pdf

[29] Hu, Y., Xie, X.R., Han, Y.D., *et al.* (2005) A Survey to Design Method of Security Architecture for Power Information Systems. *Power System Technology*, **29**, 35-39.

[30] Greenberg, A. (2011) Congress Alarmed at Cyber-Vulnerability of Power Grid.
http://www.forbes.com/2008/05/22/cyberwar-breach-government-tech-security_cx_ag_0521cyber.html

[31] Meserve, J. (2011) Staged Cyber Attack Reveals Vulnerability in Power Grid.
http://articles.cnn.com/2007-09-26/us/power.at.risk_1_generator-cyber-attack-electric-infrastructure?_s=PM:US

[32] Poulsen, K. (2011) Report: Cyber Attacks Caused Power Outages in Brazil.
http://www.wired.com/threatlevel/2009/11/brazil/

[33] Oman, P., Schweitzer, E. and Roberts, J. (2011) Protecting the Grid from Cyber Attack Part I: Recognizing Our Vulnerabilities.
http://www.elp.com/index/display/article-display/130136/articles/utility-automation-engineering-td/volume-6/issue-7/features/protecting-the-grid-from-cyber-attack-part-i-recognizing-our-vulnerabilities.html

[34] Wikipedia (2011) Stuxnet. http://en.wikipedia.org/wiki/Stuxnet

6

A Comprehensive Analysis of Plug in Hybrid Electric Vehicles to Commercial Campus (V2C)

Andrew D. Clarke, Elham B. Makram

Department of Electrical and Computer Engineering, Clemson University, Clemson, USA
Email: adclark@g.clemson.edu, makram@clemson.edu

Abstract

Vehicle to grid is an emerging technology that utilizes plug in hybrid electric vehicle batteries to benefit electric utilities during times when the vehicle is parked and connected to the electric grid. In its current form however, vehicle to grid implementation poses many challenges that may not be easily overcome and many existing studies neglect critical aspects such as battery cost or driving profiles. The goal of this research is to ease some of these challenges by examining a vehicle to grid scenario on a university campus, as an example of a commercial campus, based on time of use electricity rates. An analysis of this scenario is conducted on a vehicle battery as well as a stationary battery for comparison. It is found that vehicle to campus and a stationary battery both have the potential to prove economical based on battery cost and electricity rates.

Keywords

Electric Vehicles, Power Distribution, Economic Analysis, Vehicle to Building (V2B), Battery Storage

1. Introduction

Plug in hybrid electric vehicles (PHEVs) are becoming more prevalent. Many studies seek to examine possible services these vehicles may provide to electric utilities. In order to understand the implications of introducing PHEVs to a distribution system, one must first define what constitutes classification of a PHEV. Hybrid electric vehicles (HEVs) use both an internal combustion engine and an electric system usually consisting of a battery and a motor to meet the requirements of propulsion. A vehicle must meet the following three requirements beyond those of a typical HEV in order to be classified as a PHEV by the IEEE [1].

1) A 4 kWh battery that is used for propulsion must be installed in the vehicle.

2) A connection to the electric grid must be present to recharge the battery.

3) A vehicle must be able to drive 10 or more miles solely using electric power.

Each PHEV used for the analysis presented in this paper meets these requirements with a battery capacity of 16.5 kWh and all electric range of 38 miles [2]. AC PHEV chargers are typically broken up into three categories. These categories are listed in [3] and summarized in **Table 1**. Based on the charging stations recommended for the PHEV used in this study by [4], only Levels 1 and 2 chargers are considered. A shorter charge time for the PHEV used in this study is listed when a Level 2 charger is used [2]. It is assumed that most consumers will choose the faster charge time if available, so a maximum charge/discharge rate is chosen to correspond to a Level 2 charger.

One of the services that PHEVs may be able to provide to an electric utility is peak shaving through vehicle to grid (V2G). In V2G, PHEV batteries are charged during off peak energy usage times and discharged during peak energy usage. One method of implementing this type of scheme involves using time-of-use (TOU) rates. TOU rates charge more for energy used during peak usage periods than for energy used during off peak usage periods [5]. Using TOU rates in a V2G scheme, a PHEV owner would charge the vehicle battery during off peak pricing periods and discharge the battery back to the electric grid during peak pricing periods.

V2G schemes typically require a communication network to be installed to allow utilities to communicate with PHEV chargers. Also, adding PHEVs utilizing V2G to a radial distribution system that typically sees only unidirectional power flows may create bidirectional power flows on the system. Both the additional communication network and bidirectional power flows lead to a high cost associated with V2G [6]. Furthermore, PHEVs are only capable of supplying a small amount of power in comparison to the overall system load of a large utility. Due to these factors, using PHEVs for peak shaving through V2G will likely face many challenges. Some existing studies attempt to conduct an economic analysis of V2G, however they either do not consider actual vehicle driving cycle data, fail to include the costs of negative impacts on the battery, or are conducted on large scale systems where the implementation of V2G is expected to be slow to gain traction [6]-[8]. Also, studies fail to show the breakeven point for PHEV owners in terms of TOU rates.

This research aims to implement V2G on a university campus system or other large commercial campus, henceforth referred to as V2C. This type of system will alleviate some of these concerns while still allowing the potential for both the consumer and utility to benefit. In the system being studied, which is the Clemson University distribution system, the campus is seen as a large net consumer by the electric utility, similar to a large commercial building. Therefore, a large number of vehicles can be connected before all of the campus loads would be supplied by sources other than the electric utility. This would eliminate the issue of two way power flows on the electric utility system caused by the campus or other large commercial building supplying power back to the electric utility. Also, by using TOU rates, a decentralized method of control is used to command vehicle charging and discharging in order to greatly reduce or eliminate the need for communication. The electric utility also benefits through a reduction of load power drawn by the university campus during peak energy usage times. The economic analysis on such a scheme is conducted to show at what point it becomes economically beneficial for PHEV owners to participate. An economic comparison with stationary battery energy storage is also conducted.

The structure of this paper is as follows: Section 2 contains details of spatial and temporal PHEV models used for this study; Section 3 contains specifics of the vehicle to campus algorithm; Section 4 has details and the results for the economic analysis; Section 5 shows the peak shaving impacts of the vehicle to campus algorithm; and Section 6 contains the final conclusion of this paper. Notation may be found at the end of the paper.

Table 1. PEV charging levels [3].

Charging Level	Supply Voltage	Maximum Current	Real Power
1	120 V, 1 ph	12 A	1.44 kW
2	208 - 240 V, 1 ph	32 A	6.66 - 7.68 kW
3	208 - 600 V, 3 ph	400 A	>7.68 kW

2. Vehicle Spatial and Temporal Distributions

In order to complete an accurate analysis of V2C, the locations of PHEVs during different times of the day are needed. Both temporal and spatial distributions of PHEVs must be determined. The spatial distribution is used to determine how far a PHEV has traveled when it arrives on campus from home and how far it must travel to return home. The temporal distribution is used to determine what time PHEVs arrive on campus and what time they leave to return home. In order to develop accurate distributions, data from the 2009 National Household Travel Survey is used along with the distribution fitting command in MATLAB [9]. For the spatial distribution, the responses to how far a worker's job is from home are used. These responses correspond to the variable DISTTOWK in [9]. Based on the responses, an exponential distribution with the probability density function shown in **Figure 1** is chosen for the distance driven between work and home. The average distance driven between work and home is determined to be 14.1 miles.

The times PHEVs arrive on campus as well as when they leave must also be approximated. For the temporal distribution of when vehicles arrive on campus, the responses to the end time of all trips with the destination of work are used. For the temporal distribution of when vehicles leave campus to return home, the responses corresponding to the start time of all trips with the destination of home are used. These responses correspond to the variables ENDTIME, STRTTIME, and WHYTO for the arrival time, leaving time, and purpose of the trip, respectively. Based on the responses, normal distributions with the probability density functions shown in **Figure 2** are chosen for the arrival and departure times. The average arrival time of PHEVs on campus is determined to be 9:07 am and the average departure time of PHEVs from campus is determined to be 3:23 pm.

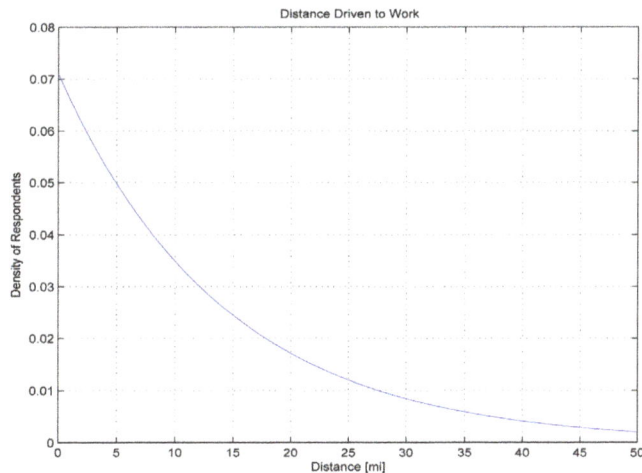

Figure 1. Spatial probability density function.

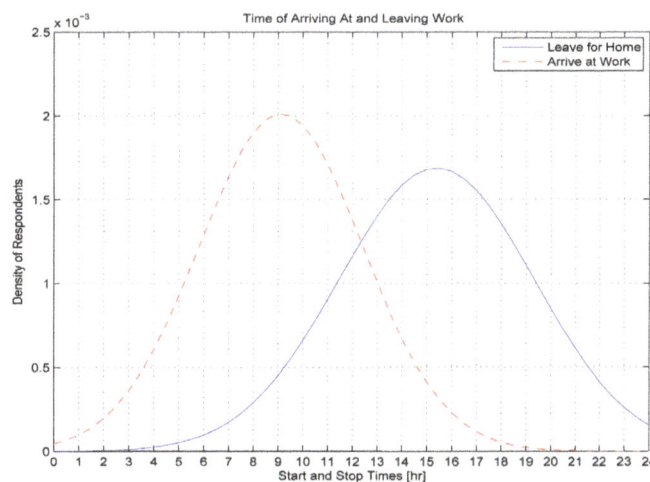

Figure 2. Temporal probability density function.

3. Vehicle to Campus Algorithm

In order to implement V2C, a decentralized algorithm is implemented. The algorithm requires the PHEV owner to estimate how far the vehicle will be driven after departure from campus and also input the time the PHEV will depart. The goal of the algorithm is to utilize V2C to supply as much energy as possible during the peak pricing period while also maintaining enough charge upon departure that the distance input by the PHEV owner can be traveled solely on electric power. The necessary state of charge (SOC) is calculated using the estimated distance input by the user as well as the 98 MPGe and 16.5 kWh battery capacity given in [2]. Any charging necessary is performed during off peak pricing if possible. The only time on peak charging would occur is the case where a vehicle is not connected long enough during off peak times to reach the necessary SOC required by the PHEV owner.

The algorithm attempts to charge the PHEV as much as possible before the on peak period, limited by a maximum charging rate of 7 kW, which is chosen based on a level 2 charger [3]. The algorithm also attempts to discharge the PHEV as much as possible during on peak periods, limited by maintaining enough energy in the battery at the time of departure and a maximum discharging rate of 7 kW. If the PHEV remains connected after the peak period ends, off peak charging is also allowed during this time. An SOC operating range of 10% - 89.2% is chosen based on 98 MPGe, 38 mile all electric range, and 16.5 kWh battery capacity [2]. An efficiency of 90% is assumed for both charging and discharging operations [8]. Charging and discharging rates are reduced from 7 kW wherever possible such that the rate is the lowest that will allow completion of the energy transfer in the allotted time.

For the economic analysis, it is assumed that the PHEV is driven the average of 14.1 miles to campus and arrives at 9:07 am. It is also assumed that the vehicle will be driven the average of 14.1 miles from campus to home upon departure at 3:23 pm. On peak pricing is assumed to start at 2:00 pm and to end at 8:00 pm [10]-[16]. Based on these parameters, the SOC profile of the PHEV battery while the vehicle is parked in campus is shown in **Figure 3**. The blue line in **Figure 3** represents the energy stored in the PHEV battery over time, starting at the arrival time and ending at the departure time. The vertical red lines show the beginning and end of the on peak pricing period.

4. Economic Analysis

In order to fully examine the economic analysis of a V2C scheme, a comparison with a stationary battery that has the same capacity as the PHEV battery is conducted. The same SOC operating range and efficiency for the PHEV battery are also assumed for the stationary battery. **Figure 4** shows the energy stored in the stationary battery throughout the day.

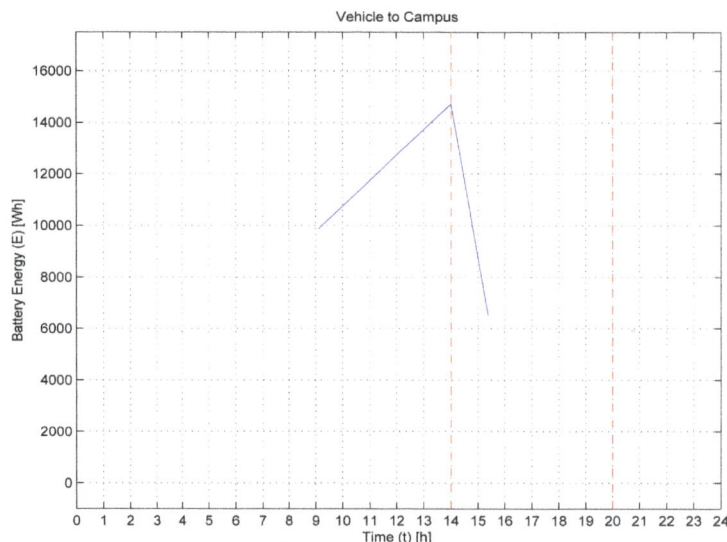

Figure 3. Energy profile of the PHEV battery throughout the day while parked on campus.

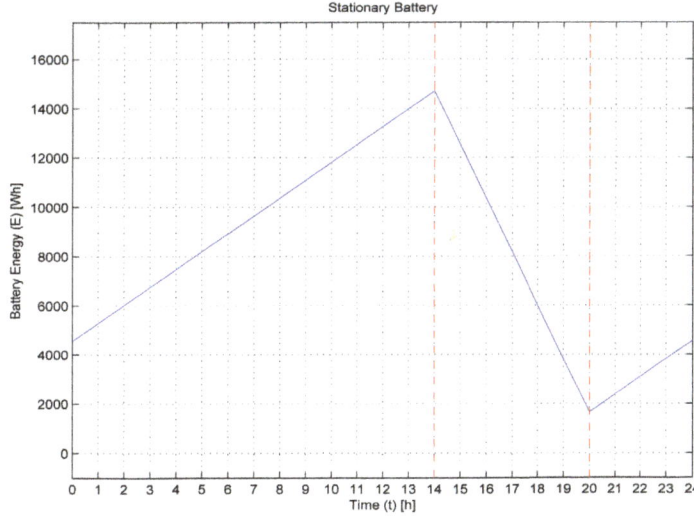

Figure 4. Energy profile of the stationary battery throughout the day.

To produce a meaningful comparison for V2C operation, a baseline cost for operating a PHEV without V2C must be considered. For the baseline, the cost of charging the PHEV only during off peak is considered. Also, only the energy necessary to drive to work and make the return trip home is necessary to consider as no other energy is depleted from the PHEV battery. Charging efficiency is also included in the calculation. The baseline cost is calculated using Equation (1). For operation without V2C, there is no accelerated degradation of the battery, so no additional cost to compensate for that is necessary.

$$Cost_{BL} = (E_W + E_H) * \text{Rate}_{Off} \tag{1}$$

Whenever V2C is employed, an accelerated degradation of the PHEV battery occurs compared to without V2C due to the increased cycling of the battery. After repeated cycling, a battery loses some of its usable storage capacity. For this study, a cycle life of 2500 cycles was chosen based on [2] [8]. This must be accounted for in order to give PHEV owners an accurate estimate of whether or not it is economical to participate in V2C. This cost is accounted for using Equation (2). This compensates the PHEV owner for a partial cycling of the battery based on how much energy is discharged from the battery for V2C operations [8].

$$Cost_C = \frac{Cost_B}{Cyc_R} * \frac{E_D}{E_B} \tag{2}$$

It is stated in [17] that the goal for battery price in PHEVs varies from \$150 - \$500 per kWh. For this study, three different battery costs are studied including the two extremes and one halfway between these estimates. These costs include \$150 per kWh, which corresponds to a battery cost of \$2475, \$325 per kWh, which corresponds to a battery cost of \$5362.50, and \$500 per kWh, which corresponds to a battery cost of \$8250.

The total cost of energy exchange during V2C is given by Equation (3). The total savings that are seen from using V2C are calculated using Equation (4).

$$Cost_E = E_C * \text{Rate}_{Off} - E_D * \text{Rate}_{On} \tag{3}$$

$$Sav_{V2C} = Cost_{BL} - (Cost_C + Cost_E) \tag{4}$$

The results are shown in **Figure 5** for the three battery costs. The axes show the off peak price, the price difference between peak and off peak prices, and the total savings. In **Figure 5(a)**, **Figure 5(c)**, and **Figure 5(e)** any points above the black surface represent an economic benefit from utilizing V2C while those below it represent a loss of money through utilizing V2C. **Figure 5(b)**, **Figure 5(d)**, and **Figure 5(f)** give an overhead view of these surfaces, where the magenta line represents the intersection with the black surface. All points above the line represent an economic benefit from utilizing V2C while those below it represent a loss of money through utilizing V2C.

It can be seen that as battery costs increase, the difference between peak and off peak price must increase in

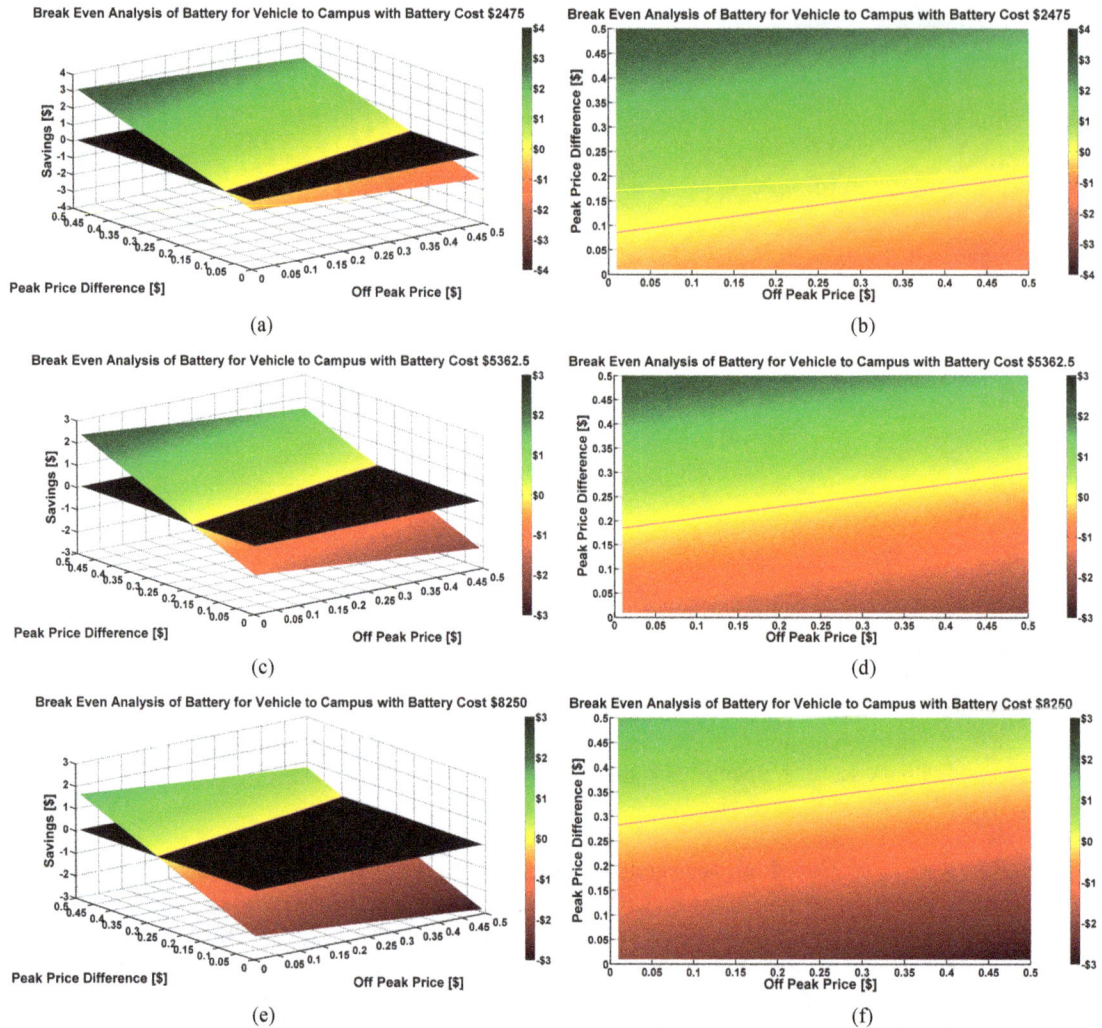

Figure 5. Break even analysis of vehicle to campus using the average cycle. (a) Battery cost—$2475; (b) Battery cost—$2475—overhead view; (c) Battery cost—$5362.50; (d) Battery cost—$5362.50—overhead view; (e) Battery cost—$8250; (f) Battery cost— $8250—overhead view.

order to realize an economic benefit from V2C. In all cases however, an economic benefit is possible if the difference between peak price and off peak price is large enough. It can also be seen that at the high off peak prices, the difference between peak and off peak prices must be larger than the low off peak prices in order to see an economic benefit. This is due to the charging and discharging efficiencies, both of which are below 100%. A PHEV owner may have to pay for energy losses, in which case this must be included in the analysis for accuracy. At high off peak prices, the cost of energy losses is higher than during low off peak prices.

A similar economic analysis is conducted for a stationary battery. **Figure 4** shows a stationary battery's SOC profile and the cycle compensation is again given by Equation (2). In this case however, ED represents the full usable battery capacity, EB. For the stationary battery, the savings are calculated using Equation (5).

$$Sav_{SB} = -\left(Cost_C + Cost_E\right) \tag{5}$$

The results of the economic analysis for the stationary battery can be found in **Figure 6**. Again, all points above the black surface or magenta line represent an economic benefit from utilizing stationary storage while those below represent a loss of money. It can be seen that the break even line for the PHEV battery and stationary battery is the same. This is due to using the same battery parameters for both comparisons. However, the amount of savings is different between the two analyses. This is due to the different charging and discharging profiles of the PHEV and stationary batteries.

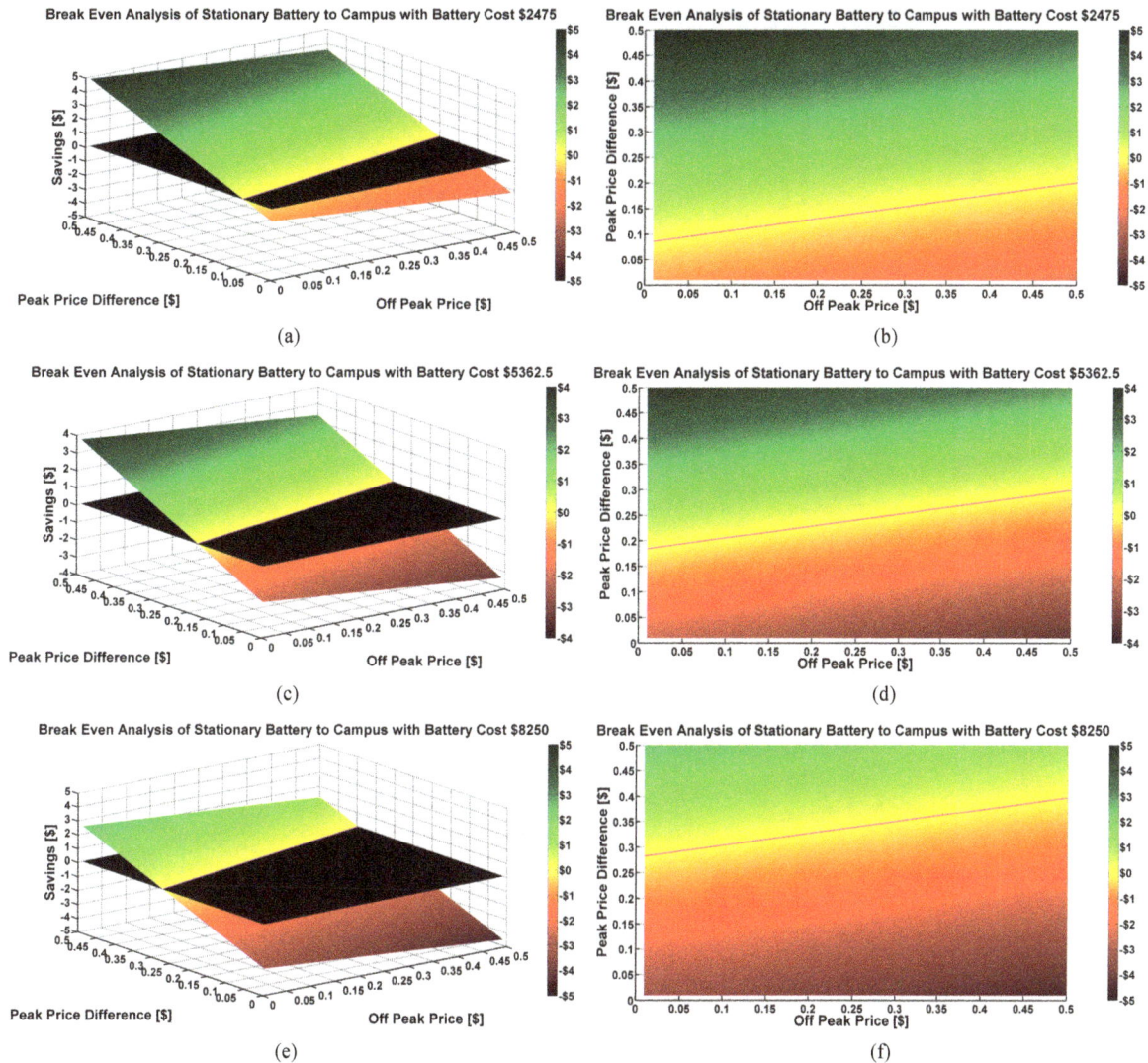

Figure 6. Break even analysis of stationary battery to campus. (a) Battery cost—$2475; (b) Battery cost—$2475—overhead view; (c) Battery cost—$5362.50; (d) Battery cost—$5362.50—overhead view; (e) Battery cost —$8250; (f) Battery cost—$8250—overhead view.

5. Peak Shaving Impacts

Based on the ability of V2C to prove profitable, it has been shown that V2C has the potential to benefit consumers without impacting required driving behavior. This section focuses on showing the benefit of V2C to the electric utility feeding campus. The V2C algorithm previously described works by charging PHEVs during off peak rate times and discharging them during peak rate times. The peak rate times correspond to the peak load values seen from campus loads. **Figures 7-9** show the campus load profile on the 12.47 kV circuit for a spring, summer, and winter day based on a base case simulation, respectively. These load profiles do not include V2C or stationary battery energy storage. It can be seen that the peak load for the spring day occurs between 2 and 3 pm. For the summer day, the peak load also occurs between 2 and 3 pm. The winter day's peak occurs between 12 and 1 pm. It can also be seen that the highest peak load occurs during the spring season.

In order to examine the impacts of adding PHEVs to the campus system while utilizing V2C, a total of 300 PHEVs are added in a total of 8 car parks around campus. The car parks are located based on currently available parking lots on campus. Six of the car parks contain thirty PHEV charging stations each, while the other two contain sixty PHEV charging stations each. The distance driven to campus, arrival time, and departure times are all chosen using random variables corresponding to the distributions described previously, with the constraint

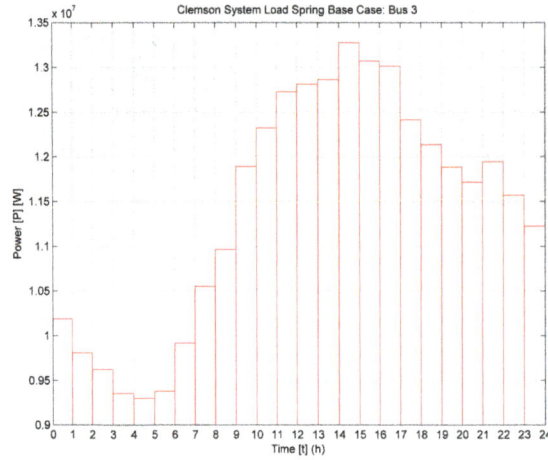

Figure 7. Clemson University 12.47 kV system load—spring-base case.

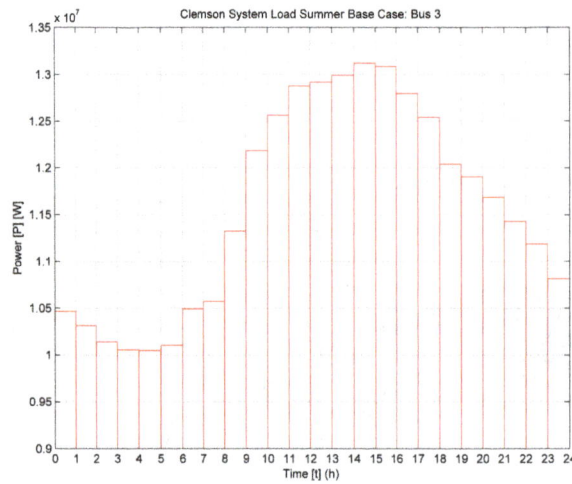

Figure 8. Clemson University 12.47 kV system load—summer-base case.

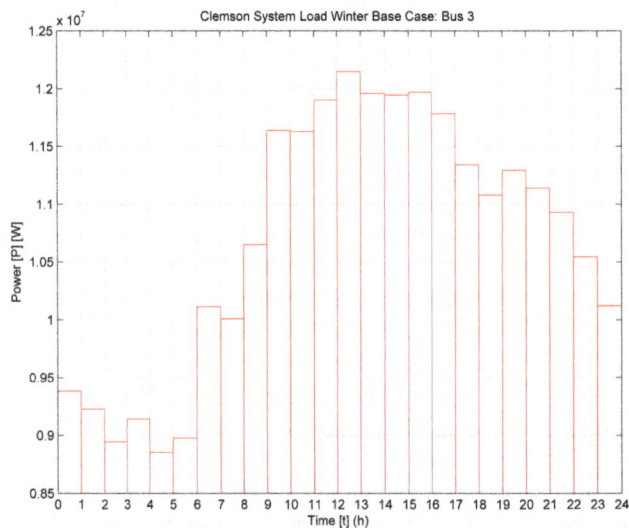

Figure 9. Clemson University 12.47 kV system load—winter-base case.

that departure time must be later than arrival time. Arrival and departure times are rounded to the nearest minute. Based on these parameters, the V2C algorithm is applied to each vehicle to determine the charge and discharge rates. In order to examine the maximum effects of peak shaving due to PHEVs, it is assumed for this section that the difference between on peak and off peak rates is high enough that V2C proves profitable. The charge and discharge rates from the V2C algorithm are then applied to the 300 PHEVs. Due to the arrival and departure times, not all PHEVs are present on campus during the peak rate period, meaning not all PHEVs participate in peak shaving. Based on the temporal distributions however, it can be seen that the majority of PHEVs that park on campus regularly will be present during the beginning of the peak rate period, with the availability decreasing as the time gets later.

Figures 10-12 show the campus 12.47 kV system load with PHEVs participating in V2C for a spring, summer, and winter day, respectively. It can be seen that the peak system load is reduced for both the spring and summer cases. For the winter case, the load is reduced during the utility peak pricing period, however this does

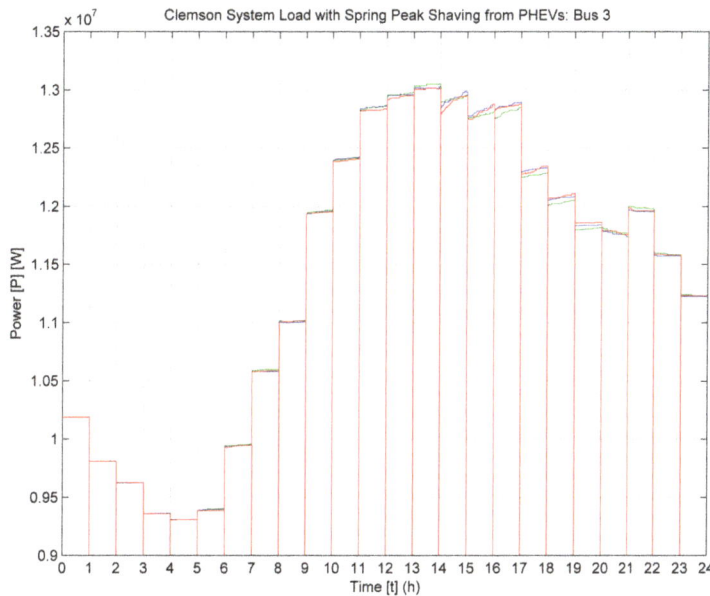

Figure 10. Clemson University 12.47 kV system load—spring-V2C.

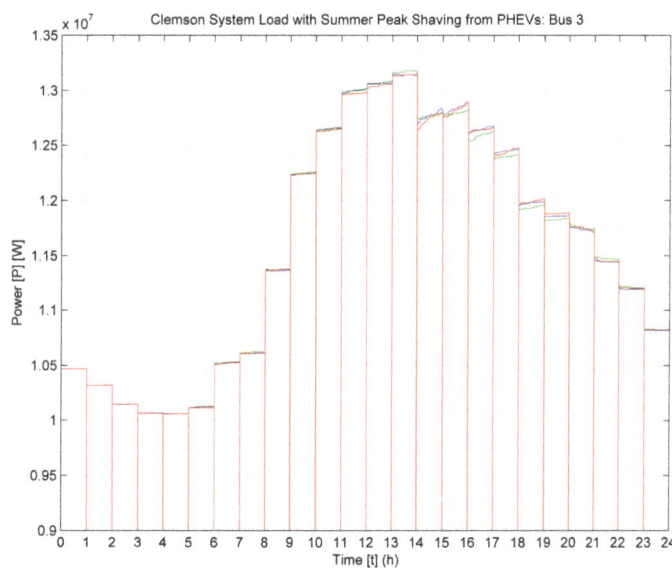

Figure 11. Clemson University 12.47 kV system load—summer-V2C.

not overlap with the system peak load, so the peak is not reduced. Due to the use of PHEVs in a V2C campus scenario, the electric utility providing electric power to campus can expect a significant reduction of campus load during the peak pricing periods. By drawing less energy during peak rate periods, the campus will also save money on the purchase of electricity.

In order to show the viability of using V2C as an acceptable source for peak shaving, a comparison with stationary battery energy storage is again conducted. In order to draw a meaningful comparison, each PHEV is replaced with an equivalent sized stationary battery of the same capacity at the same location. The profile of the stationary battery energy is shown in **Figure 4**. The load profile for the campus with stationary energy storage is shown in **Figures 13-15** for a spring, summer, and winter day, respectively.

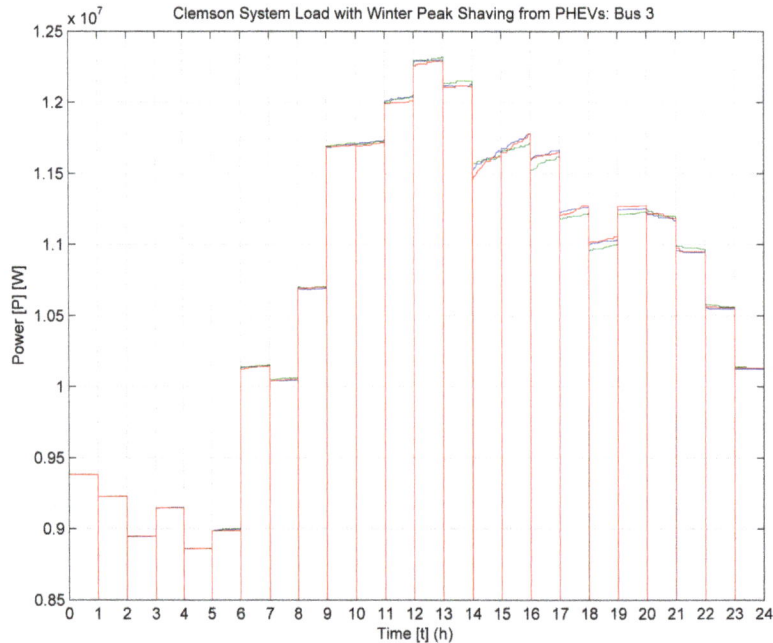

Figure 12. Clemson University 12.47 kV system load—winter-V2C.

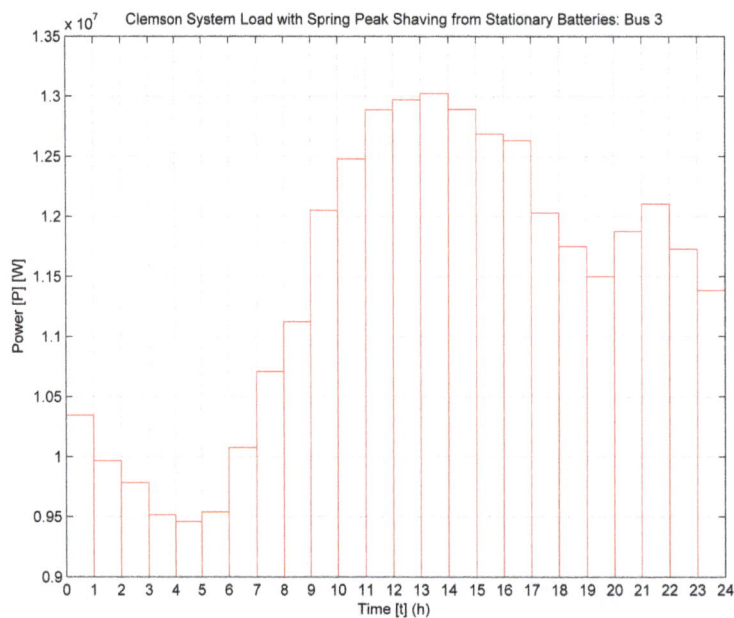

Figure 13. Clemson University 12.47 kV system load—spring-stationary battery.

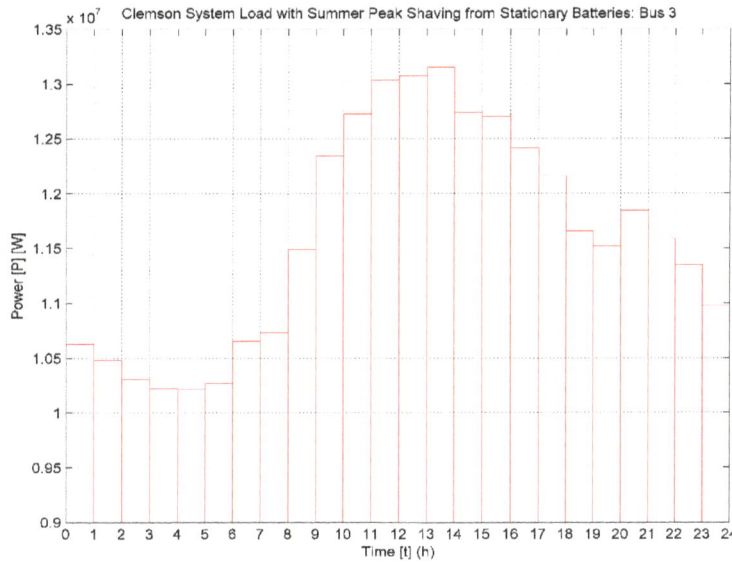

Figure 14. Clemson University 12.47 kV system load—summer-stationary battery.

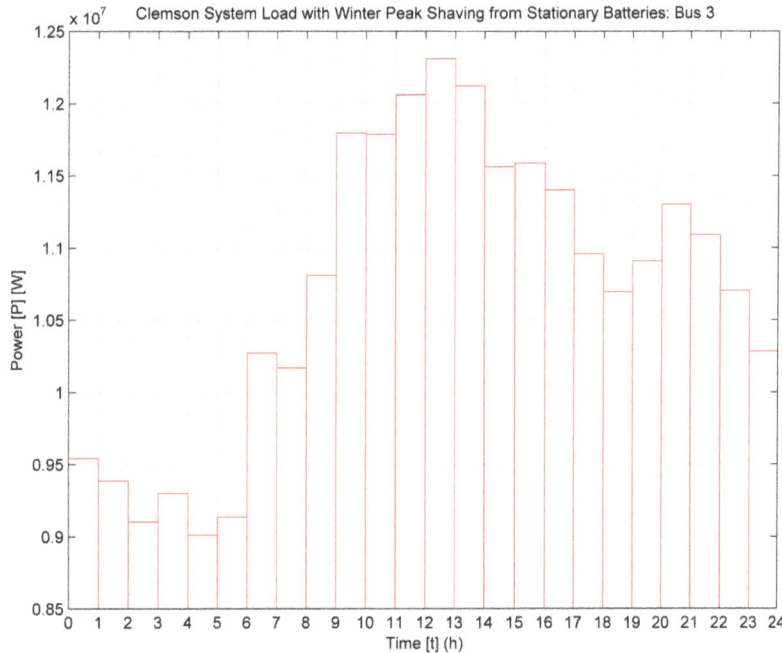

Figure 15. Clemson University 12.47 kV system load—winter-stationary battery.

It can be seen that the peak shaving impacts of the stationary battery energy storage are similar to that of the PHEVs using V2C. The main difference is that more peak shaving occurs with the stationary battery energy storage than with PHEVs using V2C. This is especially apparent during the latter portion of the peak rate time. This is due to the disconnection of PHEVs based on the temporal distributions and the minimum SOC requirement placed on individual PHEVs by the spatial distribution.

6. Conclusion

An economic analysis and peak shaving analysis for both V2C and stationary storage are conducted. It is shown that V2C and stationary storage can both be cost effective and allow for peak shaving on a university campus. The cost effectiveness and the amount of savings are dependent on the battery cost and electricity pricing during both on and off peak times. Under the V2C scenario, the PHEV owner is left with enough energy to travel a desired distance solely under electric power while still participating in V2C. Implementation of V2C reduces the load of the university campus, as seen by the electric utility, during peak usage times, thus benefiting the electric utility without the concern of two way power flows occurring on the utility system. While a stationary battery has the potential to save more than a PHEV battery, the break even point between the two is the same. Thus, utilizing V2C will allow a university campus to achieve similar benefits without the initial high cost and space requirements of a stationary battery.

Acknowledgements

The authors would like to thank the members of the Clemson University Electric Power Research Association (CUEPRA) for their financial support and valuable discussion.

References

[1]   Anumolu, P., Banhazl, G., Hilgeman, T. and Pirich, R. (2008) Plug-In Hybrid Vehicles: An Overview and Performance Analysis. 2008 *IEEE Long Island Systems, Applications and Technology Conference*, Farmingdale, 2 May 2008, 1-4. http://dx.doi.org/10.1109/LISAT.2008.4638946

[2]   Chevrolet (2014) 2014 Volt: Electric Car—Hybrid Car Chevrolet Web. http://www.chevrolet.com/volt-electric-car.html

[3]   Kisacikoglu, M.C., Ozpineci, B. and Tolbert, L.M. (2010) Examination of a PHEV Bidirectional Charger System for V2G Reactive Power Compensation. 2010 *25th Annual IEEE Applied Power Electronics Conference and Exposition (APEC)*, Palm Springs, 21-25 February 2010, 458-465. http://dx.doi.org/10.1109/APEC.2010.5433629

[4]   Bosch. (2014) Charging Stations Bosch Electric Vehicle Solutions Web. http:/www.pluginnow.com/charging_stations

[5]   Davis, B.M. and Bradley, T.H. (2012) The Efficacy of Electric Vehicle Time-of-Use Rates in Guiding Plug-In Hybrid Electric Vehicle Charging Behavior. *IEEE Transactions on Smart Grid*, **3**, 1679-1686. http://dx.doi.org/10.1109/TSG.2012.2205951

[6]   Yilmaz, M. and Krein, P.T. (2013) Review of the Impact of Vehicle-to-Grid Technologies on Distribution Systems and Utility Interfaces. *IEEE Transactions on Power Electronics*, **28**, 5673-5689. http://dx.doi.org/10.1109/TPEL.2012.2227500

[7]   Pang, C., Dutta, P. and Kezunovic, M. (2012) BEVs/PHEVs as Dispersed Energy Storage for V2B Uses in the Smart Grid. *IEEE Transactions on Smart Grid*, **3**, 473-482. http://dx.doi.org/10.1109/TSG.2011.2172228

[8]   Das, R., Thirugnanam, K., Kumar, P., Lavudiya, R. and Singh, M. (2014) Mathematical Modeling for Economic Evaluation of Electric Vehicle to Smart Grid Interaction. *IEEE Transactions on Smart Grid*, **5**, 712-721. http://dx.doi.org/10.1109/TSG.2013.2275979

[9]   US Department of Transportation, Federal Highway Administratio (2014) 2009 National Household Travel Survey. http://nhts.ornl.gov

[10] PG & E (2014) Time-of-Use. http://www.pge.com/en/mybusiness/rates/tvp/toupricing.page?WT.mc_id=Vanity_tou

[11] PPL Electric Utilities (2014) Time-of-Use Rate Option. https://www.pplelectric.com/at-your-service/electric-rates-and-rules/time-of-use-option.aspx

[12] Southern California Edison (2014) Residential Rate Plans. https://www.sce.com/wps/portal/home/residential/rates/residential-plan/tou/

[13] Portland General Electric (2014) Time of Use: Pricing. http://www.portlandgeneral.com/residential/your_account/billing_payment/time_of_use/pricing.aspx

[14] Con Edison (2014) Voluntary Time-of-Use. http://www.coned.com/customercentral/energyresvoluntary.asp

[15] NV Energy (2014) Residential Time of Use for Southern Service Territory. Time of Use Rate for Home. https://www.nvenergy.com/home/paymentbilling/timeofuse.cfm

[16] BGE (2014) Time of Use Pricing. http://www.bge.com/waystosave/manageyourusage/pages/time-of-use-pricing.aspx

[17] Anderson, D. (2009) An Evaluation of Current and Future Costs for Lithium-Ion Batteries for Use in Electrified Vehicle Powertrains. Master's Thesis, Environmental Management, Duke University, Durham.

Notation

$Cost_B$: The cost of the battery;

$Cost_{BL}$: The baseline charging cost without V2C;

$Cost_C$: The compensation that is necessary for each instance of V2C;

$Cost_E$: The total energy cost;

Cyc_R: The number of cycles the battery is rated for;

E_B: The usable capacity of the battery;

E_C: The total energy supplied to the vehicle during charging including losses;

E_D: The total energy supplied from the vehicle to the campus, after losses are subtracted;

E_H: The energy used to home from work;

E_W: The energy used to travel from home to work;

$Rate_{Off}$: The price of electricity during off peak times;

$Rate_{On}$: The price of electricity during peak times;

Sav_{SB}: The amount of money saved using a stationary battery;

Sav_{V2C}: The amount of money saved using V2C.

7

An Innovative Approach in Balancing Real Power Using Plug in Hybrid Electric Vehicles

Andrew D. Clarke, Elham B. Makram

Department of Electrical and Computer Engineering, Clemson University, Clemson, USA
Email: adclark@g.clemson.edu, makram@clemson.edu

Abstract

Many distribution systems operate under unbalanced loading conditions due to the connection of single phase loads to a three phase system. As PHEVs become more prevalent, it is expected that this unbalance will be further exacerbated due to the power draws from single phase chargers. Unbalanced loads reduce the overall system operating efficiency and power transfer capability of assets. In this paper, a new method of mitigating real power unbalance is suggested. It works by selecting which of the three phases that each single phase PHEV charger in a car park should be connected to. The algorithm is then tested on a simulation model of a real world distribution system. Using the balancing algorithm, balancing of the real power flowing through the feeder to the car park is accomplished.

Keywords

Unbalance, Electric Vehicles, Power Distribution, Power System Control, Power System Modeling

1. Introduction

Many distribution systems are inherently unbalanced due to their configuration. Balancing of distribution systems is often attempted by spreading single phase loads across the three phases by electric utilities however unbalance is still common due to variation of the single phase loads. This unbalance is expected to be further amplified by the connection of single phase Plug In Hybrid Electric Vehicle (PHEV) chargers which has the potential to negatively impact existing assets [1]. This is due in part to the unequal connection of PHEVs to the three phases of the electric distribution system [2]. Many current studies attempt to examine the effects of PHEVs on distribution systems, however most neglect the important impacts PHEVs have on unbalance in a three phase system and the potential of using PHEVs to reduce unbalance. The contribution of this paper is using PHEVs to balance phases. Unbalance in a system may not necessarily harm the system by itself, but it is certainly not the most efficient way to operate. The unbalance conditions discussed in this paper deal with the unequal real power consumption of loads throughout the system. Whenever unbalance exists in a distribution system, negative and zero sequence currents are generated. Neither of these are usually desired in power systems however. Unbalance in a system can reduce power transfer capability of equipment, increase losses, interfere with fault protection devices, and produce overheating in motors and transformers [3] [4].

In a distribution system, if this unbalance condition is mitigated, potential problems can be avoided. The goal of this paper is to balance the real power while allowing PHEV charging and discharging operations to continue virtually unaffected. PHEV chargers represent a large, controllable, single phase load. In a large car park, it is expected that a three phase supply will be available, with vehicles distributed between the phases. Currently, it is not possible to force users to charge from a particular phase. A new idea that compensates for unbalance by allowing PHEV chargers in a car park to automatically change the phase they are drawing power from based on power flows in a distribution system is presented in this paper. An algorithm is then tested using a simulated unbalanced distribution system with a car park containing PHEV chargers.

The structure of this paper is as follows: Section 2 contains details of the PHEV model used for this study; Section 3 contains specifics of the distribution system being simulated, which in this case represents the Clemson University distribution system; Section 4 has details on the real power balancing using PHEVs; Section 5 has the results of simulations; and Section 6 contains the final conclusion of this paper.

2. PHEV Model

With the high cost of fossil fuels and the recent push to make automobiles more environmentally friendly, many manufacturers have started making PHEVs. According to [5], the IEEE places three criteria on a vehicle in order for it to fall under the classification of a PHEV.

1) The vehicle must have a battery with a capacity of at least 4 kWh that is used to drive the vehicle.

2) The vehicle must be able to be recharged from the electric grid.

3) The vehicle must be capable of traveling at least 10 miles on battery power alone.

Most AC PHEV chargers fall into one of the three categories listed in **Table 1** [6]. Based on the expectation that level 1 and 2 chargers will utilize commonly available building wiring, level 3 charging will not be considered in this paper. Due to the higher power transfer available with level 2 charging, which in turn typically leads to a faster recharge than level 1 charging, it is expected that most consumers will desire this type of charger if available. Based on [7], it is shown that a single phase PHEV charger can be developed with the capability of operating at a commanded real and reactive power. In order to decrease simulation time compared with a full switching model, which requires a very small time step, it is assumed that the desired current can be obtained using a PHEV charger. Therefore, a current source is used in place of a PHEV charger. The current is calculated based on the measured terminal voltage and commanded real and reactive powers using Equation (1), where I_{peak} is the peak phasor value of current solved for, V_{peak} is the measured peak voltage phasor, P is the real power, and Q is the reactive power.

$$I_{peak} = \frac{2(P + jQ)^*}{V_{peak}^*} \tag{1}$$

The model used for simulation purposes is allowed to consume or supply power up to 7 kW, which is representative of a level 1 or level 2 PHEV AC charger [6]. The power for each PHEV is chosen between 0 - 7 kW charging or discharging power using a pseudorandom function in MATLAB to simulate volatility in PHEV availability and charging patterns.

3. Clemson University Distribution System

For this study, the Clemson University electric distribution system is modeled in Sim Power Systems, an add-on toolbox for MATLAB. A system diagram can be found in **Figure 1**. Parameters used were provided by Clemson University Utility Services. Due to the large amount of data corresponding to a system this size, inclusion of all system parameters is not feasible in this paper. The system is a primarily three phase, 12.47 kV distribution system connected by underground tape shielded cables. The system is fed by a connection to a 44 kV transmission system. At peak load for 2013, the system was drawing 22.03 MVA of three phase apparent power. Most buildings on campus are fed by three phase distribution transformers that step the voltage down to either 480 V or

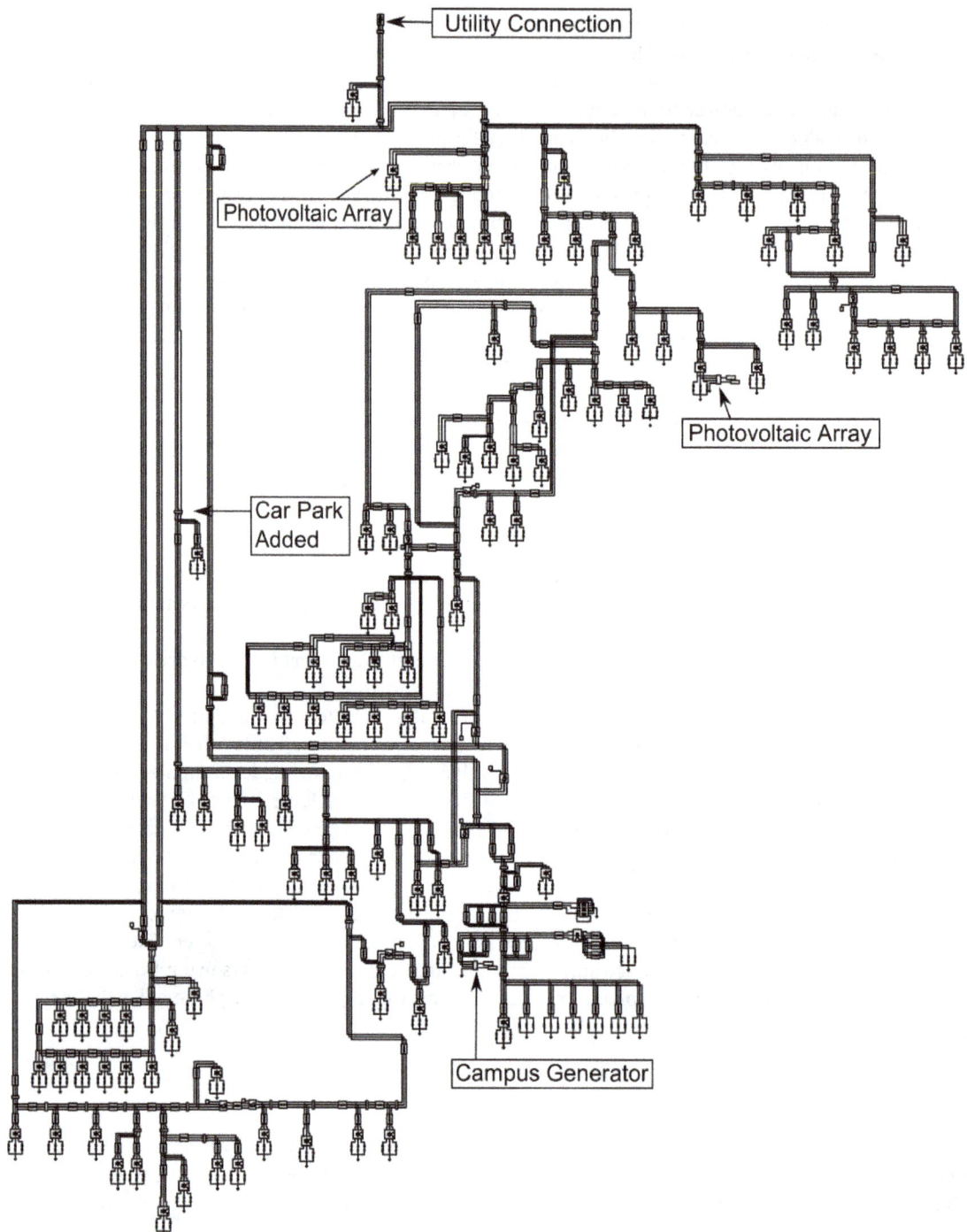

Figure 1. Clemson university electric distribution system.

Table 1. PHEV charging levels [6].

Charging level	Nominal supply voltage	Maximum current	Continuous input power
1	120 V, 1 phase	12 A	1.44 kW
2	208 - 240 V, 1 phase	32 A	6.66 - 7.68 kW
3	208 - 600 V, 3 phase	400 A	>7.68 kW

208 V. There are also two 15 kW photovoltaic arrays and a 5.5 MVA gas turbine used for peak load shaving.

In order to test the balancing algorithm, an unbalanced load is applied to the system. The powers of all the loads connected to phase A are increased by 2%, the powers of all the loads connected to phase B are increased by 1%, and the powers of all the loads connected to phase C are decreased by 3%, all compared to balanced conditions. It is important to note however that these loads are connected on the secondary sides of the distribution transformers, some of which are Delta-Wye connected. This causes the unbalance in the system to appear on different feeders compared to the unbalance in the loads.

A PHEV car park is added to the campus system simulation at one of the major parking lots on campus to show the capability of using PHEVs to balance phases. A total of up to 30 PHEV charging stations are connected to the system in the car park at a given time, based on the power drawn by each PHEV and the algorithm described in this paper. The car park is connected to the system using a Wye Grounded-Wye Grounded three phase transformer. This transformer type is chosen because the balancing algorithm described in this paper intentionally unbalances the PHEV load on the secondary of this transformer in order to balance the real power flowing elsewhere in the system. Using a Wye Grounded-Wye Grounded three phase transformer allows the unbalanced power on one phase of the secondary to not influence another phase on the primary side of the transformer.

4. Real Power Balancing Using PHEVs

The goal of this idea is to automatically change the phase that PHEV chargers are connected to based on measured power flows in the system. Some vehicle chargers that are drawing power will be switched to the less heavily loaded of the three phases and some vehicle chargers that are supplying power will be switched to more heavily loaded phases. The effect of this is shifting some power away from heavily loaded phases. This in turn increases power on lightly loaded phases in order to balance the three phases as necessary. To accomplish this, the algorithm described in this section works by taking in information from each charger about how much real power it is drawing and responding with what phase each PHEV charger should be connected to. It also requires an input of the unbalanced real power measured where the balancing occurs. In order to be successful, the PHEVs must draw their power from where the algorithm will balance power.

In addition to the description given in this section, **Figure 2** shows a flowchart of this algorithm. Once a specified time delay between calculations has passed, an approximation of the real power that would flow through each phase without the connection of PHEVs is calculated using Equation (2). The real power measured from the system is required by the algorithm. The real power consumed by each charger and memory of which phase each PHEV charger is connected to from the previous time step are also required. By neglecting the feeder losses caused by the addition of the PHEV to a phase, Equation (2) can be used. However, the losses caused by the addition of a PHEV to a phase are anticipated to be much smaller than the power drawn by the PHEV charger.

$$P_{\text{Estimate}_{\text{Phase}}} = P_{\text{Measured}_{\text{Phase}}} - \sum_{x \in \text{Phase}} P_{\text{PHEV}}(x) \tag{2}$$

The estimated values of real powers for each phase without PHEVs connected are compared with the values from the previous calculation. If the estimated real power has not changed by a certain percentage and no PHEVs have been connected or disconnected since the last placement, the algorithm exits and waits for the specified time before starting over. If the estimated real power has changed or a PHEV has been connected or disconnected since the last placement, the algorithm is allowed to continue and the PHEVs are then sorted into a list based on the absolute value of the power consumed or supplied by each charger, with the largest listed first. Starting with the first power in the sorted list, the real power of the PHEV charger associated with it is checked to determine if it is drawing real power, supplying real power, or neither. If the PHEV charger is drawing real power, it is assigned a connection to the phase with the minimum estimated real power. This will cause an increase in the observed real power load on that phase. If the PHEV charger is supplying real power, it is assigned a connection to the phase with the maximum estimated real power. This will cause a decrease in the observed real power load on that phase. If the PHEV charger is neither drawing nor supplying real power, it is disconnected from the system. Based on the assigned phase connection, the estimated real power is recalculated using Equation (3).

$$P_{\text{EstimateNew}_{\text{Phase}}} = P_{\text{Estimate}_{\text{Phase}}} + P_{\text{PHEV}}(x) \tag{3}$$

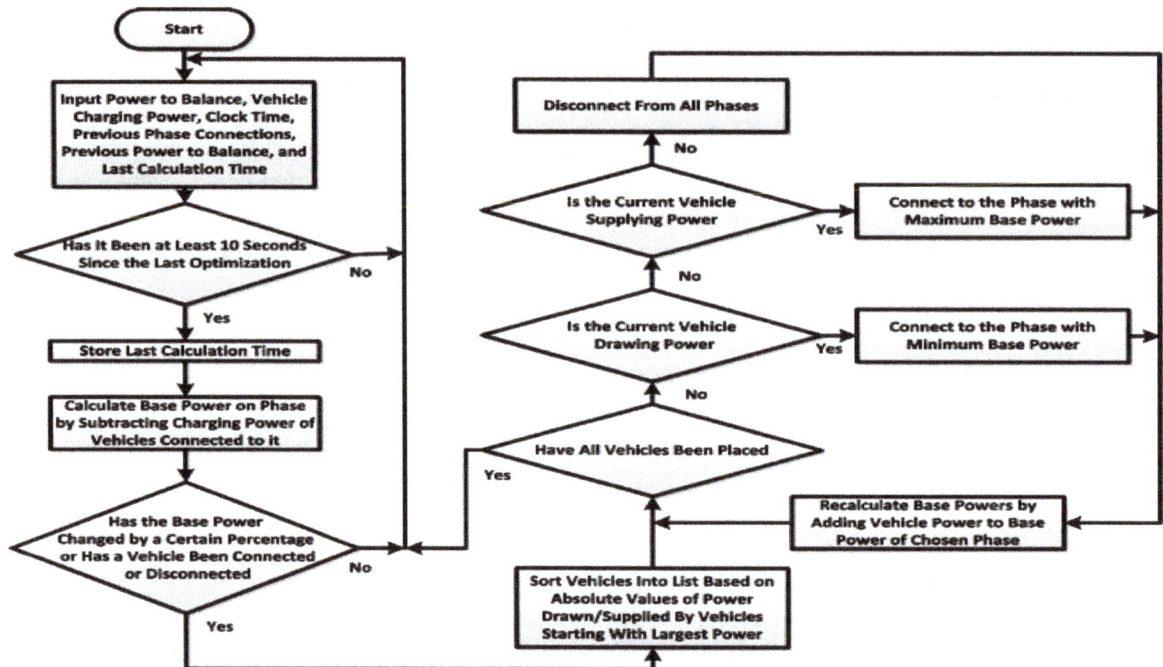

Figure 2. Real power balancing algorithm.

This assignment will continue until all PHEVs in the sorted list are placed. The switching elements connected to the PHEV chargers are then commanded to automatically switch without user interaction. As a result, this algorithm will exit and wait for the specified time before starting over. The time before the next calculation is based on a tradeoff. By increasing the time before the next calculation, wear on the switching elements will be reduced. By decreasing the time before the next calculation, the accuracy will be increased and PHEVs will be allowed to begin charging faster after connection to a charger. Due to the automatic switching used, the vehicles will continue to charge or discharge even if they are commanded to change phases multiple times during connection. In this case, the switching elements are represented through three single phase breakers added at the terminals of each PHEV charger, which allows each individual PHEV charger to be automatically connected to any of the three phases based on the commanded phase. When a commanded phase is sent from the algorithm to the set of three single phase breakers connected to a PHEV charger terminals, all of the breakers open except for the one attached to the commanded phase, which is triggered to close. Switches consisting of power electronics could also be used in place of breakers.

5. Results

Using the system described in Section 3, the real power balancing algorithm is applied to the connected car park. The real power measured that the algorithm will attempt to balance is flowing into the bus where the car park is connected. This bus feeds both the car park as well as other downstream loads, which are the original source of unbalance in the feeder. **Figure 3** shows the real power flowing into the bus both before and after the balancing algorithm is applied. For the first half of the time shown, the balancing algorithm is not applied and PHEVs are connected according to **Table 2**. After the first half of the time shown, the balancing algorithm is applied and PHEVs are assigned to phases in such a way as to balance the real power flowing into the bus. As can be seen in **Figure 3**, applying the real power balancing algorithm to the car park almost perfectly balances the real power flowing into the bus. This balancing is accomplished by switching vehicles from the heavily loaded phases, A and B, to the lightly loaded phase, C. **Table 3** shows the powers for each phase and maximum percent difference both before and after balancing occurs, as measured at the car park bus. This in turn has the effect of pushing the entire system towards a more balanced operation, as can be seen in **Figure 4**. At this bus, the load on the two heavily loaded phases, A and B, is again reduced and the load on the lightly loaded phase, C, is increased by approximately the same amount of the reductions of the other two phases. **Table 4** shows the powers for each

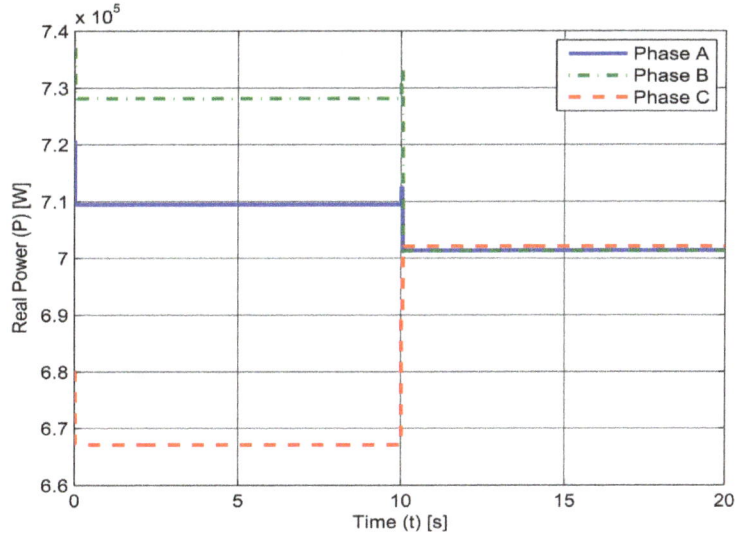

Figure 3. Results of real power balancing algorithm—car park bus.

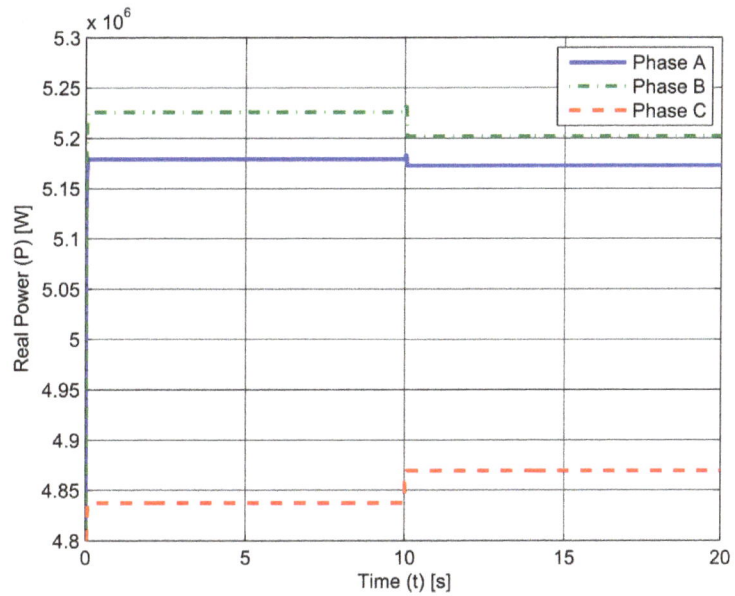

Figure 4. Results of real power balancing algorithm—utility feed.

Table 2. Initial PHEV phase connections.

PHEV number	Connected phase
1, 4, 7, 10, 13, 16, 19, 22, 25, 28	A
2, 5, 8, 11, 14, 17, 20, 23, 26, 29	B
3, 6, 9, 12, 15, 18, 21, 24, 27, 30	C

Table 3. Results of real power balancing algorithm—car park bus.

Case	Phase A power	Phase B power	Phase C power	Maximum percent difference
Before balancing	709.5 kW	728.2 kW	667.1 kW	8.758%
After balancing	701.4 kW	701.3 kW	702.2 kW	0.128%

phase and maximum percent difference both before and after balancing occurs, as measured at the utility feed.

During normal operation of distribution systems, many loads may vary frequently. In order to be practical for real world applications, the algorithm must be able to adapt to load variations. The algorithm's ability to keep up with these variations can be seen in **Figure 5**. In this figure, the first half of the time shown is the same power as is shown in the second half of **Figure 3**. During this period, the real power balancing algorithm has been applied. At halfway through the time shown, the load is instantaneously changed. At this point, the balancing algorithm keeps the PHEVs connected to the same phases as before the load change for 10 seconds. This is the time between balancing calculations selected for this algorithm. Therefore this represents the longest time between a system change and recalculation to assign a new phase configuration. After this point, the balancing algorithm runs again and the phase assignments of the PHEVs are switched in order to better balance the system. **Figure 6** extends the changing load profile for all 24 hours in the day, where each hour is represented by the first 20 seconds of data for that hour's simulation. It should be noted that **Figure 5** and **Figure 6** are the results of multiple simulations plotted one after the other due to limitations on changing load values in Sim Power Systems during a simulation. After the first simulation, the initial values for PHEV phase connections and powers in the simulations are taken from the results of the prior simulation however.

6. Conclusion

It has been successfully shown that this new idea will enable PHEV charging and discharging to have a positive impact on a distribution system by reducing real power unbalance. The phase balancing algorithm assigns each PHEV to an individual phase based on the real power that the charger draws or supplies. The presented algorithm correctly operates for both charging and discharging PHEVs. By changing the phase without altering the power of each charger, the system is able to benefit while PHEV owners will see almost no change in operation compared with a charger that has a static phase connection. It is also possible to connect or disconnect PHEVs from the system at any time and it will recalculate a new phase assignment to move towards a balanced system.

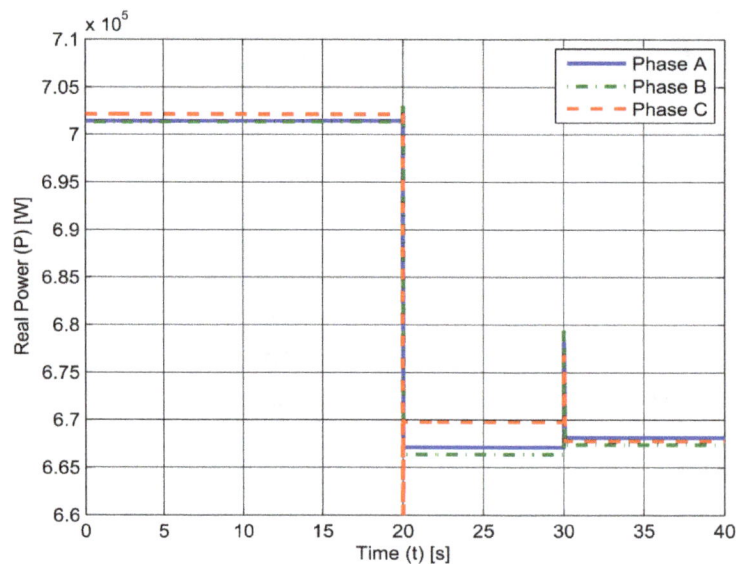

Figure 5. Results of real power balancing algorithm under changing load—car park bus.

Table 4. Results of real power balancing algorithm—utility feed.

Case	Phase A power	Phase B power	Phase C power	Maximum percent difference
Before balancing	5.18 MW	5.23 MW	4.84 MW	7.711%
After balancing	5.17 MW	5.20 MW	4.87 MW	6.593%

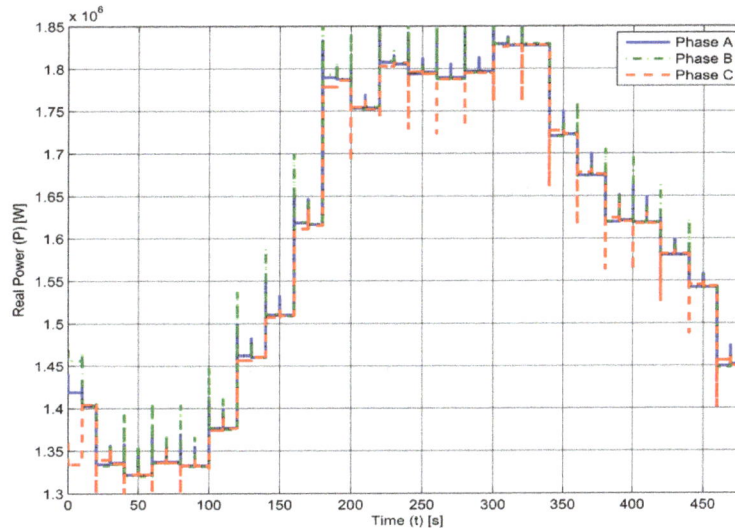

Figure 6. Results of real power balancing algorithm under changing load, daily load profile—car park bus.

Acknowledgements

The authors would like to thank the members of the Clemson University Electric Power Research Association (CUEPRA) for their financial support and valuable discussion.

References

[1]   Oe, S.P., Christopher, E., Sumner, M., Pholboon, S., Johnson, M. and Norman, S.A. (2013) Microgrid Unbalance Compensator—Mitigating the Negative Effects of Unbalanced Microgrid Operation. *4th IEEE/PES Innovative Smart Grid Technologies Europe (ISGT EUROPE)*, 6-9 October 2013, 1-5.

[2]   Dharmakeerthi, C.H., Mithulananthan, N. and Saha, T.K. (2011) Overview of the Impacts of Plug-In Electric Vehicles on the Power Grid. *IEEE PES Innovative Smart Grid Technologies Asia (ISGT)*, 13-16 November 2011, 1-8.

[3]   Chindris, M., Cziker, A., Miron, A., Balan, H., Iacob, A. and Sudria, A. (2007) Propagation of Unbalance in Electric Power Systems. *9th International Conference on Electrical Power Quality and Utilisation*, 9-11 October 2007, 1-5.

[4]   Makram, E., Zambrano, V., Harley, R. and Balda, J. (1989) Three Phase Modeling for Transient Stability of Large Scale Unbalanced Distribution Systems. *IEEE Transactions on Power Systems*, **4**, 487-493.
       http://dx.doi.org/10.1109/59.193820

[5]   Anumolu, P., Banhazl, G., Hilgeman, T. and Pirich, R. (2008) Plug-In Hybrid Vehicles: An Overview and Performance Analysis. *IEEE Long Island Systems, Applications and Technology Conference*, 2 May 2008, 1-4.

[6]   Kisacikoglu, M.C., Ozpineci, B. and Tolbert, L.M. (2010) Examination of a PHEV Bidirectional Charger System for V2G Reactive Power Compensation. *25th Annual IEEE Applied Power Electronics Conference and Exposition (APEC)*, 21-25 February 2010, 458-465.

[7]   Clarke, A.D., Bihani, H.A., Makram, E.B. and Corzine, K.A. (2014) Analysis of the Impact of Different PEV Battery Chargers during Faults. *Journal of Power and Energy Engineering*, **2**, 31-44.
       http://dx.doi.org/10.4236/jpee.2014.28004

8

Analysis of the Impact of Different PEV Battery Chargers during Faults

Andrew D. Clarke, Himanshu A. Bihani, Elham B. Makram, Keith A. Corzine

Department of Electrical and Computer Engineering, Clemson University, Clemson, USA
Email: adclark@g.clemson.edu, hbihani@g.clemson.edu, makram@clemson.edu, corzine@clemson.edu

Abstract

With a high penetration of Plug-In Electric Vehicles (PEVs) in the electric grid, utilities will have to face the challenges related to them. Considerable research is being done to study and mitigate the impact of PEVs on the electric grid and devise methodologies to utilize them for energy storage and distributed generation. In this paper, the impact of PEVs in a smart car park, placed in an unbalanced distribution system, during a single line to ground fault with auto-recloser operation is studied. Level-2, bidirectional battery chargers with current-controlled and voltage-controlled voltage source converters are modeled for the battery charging systems of the PEVs. A smart car park, with 16 vehicles connected to each of the three phases is simulated at one of the buses in the IEEE 13 Bus Test Feeder. The impacts observed during the fault are analyzed and a method to mitigate them is suggested.

Keywords

Electric Vehicles, Battery Chargers, Voltage Source Converters, Current Control, Voltage Control, Power Distribution Faults

1. Introduction

Many countries in the world rely on imports to satisfy their petroleum needs. Heavy reliance of the transportation sector on petroleum-based fuel is a matter of concern for most of the world governments. These concerns are mainly associated with:

1) Rising level of carbon dioxide in the atmosphere which is considered a contributor to global warming and climate change.

2) Threat to energy security due to dependence on a fuel source controlled by a different entity or government.

3) Rising cost of petroleum which has a tremendous negative impact on the economy.

The Energy Independence and Security Act of 2007 as well as the Energy Improvement and Extension Act of 2008 reflect the inclinations of the political faction of the USA to alter the energy consumption pattern away from petroleum [1]. Developing alternative fuel sources for vehicles is a step in that direction. Electric propulsion technology used by PEVs has gained ground in recent times. Battery Electric Vehicles (BEVs), Plug-In Hybrid Electric Vehicles (PHEVs), and Extended Range Electric Vehicles (EREVs), which all have the capability to be charged from the distribution system, come under the gamut of PEVs. The entire energy requirement for BEVs is satisfied by its large battery pack alone, while PHEVs and EREVs rely on gasoline as well as a battery pack to meet their energy needs [2]. Almost every major automobile manufacturer has a PEV on road or plans to have one in the near future [1]. One of the biggest constraining factors thwarting the large scale adoption of the PEVs is their high cost. The high cost is mainly associated with the battery pack which is an integral component of PEVs. Breakthroughs in battery technology may have positive impacts on the PEV market [3].

Though PEVs provide solutions to the concern associated with petroleum-based fuel, a high penetration of the PEVs will pose different kinds of challenges to the electric utility industry. In terms of load, PEVs will have a very different characteristic and utilities will have to devise methodologies to adapt their system to it. On the other hand, the capability of PEVs as a source of distributed generation and energy storage can be used by the utility industry for regulation services and as an enabler for high penetration of intermittent renewable energy sources. Considerable amount of research has already been done to study the impacts of PEVs on the electric grid and concocting methodologies to alleviate them. In [4], the aggregated load profile of PHEVs based on the data from National Household Travel Survey (NHTS), All Electric Range (AER) of PHEVs and its charging level are analyzed. Based on the analysis, different smart charging policies are suggested to shift the peak of the load profile to a desirable time of the day. Research has also focused on studying the impact of PEVs on the assets of the distribution system. Distribution transformers are considered critical assets in a distribution system. Estimating the remaining life of a transformer is useful for the reliability and planning of the system. Aging of a transformer is mainly associated with degradation of its insulation, which is very sensitive to the temperature. In [5], the hot spot temperature in the transformer is estimated using NHTS data and the IEEE standard dynamic thermal model. It concludes that with PEVs in the system, ambient temperature plays a critical role in accelerated aging of a transformer and that some of the proposed smart charging algorithms may actually have an adverse effect on its life. Research in the domain of PEVs has also focused on developing smart charging algorithms to minimize losses and investment costs in distribution systems [6] [7] and to optimally utilize PEVs for regulation and ancillary services [8] [9].

Faults on a distribution system are a common phenomenon and it being a Single Line to Ground (SLG) fault has the highest probability. Work done in the domain of analyzing the impacts which PEVs in a smart car park may have on the unbalanced distribution system during faults is limited. In this paper a smart car park of PEVs in the IEEE 13 Node Test Feeder is simulated. A SLG fault with auto-reclosure operation is created and the response of the smart car park, in terms of impacts on system voltages, currents and power flows is analyzed. In this study, a high-magnitude switching voltage, which would negatively impact the grid, is observed during the time when the recloser of the faulted phase is open and the fault on the system has been cleared. This high-magnitude switching voltage may cause severe damage to the insulation and the equipments installed in the distribution system. Extra control logic in the control of the battery chargers of PEVs is added to isolate the vehicles from the system during fault and fault recovery to mitigate the problem of this high-magnitude switching voltage. The rest of the paper is organized in the following manner. Detailed modeling of the battery charging system of PEVs is explained in Section 2. Section 3 gives an overview of the IEEE 13 Node Test Feeder and details regarding the fault analysis performed. Section 4 analyzes the impacts which PEVs have on the system during a fault and explains the methodology used to mitigate them. Conclusions from the studies are discussed in Section 5, and future work is described in Section 6.

2. Battery Charging System

The equipment that forms the link between PEVs and the electric grid is the Battery Charging System (BCS). A BCS is mainly comprised of the battery and power electronic converters. Power electronic converters enable the battery to be charged from the grid as per its specific requirements.

Lithium-ion (Li-ion) batteries are expected to be used for energy storage for the coming generation of PEVs

[10]. Constant Current (CC) and Constant Voltage (CV) charging is the most common charging profile used for Li-ion batteries [10]. Li-ion batteries with Lithium-Iron-Phosphate (LFP) cathode composition find wider usage for automotive applications [10]. In the case of LFP, CC charging consumes 75% of the total charging time [10]. A typical CC-CV characteristic of an LFP cell is shown in **Figure 1**. Since the State Of Charge (SOC) of the battery remains almost constant during a short time period, only constant current charging is considered.

The different charging methods in North America are shown in **Table 1** [11]. Due to its standardization and higher charging speed than AC Level 1, an AC Level 2 battery charger is expected to find a wider adoption and hence is used as the basis for modelling a bi-directional battery charger for this paper. A typical PEV battery charger consists of an AC/DC stage and a DC/DC stage. In this bidirectional battery charger, the AC/DC stage is an active converter while DC/DC stage can act as a buck converter or a boost converter. Thus with this bi-directional battery charger, the battery can be charged from the grid as well as discharged to the grid.

Power flow can be controlled using inherently efficient and compact Voltage Source Converters (VSCs) [12]. VSCs can be classified as Current Controlled Voltage Source Converters (CCVSC) and Voltage Controlled Voltage Source Converters (VCVSC). A CCVSC directly controls the current flowing into the VSC. This is achieved by generating switching signals based on the error in input current with respect to the reference value. In the case of a VCVSC, the magnitude and the angle of the converter side voltage with respect to the grid side voltage phasor is controlled to obtain the desired power flow [12]. Two types of battery charging system controllers based on CCVSC and VCVSC are developed. This helps in comparing the similarities and differences in the impacts the control methodologies of the battery chargers for PEVs have on the distribution system.

2.1. CCVSC Type Battery Charger

The topology of a CCVSC type battery charger with two different types of filters is as shown in **Figure 2** and **Figure 3**, respectively. The topology and the parameters of the charger are adopted from [10]. The value of the

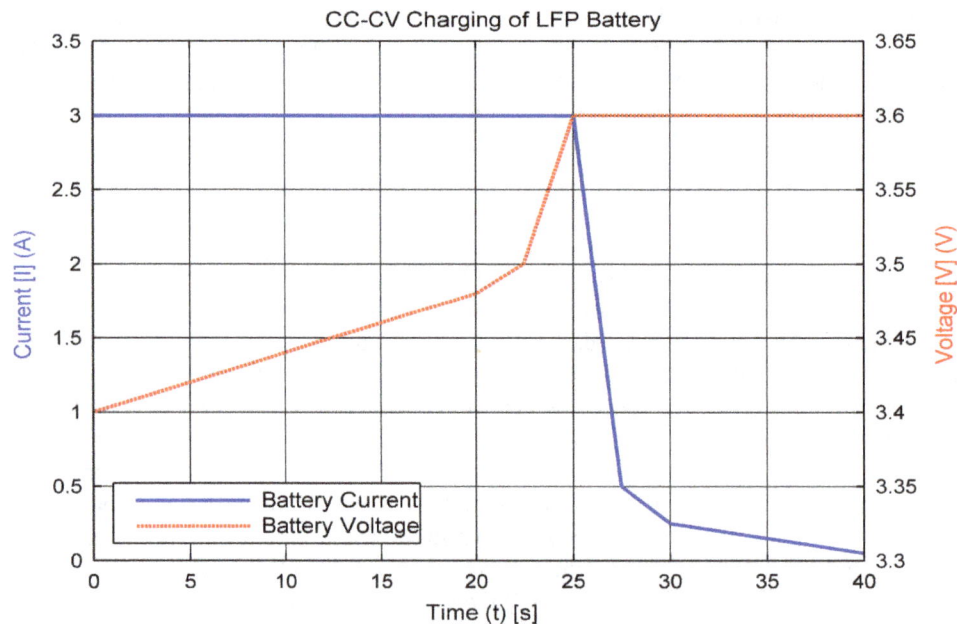

Figure 1. CC-CV charging characteristic of LFP type Li-ion battery cell [10].

Table 1. Charging methods in North America [11].

Charging Method	Nominal Supply Voltage	Maximum Current	Continuous Input Power
AC Level 1	120 V, 1 Phase	12 A	1.44 kW
AC Level 2	208 - 240 V, 1 Phase	32 A	6.66 - 7.68 kW
AC Level 3	208 - 600 V, 3 Phase	400 A	>7.68 kW
DC Charging	600 V Maximum	400 A	<240 kW

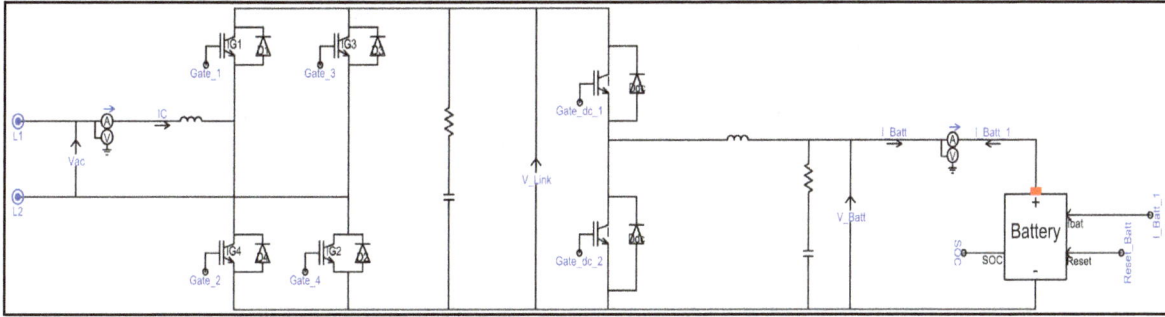

Figure 2. Topology of CCVSC type charger with L type filter.

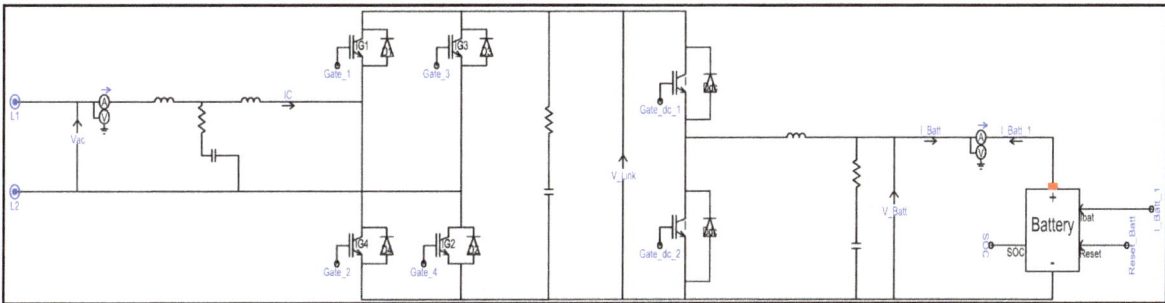

Figure 3. Topology of CCVSC type charger with LCL type filter.

coupling inductor and DC/DC converter filter inductor are modified by observing the steady state and transient response of the charger. The negative impacts of a VSC are suppressed by connecting a filter between the converter and the grid [13]. L and LCL type of filters are used for analysis in this paper. For the L type filter, a 6 mH inductor is used and for the LCL type filter, two 3 mH inductors are used. A switching frequency of 3 kHz is selected for both the AC/DC and DC/DC converters. The cut-off frequency of 1.5 kHz is selected for LCL filter. Based on the information given in [13] and using (1), the value of the capacitor for the LCL filter obtained is 7.5×10^{-6} farads. Magnification of the frequency around the cutoff frequency is avoided by selecting the value of the damping resistor in series with capacitor to be 0.5 ohms.

$$\text{Cutoff Frequency} = \left(\frac{1}{2\pi} \right) \times \sqrt{\left(\frac{L_1 + L_2}{L_1 \times L_2 \times C} \right)} \tag{1}$$

As outlined before, the operation of the charger shown in **Figure 2** and **Figure 3** can be divided into two stages; AC/DC Converter Stage and DC/DC Converter Stage. In unity power factor charging mode the AC/DC stage converts the AC input supply voltage to a DC voltage which appears across the DC-Link [14]. It also ensures that the input current is at unity power factor and the current harmonic distortion is low. The DC/DC converter modulates the DC-Link voltage to the required DC voltage based on the battery CC-CV charging algorithm [14].

The control of the charger is based on the methodology given in [14] [15] and described in [16]. The control scheme for the AC/DC converter stage is shown in **Figure 4**. In unity power factor charging mode, the voltage error signal is generated by comparing the reference DC-Link voltage with the actual value. After normalizing, the error signal is given first to a feedback Proportional-Integral (PI) controller. Faster transient response is obtained by stopping the integrator of the PI controller whenever its output hits upper or lower limits defined by the saturation block and the error is in the same direction. A reset signal is generated when the output of the PI controller comes out of saturation, which resets the integrators of all the PI controllers in the control system of the charger. The output of the first PI controller is the magnitude of the current demanded by the charger. The magnitude of current is multiplied with the AC grid voltage to get the desired wave shape for the reference input current signal. A current error signal is generated by comparing the reference input current with the actual current flowing into the charger. This error signal is given to the second feedback PI controller to generate gating signals for the AC/DC converter based on Sine Pulse Width Modulation (SPWM). Since the gating signals for the converter are determined by comparing the actual current flowing into the battery charger with the desired current, the control methodology is called CCVSC.

The controller for the DC/DC converter in charging mode is shown in **Figure 5**. Only CC charging is incorporated in the control as the SOC of the battery remains almost constant during the simulation time period. The current error signal is generated by comparing the CC set point with the actual value of the battery current. This error signal is then given to the feedback PI controller to generate a reference duty cycle for the DC/DC converter. The duty cycle is implemented by generating appropriate gating signals for the switches.

2.2. VCVSC Type Battery Charger

The topology of a VCVSC type of battery charger with L and LCL type of filter is exactly same as that of a CCVSC type of battery charger. As mentioned before, the desired power flow in VCVSC is obtained by controlling the converter side voltage phasor with respect to grid side voltage phasor. There are two primary methods of controlling a VCVSC type of battery charger: Direct Control and Vector Control. These methods are described and compared in [17]. In this paper vector control methodology is used as it provides independent control of the active and reactive power flows. The basic structure of the control methodology is based on [18] [19] with all the necessary modifications incorporated to adapt it to PEV application.

The very first step in implementing the vector control is to obtain rotating space vectors for the grid side voltage and current. In the case of a three phase system, rotating space vectors are readily available, but this is not the case with a single phase system. In a single phase system, a fictitious phase orthogonal to the original phase has to be created in order to obtain the rotating space vectors. The fictitious orthogonal phase can be created using various techniques such as a 90° phase shift, the Hilbert transformation, and a Second Order Generalized Integrator (SOGI) [18]. In this paper, a SOGI is used to generate the fictitious phase for the grid side voltage and current. The original phase is considered to be along the "α" axis, while the fictitious phase is considered along the "β" axis. One of the major advantages of using SOGI compared to other methods is that it provides filtering of the input quantity [18]. The structure of SOGI used in the control of the battery charging system is as shown in **Figure 6** [18]. **Table 2** lists the input and output signals of SOGI.

The generation of the rotating space vectors is graphically shown in **Figure 7**. Once the rotating space vectors

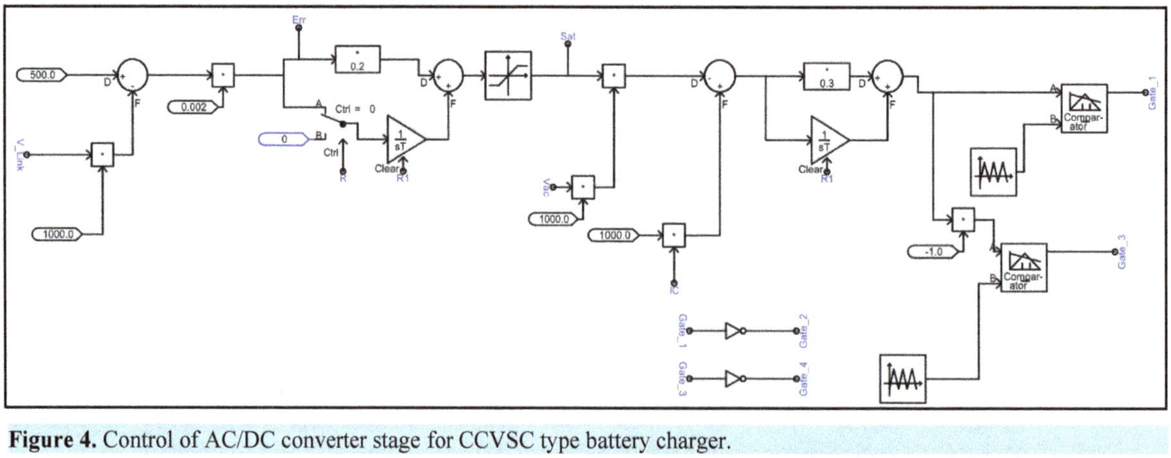

Figure 4. Control of AC/DC converter stage for CCVSC type battery charger.

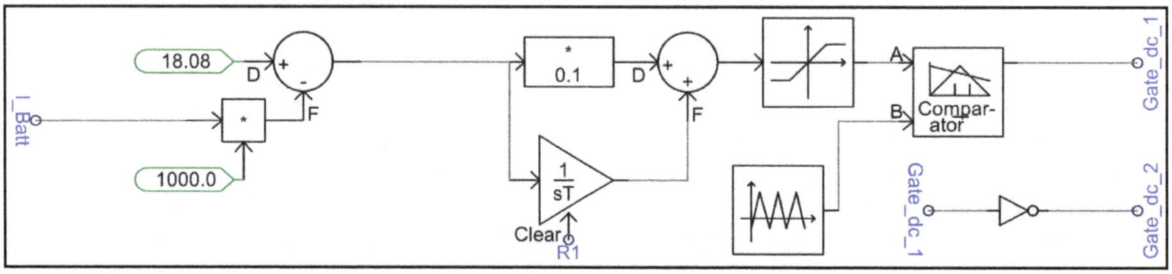

Figure 5. Control of DC/DC converter stage for CCVSC type battery charger.

for the grid side voltage and currents are obtained, the pseudo control algorithm to obtain the converter side voltage phasor for the desired power flow is as given below:

1) The synchronously rotating "d" axis is aligned with the grid side voltage space vector and lagging "q" axis orthogonal to it. In order to align the "d" and "q" axis as per the requirement, the value of angle "θ" needs to be known at all times. The value of angle "θ" is obtained by using a Phase Locked Loop (PLL) for the grid side voltage phasor. The transformation matrix "T" to map the grid side voltage space vector along the "d" and "q" axes is given in (2).

$$T = \begin{bmatrix} \sin(\theta) & \cos(\theta) \\ -\cos(\theta) & \sin(\theta) \end{bmatrix} \tag{2}$$

2) The grid side current space vector is also mapped along "d" and "q" axes using "T". The "d" axis current

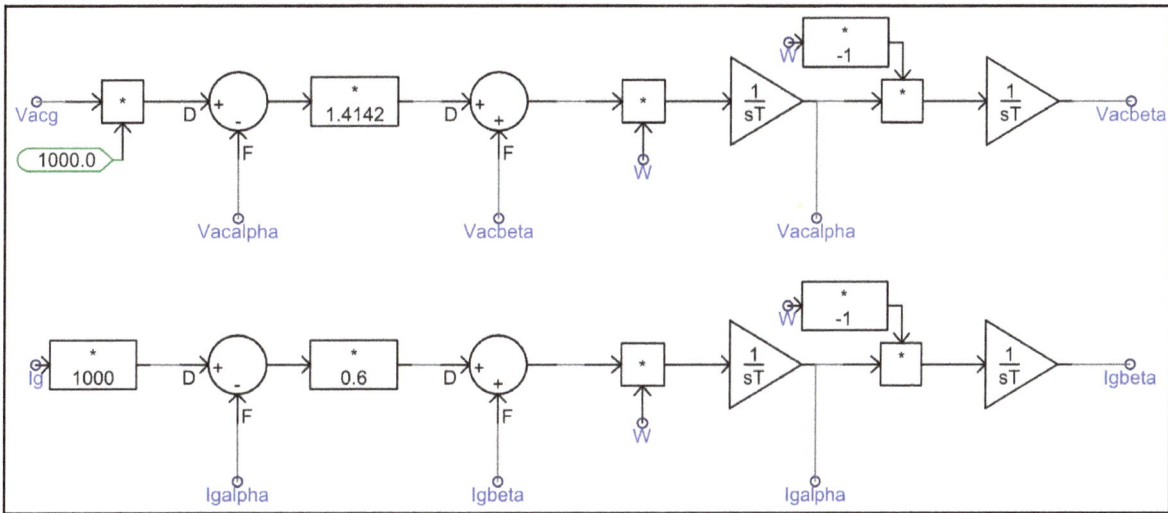

Figure 6. Structure of second order generalized integrator [18].

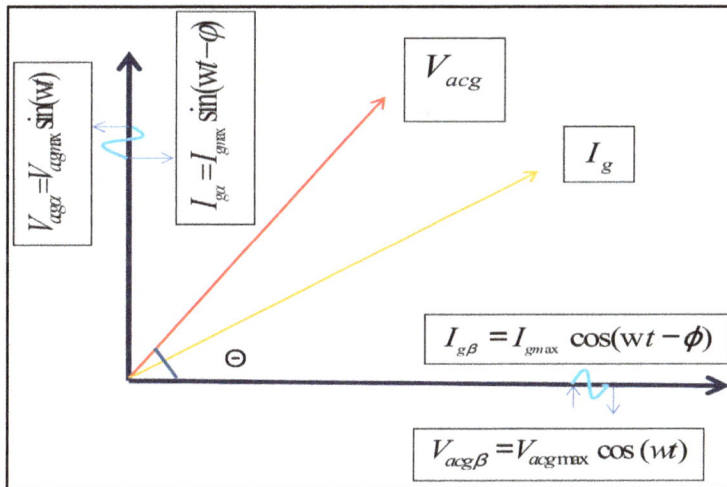

Figure 7. Rotating space vectors for grid side voltage and current.

Table 2. α-β axis values for the grid side voltage and current.

Input	α	β
$V_{acgmax} \sin(wt)$	$V_{acgmax} \sin(wt)$	$V_{acgmax} \cos(wt)$
$I_{gmax} \sin(wt - \varphi)$	$I_{gmax} \sin(wt - \varphi)$	$I_{gmax} \cos(wt - \varphi)$

is in phase with the voltage and hence controls the active power flowing between the grid and the converter while the "q" axis is 90 degrees out of phase and hence controls the reactive power. The transformed equivalent circuit between grid and converter side voltages in the "d-q" domain with the filter capacitor neglected is shown in **Figure 8** [19]. V_{acd} and V_{acq} are the grid side voltages while V_{acinvd} and V_{acinvq} are the converter side voltages in the "d-q" domain. Based on **Figure 8**, the converter side voltages in the "d-q" domain can be related to the grid side voltage as given in (3) and (4).

$$V_{acinvd} = V_{acd} - V_L - wL * L_{gq} \tag{3}$$

$$V_{acinvq} = V_{acq} - V_L + wL * L_{gd} \tag{4}$$

Equations (3) and (4) are implemented in control as described in steps 3, 4 and 5 of the algorithm and as shown in **Figure 9** and **Figure 10**.

3) The error in the DC-Link voltage with respect to its reference value is used to generate the reference value for "d" axis current.

4) The difference in the actual reactive power flowing between the grid and converter and the commanded value is used to generate a reference value of "q" axis current.

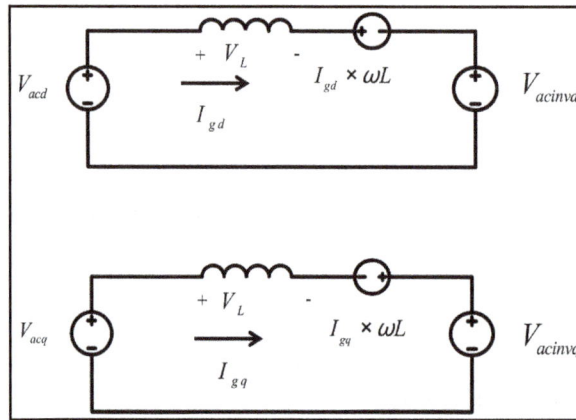

Figure 8. Transformed equivalent circuit in d-q domain.

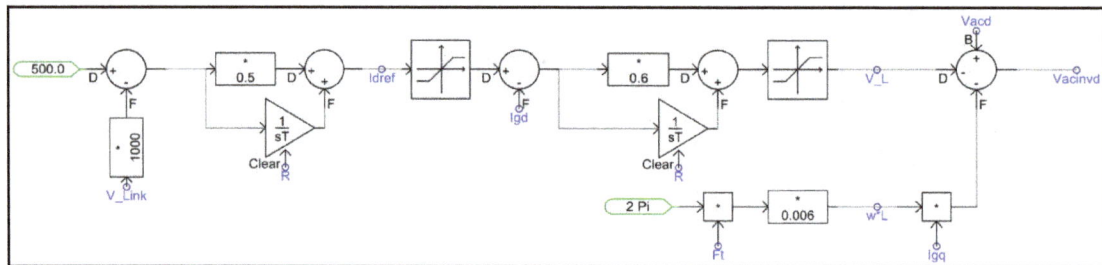

Figure 9. Control circuit to obtain converter side "d" axis voltage.

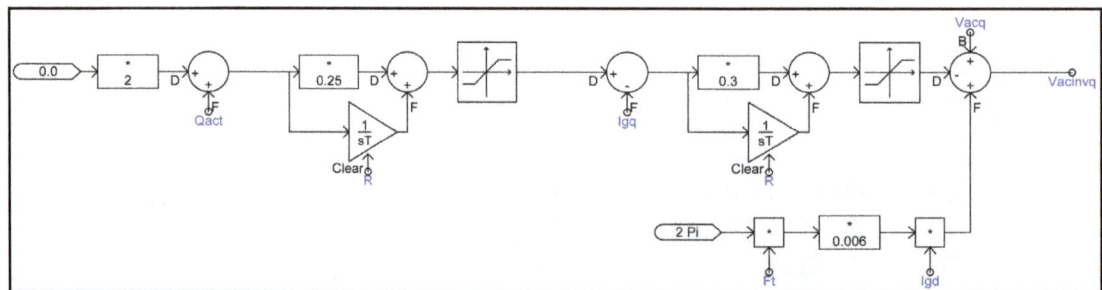

Figure 10. Control circuit to obtain converter side "q" axis voltage.

5) Errors in the "d" and "q" axis currents, based on the actual and reference values obtained from steps 2, 3, and 4, are used to generate the converter side "d" and "q" axis voltages as shown in the figures below.

6) The converter side "d" and "q" axis voltages are transformed back to the "α" and "β" axes using the inverse of the transformation "T". "β" being a fictitious axis is discarded. The "α" axis converter side voltage is the desired voltage waveform which is implemented using SPWM.

3. System Description

The IEEE 13 Node Test Feeder, shown in **Figure 11**, is used for this research. This is an unbalanced, primarily 4.16 kV distribution test system [20]. Due to the small time step required for the simulation and the short lines present in the system, all distribution lines and cables are modeled using mutually coupled lines. The mutually coupled line model neglects line charging but provides a much faster simulation by allowing for a larger solution time step compared to the detailed model found in [21].

In order to determine the effect the penetration of PEVs has on a distribution system, a 416 V bus is added to the system for the connection of PEVs. This bus is connected to the IEEE 13 Node Test Feeder through a Delta-Wye Grounded transformer attached to bus 680. This bus was chosen for the connection of PEVs because it is originally left empty in the test system. In real world systems, the connection point will be largely dependent on physical location in the system. On each phase of the secondary, sixteen PEV chargers are connected in parallel. Each vehicle draws approximately 6.2 kW of power during charging, for a total additional load of 99.2 kW per phase above the base system load. Single phase reclosing is also added to the system for fault protection. A single phase recloser with the opening and closing timings found in **Table 3** is added at bus 671 on phase A of the line connecting bus 671 to bus 680. All simulations in this study are carried out using PSCAD. A temporary single phase to ground fault is applied to phase A of bus 680. The fault is applied 5 seconds into the simulation in order to allow the system to initialize and settle into steady state operation before the disturbance. The fault is then cleared 0.275 seconds after it is applied. The clearing of the fault occurs while the single phase recloser has phase A of the line connecting buses 671 and 680 disconnected from the main system. In order to isolate the impact of adding vehicles to the distribution system at bus 680 from the impact of adding a large load at the same location, a base case is first simulated with a constant power load connected to the added bus in place of vehicle chargers. The constant power load is equivalent to the power drawn by the vehicles. All results shown

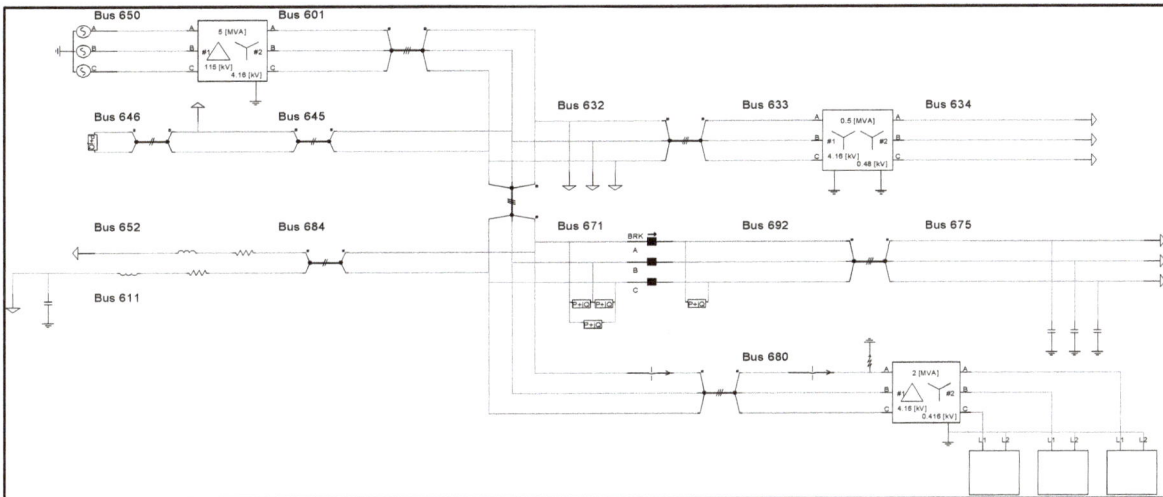

Figure 11. IEEE 13 node test feeder.

Table 3. Recloser timings.

Cycle Type	Open Time	Close Time
Fast	0.1 Second	0.01 Second
Slow	0.1 Second	0.5 Second

compare the cases of vehicle chargers connected to bus 680 with the base case with the equivalent load connected.

4. Impacts

The differences between the two control structures as well as the different components that comprise the L and LCL type filters leads the four chargers to each have a different impact on the system.

4.1. VCVSC Type Battery Charger with L Filter

The VCVSC with L filter shows a negative impact on the system only when the fault is cleared from the system but the recloser is still open during a slow cycle. The fault clears from the system at 5.275 s and the recloser remains open until 5.33 s. During this period, a high magnitude, high frequency switching voltage is fed back into the system on phase A of bus 680 from the vehicle chargers. This distortion in the voltage on phase A of bus 680 comes from the vehicles connected to phases A and C on the secondary side of the Delta-Wye Grounded transformer. The vehicles on phase B remain largely unaffected due to the connection of the transformer. The peak magnitude of this voltage is over two times the steady state peak voltage. Once the recloser reconnects phase A of bus 680 to the rest of the distribution system, the voltages regain their steady state values. **Figure 12** shows the voltage waveforms on phase A of bus 680 during the fault and recovery.

After the recloser reconnects phase A of the distribution system to bus 680, a decaying transient is seen in the current while the vehicle chargers resume steady state operation. **Figure 13** shows the current waveform of phase B of bus 680 during the fault and recovery. A significant impact was not seen at other buses or on other phases in the system during fault recovery due to the recloser being in the open position.

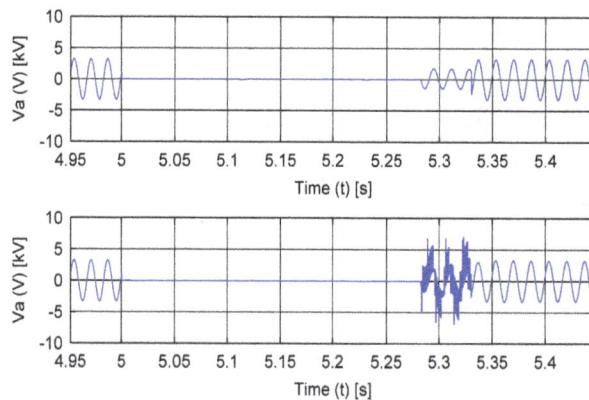

Figure 12. Bus 680 phase A voltages-base case with equivalent load and VCVSC type battery charger with L filter.

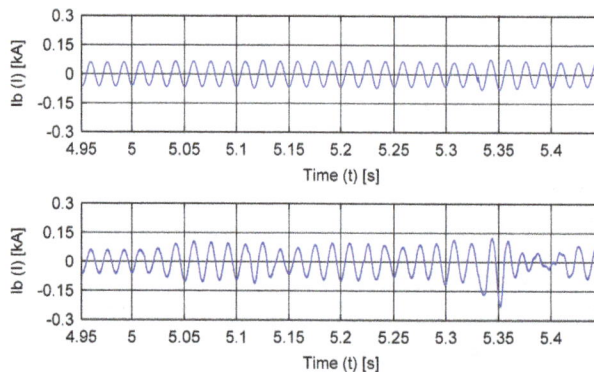

Figure 13. Bus 680 phase B currents-base case with equivalent load and VCVSC type battery charger with L filter.

4.2. VCVSC Type Battery Charger with LCL Filter

The VCVSC with LCL filter also shows a negative impact on the system during the same period as the VCVSC with L filter. With this charger model, a similar high magnitude, high frequency switching voltage as seen from the VCVSC with L filter is again fed back into the system on phase A of bus 680. In this case however, the peak magnitude is over nine times the steady state peak value compared to the much smaller magnitude seen with the L filter. **Figure 14** shows the voltage waveforms on phase A of bus 680 during the fault and recovery.

A transient is again seen in the current after the recloser reconnects phase A of bus 680 to the rest of the distribution system. **Figure 15** shows the current waveform of phase B of bus 680 during the fault and recovery.

4.3. CCVSC Type Battery Charger with L Filter

The CCVSC with L filter shows a negative impact on the system during the same period as both of the VCVSC models. The CCVSC with L filter feeds a high magnitude, high frequency switching voltage back onto phase A of bus 680 in a similar way to the VCVSC chargers. **Figure 16** shows the voltage waveform of phase A at bus 680 during the fault and recovery. The peak magnitude of this waveform is similar to that of the VCVSC with L filter however it maintains this peak value for a longer period.

The CCVSC does not produce the same current transient problems as observed in the VCVSC. **Figure 17** shows the current on phase B of bus 680 during the fault and recovery. The transient for this charger control is

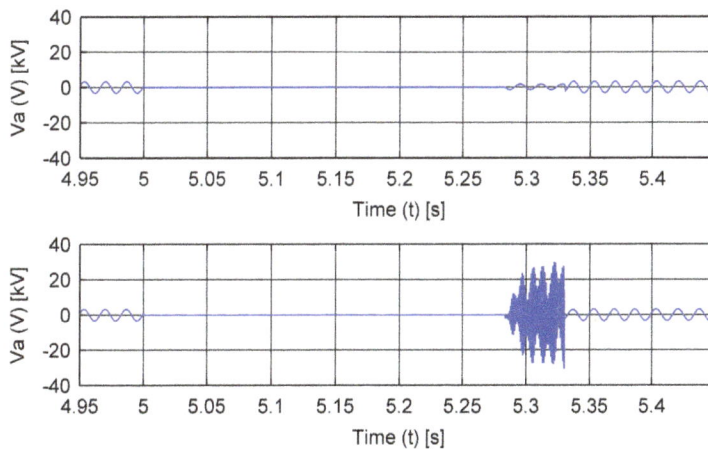

Figure 14. Bus 680 phase A voltages-base case with equivalent load and VCVSC type battery charger with LCL filter.

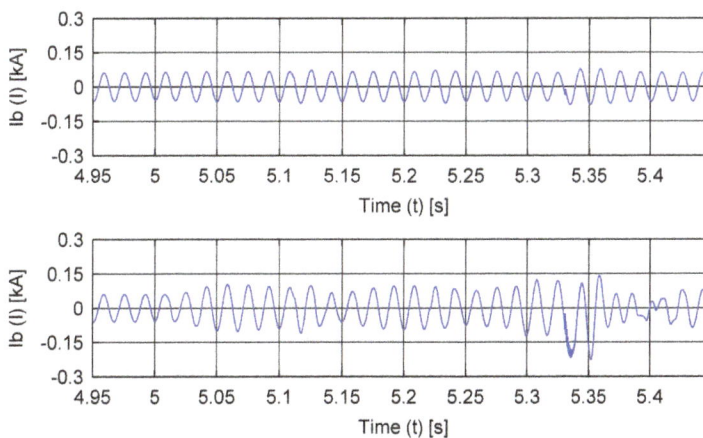

Figure 15. Bus 680 phase B currents-base case with equivalent load and VCVSC type battery charger with LCL filter.

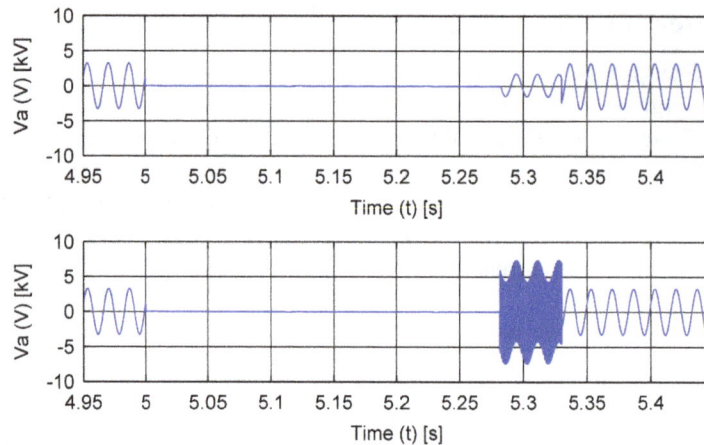

Figure 16. Bus 680 phase A voltages-base case with equivalent load and CCVSC type battery charger with L filter.

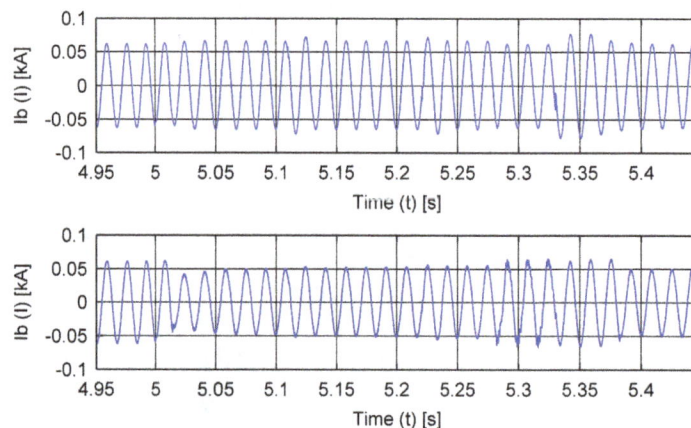

Figure 17. Bus 680 phase B currents-base case with equivalent load and CCVSC type battery charger with L filter.

much smaller and maintains a mostly sinusoidal shape.

4.4. CCVSC Type Battery Charger with LCL Filter

Since the CCVSC controls the magnitude and wave shape of the current entering the charger directly, the CCVSC with LCL filter does not show a negative impact on the distribution system during the fault recovery period. **Figure 18** shows the voltage waveform on phase A at bus 680 during the fault and recovery.

Like the other CCVSC model, the current transient seen after the recloser reconnects phase A of bus 680 to the distribution system again is mostly sinusoidal and much less pronounced than the VCVSCs. **Figure 19** shows the current of phase B on bus 680 during the fault and recovery.

4.5. Mitigation of Negative Impacts

Based on the negative impacts of both of the VCVSCs and the CCVSC with the L filter, it is desired to implement extra control structures to mitigate the negative impacts. To accomplish this task, control was added that stops all switching in the vehicle charger when the voltage at the terminals of the charger drops below 200 V. In per unit, this voltage is well below normal voltage drop limits in a distribution system. During this period where switching is stopped, all switches are left in the open position. After the terminal voltage returns to a normal value, the switching is allowed to resume. This prevents the vehicles from feeding back to the system while the recloser has phase A of bus 680 disconnected from the rest of the distribution system.

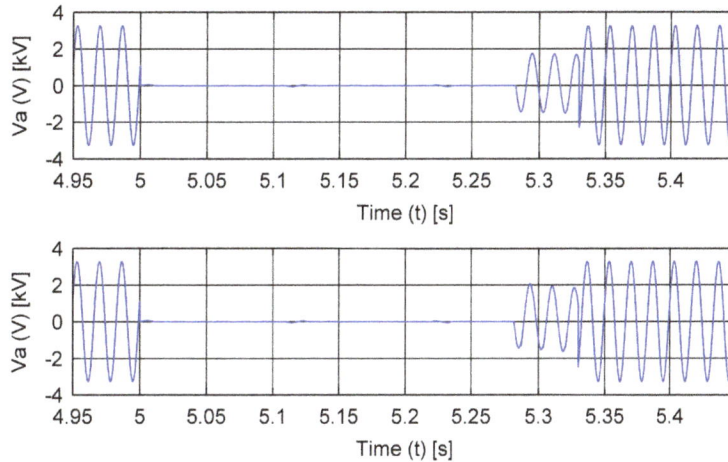

Figure 18. Bus 680 phase A voltages-base case with equivalent load and CCVSC type battery charger with LCL filter.

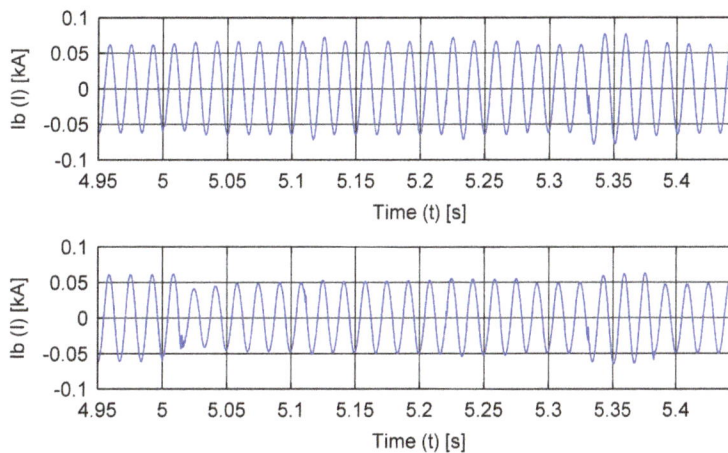

Figure 19. Bus 680 phase B currents-base case with equivalent load and CCVSC type battery charger with LCL filter.

5. Conclusions

It is found that even a relatively small penetration of PEVs has the potential to significantly impact a distribution system in a negative way. The main negative impact observed was a high-magnitude, high-frequency switching voltage on the distribution system due to recloser operation and continued vehicle charger switching during fault recovery.

- The topology of the charger control and filter type determine what this impact will be.
- The worst case occurred with the VCVSC with LCL filter.
- The CCVSC with LCL filter was the best case and did not produce a negative impact on the distribution system.

In order to prevent the potential negative impacts, extra control such as the under voltage control logic described in the previous section needs to be included in PEV chargers.

6. Future Work

In this case, the negative impacts of PEVs ceased shortly after the connection to the infinite source used to model the transmission system was restored. A high penetration of distributed generation will be added to the IEEE 13 Node Test Feeder in order to determine the impact PEVs have when the majority of the system load is served by sources other than the high-voltage transmission system.

Acknowledgements

The authors wish to thank the Clemson University Electric Power Research Association (CUEPRA) for their financial support and their valuable discussion.

References

[1] Rotering, N. and Ilic, M. (2011) Optimal Charge Control of Plug-In Hybrid Electric Vehicles in Deregulated Electricity Markets. *IEEE Transactions on Power System*, **26**, 1021-1029. http://dx.doi.org/10.1109/TPWRS.2010.2086083

[2] Bunga, S.K. (2013) Impact of Plug-In Electric Vehicle Battery Charging on a Distribution System. Thesis, The University of Tennessee at Chattanooga, Tennessee. http://search.proquest.com/docview/1356783945

[3] US Energy Information Administration (2012) Annual Energy Outlook 2012. http://www.eia.gov/forecasts/aeo/pdf/0383(2012).pdf

[4] Darabi, Z. and Ferdowsi, M. (2011) Aggregated Impact of Plug-In Hybrid Electric Vehicles on Electricity Demand Profile. *IEEE Transactions on Sustainable Energy*, **2**, 501-508. http://dx.doi.org/10.1109/TSTE.2011.2158123

[5] Hilshey, A.D., Hines, P.D.H., Rezaei, P. and Dowds, J.R. (2013) Estimating the Impact of Electric Vehicle Smart Charging on Distribution Transformer Aging. *IEEE Transactions on Smart Grid*, **4**, 905-913. http://dx.doi.org/10.1109/TSG.2012.2217385

[6] Sortomme, E., Hindi, M.M., MacPherson, S.D.J. and Venkata, S.S. (2011) Coordinated Charging of Plug-In Hybrid Electric Vehicles to Minimize Distribution System Losses. *IEEE Transactions on Smart Grid*, **2**, 198-205. http://dx.doi.org/10.1109/TSG.2010.2090913

[7] Pieltain Fernández, L., Román, T.G.S., Cossent, R., Domingo, C.M. and Frías, P. (2011) Assessment of the Impact of Plug-In Electric Vehicles on Distribution Networks. *IEEE Transactions on Power Systems*, **26**, 206-213. http://dx.doi.org/10.1109/TPWRS.2010.2049133

[8] Han, S., Han, S. and Sezaki, K. (2010) Development of an Optimal Vehicle-to-Grid Aggregator for Frequency Regulation. *IEEE Transactions on Smart Grid*, **1**, 65-72. http://dx.doi.org/10.1109/TSG.2010.2045163

[9] Sortomme, E. and El-Sharkawi, M.A. (2012) Optimal Scheduling of Vehicle-to-Grid Energy and Ancillary Services. *IEEE Transactions on Smart Grid*, **3**, 351-359. http://dx.doi.org/10.1109/TSG.2011.2164099

[10] Kisacikoglu, M.C., Ozpineci, B. and Tolbert, L.M. (2011) Reactive Power Operation Analysis of a Single-Phase EV/PHEV Bidirectional Battery Charger. 2011 *IEEE 8th International Conference on Power Electronics and ECCE Asia* (*ICPE & ECCE*), Jeju, 30 May-3 June 2011, 585-592.

[11] Kisacikoglu, M.C., Ozpineci, B. and Tolbert, L.M. (2010) Examination of a PHEV Bidirectional Charger System for V2G Reactive Power Compensation. 2010 *25th Annual IEEE Applied Power Electronics Conference and Exposition* (*APEC*), Palm Springs, 21-25 February 2010, 458-465. http://dx.doi.org/10.1109/APEC.2010.5433629

[12] Ko, S.H., Lee, S.R., Dehbonei, H. and Nayar, C.V. (2006) Application of Voltage- and Current-Controlled Voltage Source Inverters for Distributed Generation Systems. *IEEE Transactions on Energy Conversion*, **21**, 782-792. http://dx.doi.org/10.1109/TEC.2006.877371

[13] Lettl, J., Bauer, J. and Linhart, L. (2011) Comparison of Different Filter Types for Grid Connected Inverter. *Progress in Electromagnetics Research Symposium*, Marrakesh, 20-23 March 2011, 1426-1429.

[14] Fain, D. (2009) A Dual Input Bidirectional Power Converter for Charging and Discharging a PHEV Battery. Master's Thesis, Department of Electrical and Computer Engineering, Clemson University, Clemson.

[15] Verma, A.K., Singh, B. and Shahani, D.T. (2011) Grid to Vehicle and Vehicle to Grid Energy Transfer Using Single-Phase Bidirectional AC-DC Converter and Bidirectional DC-DC Converter. 2011 *International Conference on Energy, Automation, and Signal* (*ICEAS*), Bhubaneswar, 28-30 December 2011, 1-5. http://dx.doi.org/10.1109/ICEAS.2011.6147084

[16] Clarke, A., Bihani, H., Makram, E. and Corzine, K. (2013) Fault Analysis on an Unbalanced Distribution System in the Presence of Plug-In Hybrid Electric Vehicles. *Clemson Power System Conference*. http://rtpis.org/psc13/files/PSC2013_final_1358371035.pdf

[17] Sood, V. and Patel, H. (2010) Comparison between Direct and Vector Control Strategy for VSC-HVDC System in EMTP-RV. 2010 *Joint International Conference on Power Electronics, Drives and Energy Systems* (*PEDES*) & 2010 *Power India*, New Delhi, 20-23 December 2010, 1-6. http://dx.doi.org/10.1109/PEDES.2010.5712550

[18] Monfared, M., Sanatkar, M. and Golestan, S. (2012) Direct Active and Reactive Power Control of Single-Phase Grid-Tie Converters. *IET Power Electronics*, **5**, 1544-1550. http://dx.doi.org/10.1049/iet-pel.2012.0131

[19] Samerchur, S., Premrudeepreechacharn, S., Kumsuwun, Y. and Higuchi, K. (2011) Power Control of Single-Phase Voltage Source Inverter for Grid-Connected Photovoltaic Systems. 2011 *IEEE/PES Power Systems Conference and*

Exposition (*PSCE*), Phoenix, 20-23 March 2011, 1-6. http://dx.doi.org/10.1109/PSCE.2011.5772504

[20] IEEE PES Distribution System Analysis Subcommittee. IEEE 13 Node Test Feeder. http://ewh.ieee.org/soc/pes/dsacom/testfeeders/feeder13.zip

[21] IEEE PES Distribution System Analysis Subcommittee. Radial Distribution Test Feeders. http://ewh.ieee.org/soc/pes/dsacom/testfeeders/testfeeders.pdf

9

Impact of Electric Vehicles on a Carbon Constrained Power System—A Post 2020 Case Study

Zhebin Sun[1], Kang Li[1], Zhile Yang[1], Qun Niu[2], Aoife Foley[3]

[1]School of Electronics, Electrical Engineering and Computer Science, Queen's University Belfast, Belfast BT9 5AH, UK
[2]School of Mechatronic Engineering and Automation, Shanghai Key Laboratory of Power Station Automation Technology, Shanghai University, Shanghai 200072, China
[3]School of Mechanical and Aerospace Engineering, Queen's University Belfast, Belfast BT9 5AH, UK
Email: zsun03@qub.ac.uk, k.li@qub.ac.uk, zyang07@qub.ac.uk

Abstract

Electric vehicles (EVs) offer great potential to move from fossil fuel dependency in transport once some of the technical barriers related to battery reliability and grid integration are resolved. The European Union has set a target to achieve a 10% reduction in greenhouse gas emissions by 2020 relative to 2005 levels. This target is binding in all the European Union member states. If electric vehicle issues are overcome then the challenge is to use as much renewable energy as possible to achieve this target. In this paper, the impacts of electric vehicle charged in the all-Ireland single wholesale electricity market after the 2020 deadline passes is investigated using a power system dispatch model. For the purpose of this work it is assumed that a 10% electric vehicle target in the Republic of Ireland is not achieved, but instead 8% is reached by 2025 considering the slow market uptake of electric vehicles. Our experimental study shows that the increasing penetration of EVs could contribute to approach the target of the EU and Ireland government on emissions reduction, regardless of different charging scenarios. Furthermore, among various charging scenarios, the off-peak charging is the best approach, contributing 2.07% to the target of 10% reduction of Greenhouse gas emissions by 2025.

Keywords

Carbon Emissions, Electric Vehicles, Power System, PLEXOS, Energy Forecasting

1. Introduction

In recent years, climate changes caused by carbon emissions have attracted considerable attention worldwide [1]. "WMO greenhouse gases bulletin" published by World Meteorological Organization in 2011 pointed out that the concentration of greenhouse gases (GHG) in the atmosphere has reached a high record in 2010. The average concentration of CO_2 reached 389.0 ppm, while the Nitrous oxide reached 323.2 ppb, with the increase of 39% and 20% respectively comparing with their counterparts during the industrial revolution [2]. The increased GHG emissions are largely due to the use of fossil fuels. The atmospheric pollution produced by transportation industry due to large usage of fossil fuel, oil and natural gas, has become a major issue, surpassing the traditional industrial pollution [3]. In terms of carbon emissions, transportation and electricity industry are the two key sectors [4]. In the European Union (EU) there are strict binding GHG reductions targets and renewable energy targets called 20 - 20 by 2020 [5]. In each EU member state the targets have been implemented. For example in the Republic of Ireland it is proposed that GHG emissions need to be cut by 60% - 80% by 2050 [6]. Carbon dioxide (CO_2) emission is a major part of GHG which accounted for 80% of all GHG emissions in 1990-2010. Thus the Irish government has set a target that CO_2 emissions be reduced to 20% by 2020 in keeping with EU directives [7]. Transport is one of the main sectors dependent on fossil fuels, and is thus a major source of GHG emissions. It is believed that electric vehicles (EV) can reduce heavy fossil fuel dependency, support better renewable power integration and thus reduce GHG emissions. In the Republic of Ireland, a country with a large wind power resource the Irish government set a 10% EV target by 2020 [8]. Consequently, it is thought that a better reduction in GHG emissionsis achievable and increased security of energy supply by reducing oil imports and more efficient wind power integration may be possible. However, there are many technical challenges relating to mass EV deployment, such as battery reliability, as well as the impacts of stochastic charging and potentially discharging on both the grid and the generators.

A number of studies have examined the interaction of EVs with the power system considering emissions. For example, the CO_2 emissions of battery electric vehicles (BEV) and plug-in hybrid electric vehicles (PHEV) are compared in [9] [10] provides a bottom-up model of EV carbon emissions and energy impacts in the Republic of Ireland.

In this paper, the EV technologies are briefly reviewed and the impacts of EV charged in the all-Ireland single wholesale electricity market after the 2020 deadline passes is investigated using a power system dispatch model called PLEXOS for power systems. For the purpose of this work it is assumed that a 10% electric vehicle target in the Republic of Ireland is not achieved, but instead 8% is reached by 2025 considering the slower market uptake of electric vehicles. The test system is the single wholesale electricity market (SEM) of Northern Ireland and the Republic of Ireland in 2025. Four different EV charging scenarios are analysed to determine increased renewable energy penetration, the net reduction in CO_2 emissions and the percentage contribution to the 2020 target.

2. Overview of EV Technology

2.1. Electric Vehicle Types

There are several types of EVs. The common types are BEVs and PHEVs. The BEV is powered by 100% electric energy, whereas PHEVs are powered by electric energy, as well as a downsized combustion engines. Both types have an electric motor powered by a rechargeable battery. This battery is recharged by connecting it to a power supply. A BEV does not emit any tail-pipe emissions because it operates in only electric mode. However, a PHEV produces emissions when using their combustion engines to drive the car. Nevertheless, both types may implicitly produce GHG emissions due to the need to charge the battery from the grid where thermal generation units are in use.

2.2. Electric Vehicle Battery Charging

The preferred battery type for the majority of EV manufacturers is lithium-ion due to that it has a high storage capacity. The current average battery driving range and battery capacity of commercial BEVs in the market are 130 km and 22 kWh respectively [11]. Battery performance of EVs and their ability to achieve stated driving ranges is highly dependable on the driving styles and environment conditions.

There are three types of charging options available for EVs, including home charging, public charging and fast charging. **Table 1** shows the charging options [11].

3. Model of the SEM and EV Charging Scenarios

In order to assess the impact of different EV charging scenarios on the power system, a base line power system

Table 1. Battery charging options.

Type	Electrical	Resulting Charge
Home Charging	230V 16A	100% in 6 - 8 hours
Public Charging	400V 32A	50% in 30 mins
Fast Charging	400V 63A	80% in 30 mins

model is first built. PLEXOS is an MIP-based (mixed integer programming algorithm) software used for the next-generation energy market simulation and optimization [12]. The software can realistically replicate the actual operation of generators in the physical market as all technical constraints can be modeled and obeyed. PLEXOS can be used to minimize cost or maximize profit and to integrate the analysis of variable energy resources. It could also be used for electrical market analysis.

3.1. SEM Test System

The Republic of Ireland and Northern Ireland share a synchronous power system known as the All-Island Gird (AIG), which facilitates the operation of the SEM. The SEM system was built in PLEXOS using [13]. These are based on published data from NERA Economic Consultants (2008), KEMA (2007) and Energy Regulation (2011) [14]. GDP grossing function from Gross Domestic Product Preliminary Estimate 2014 association with the growth of economy potential output and population growth curve were used to update 2007-2008 energy load demand of Ireland for the year of 2025 [15]. Using gross domestic product predictions for Ireland and the UK then load demand in each hour in a typical day by 2025 was extrapolated [16]. The totally load demand in 2025 as predicted to be 57,325 GWh.

The SEM 2025 baseline model needs a number of generation data and the generation schedule of 2025. In this research the generation data was developed based on the report of SONI 2014 and EirGrid 2014 [17]. We have updated the technical details for all the generators. The total predicted installed capacity for 2025 was estimated to be 14,983 MW. **Table 2** shows all the types of generators dispatched in the SEM by 2025 without an EV load. It is obviously that gas, coal and wind generations were the three largest contributors to generation capacity.

It is forecasted that approximately 20,032 MWh are available for dispatch via wind generators and the wind generation data was updated for EirGrid since 2009. It is assumed that the power supplied by wind generation is limited to 70% of total energy demand.

The fuel price is set and updated for generation on the basis of the fuel market form UK Fuel Price forecast report [18]. The carbon cost is set to £20/t for CO_2, which is referenced with the carbon prices usually used in PLEXOS, for instance, the carbon cost was set to £30/t in PRIMES EU-wide energy model. The fuel prices are listed in **Table 3**.

The SEM is linked to the British Electricity Trading and Transmission Arrangement (BETTA) via the 500 MW Moyle Interconnector. It is assumed that any existing flow constraints on the Moyle Interconnector have been removed by 2025. Another 500 MW interconnector is under construction between Rush, County Dublin to Bark by Beach, North Wales, which is called as the East West Interconnector (EWIC). In the present work, it was assumed that the EWIC commenced operation in 2012. Total interconnection to BETTA in PLEXOS in 2025 is represented as a single gas generator with 12 different heat rates and operating costs using the SEM validated market model produced by CER and UR (2011). This is because gas-fired generation is the predominant marginal plant in the BETTA and there is a strong correlation between the cost of gas-fired generation and the BETTA market price.

3.2. EV Charging Scenarios

An EV target has been set up by the Irish government that 10% of all vehicles fleet in all island should be replaced by EVs by 2025. It is assumed for this analysis that the 10% EV target may be difficult to achieve considering some of the existing challenges and slow EV sales. Therefore a figure of 8% EV market share was assumed in 2025. Thus the total EVs is calculated to be 343,918 vehicles. Furthermore the EV types were limited to just PHEV with a 3.3 kW and 16 kWh battery in this analysis. Four EV charging profiles were examined, including 1) off-peak and 2) peak charging, 3) Electric Power Research Institute (EPRI) [19] EV charging data

Table 2. Generation mix for 2025 without EV load.

Fuel	Generation (GWh)	Share of Generation (%)
Goal	6525.69	11.59
Distillate Oil	2.48	0.00
Gas	21797.59	37.32
Hydro	1825.78	3.16
Pumped storage	152.49	0.27
Wave	707.26	1.23
Wind	19979.85	34.71
Peat	2788.59	4.84
Interconnectors	3756.76	6.86

Table 3. Fuel prices.

Fuel type	Cost £/GJ
Gas	7.02
Oil	12.06
Coal	2.12
Peat	3.18
Hydro	0

and 4) stochastic charging as illustrated in **Figure 1** during a one day period.

1) Off-peak and peak charging profiles

Rate of EV domestic charging for PHEVs with a 16 kWh battery is shown in **Table 4**. Off-Peak charging profile is obtained from this EV data associated with the No. of EV in Ireland using 0.88 efficiency if charging from 00:00. The charging power was calculated as 847.76 MW. It is assumed that 85% of EVs charge during the week and 15% of EVs charge at the weekend when built the charging profile. Off-Peak charging data are listed in **Table 5**. This charging profile has been expanded to the whole year of 2025. Then the results were added as a purchaser load in PLEXOS 2025 baseline model. It can be seen that the EV's Peak charging profiles are similar with Off-Peak during the period of charging from 16:00 to 24:00.

2) EPRI Charging Profiles

An aggregated charge profile was created for the fleet of PHEVs in the model. 100% of the charge energy requirements are apportioned to each hour of the day. The data of EPRI charging are listed in **Table 6**. In this analysis, it was assumed that the highest charging loads occur during late night and early morning hours whereas the modest loads from daytime. The public or workplace charging presumably occurred in the middle of the day. Hours of minimal charging correspond roughly with commute times. This specific charge profile creates a scenario where 74% of the charging energy is delivered from 10:00 p.m. to 6:00 a.m. (nominally off-peak). The remaining 26% is provided between 6:00 a.m. to 10:00 p.m. It is one of the simplest scenarios among many possible scenarios and it represents an initial approximation of aggregate charging behavior in a fleet of PHEVs. The charging power was calculated as 467.73 MW.

3) Stochastic charging profiles

The stochastic charging profile is very similar to the EPRI charging profile, but it has a random percent value for each hour. Both of EPRI and stochastic charging profiles' function are number of all EVs multiplying 16 kWh for each PHEV capacity, and then multiplying the percent of each hour.

3.3. 2025 SEM Baseline Model

The power generation for the regular load demand in the SEM in 2025 is met by wind, gas and coal. As shown in **Table 2** that 34.71% of power comes from wind, 37.32% power from gas and 11.59% from coal. So the base

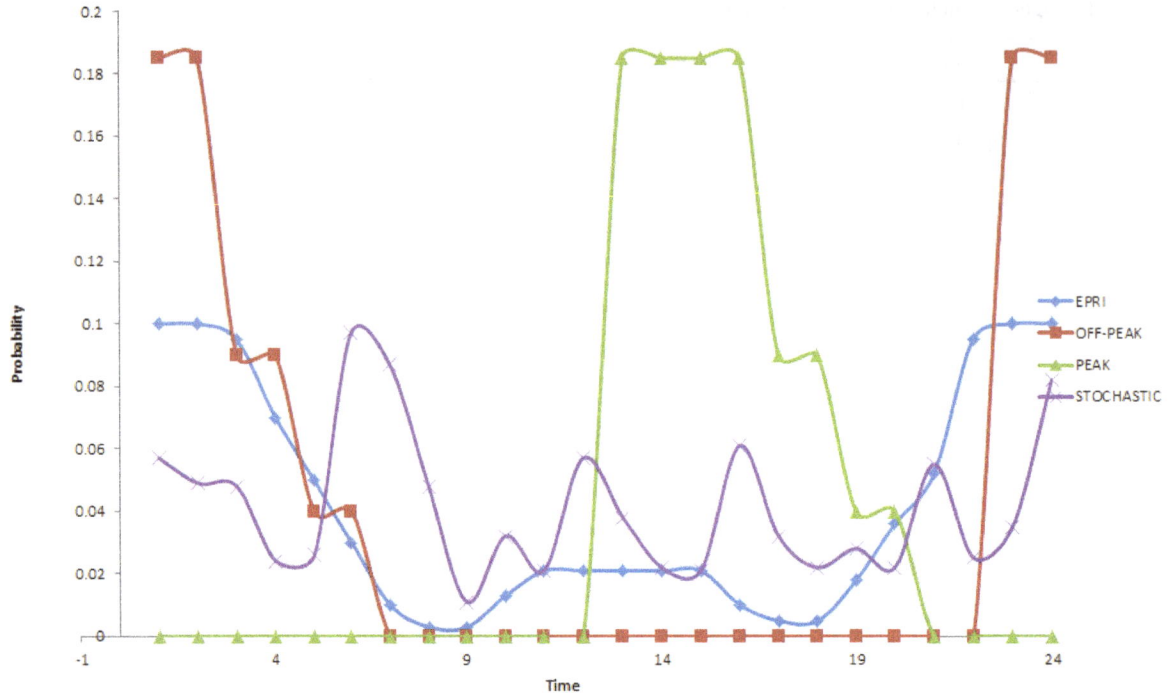

Figure 1. All type of EV charging profiles.

Table 4. Rate of EV domestic charging for PHEVs with 16 kWh battery.

Period	State of Charge	Charge Rate (kWh)
First 4 hours	73%	2.9
Intermediate 2 hours	90%	1.45
Final 2 hours	100%	0.58

Table 5. Off-Peak charging data for PHEV load demand.

Time of Charge	00:00	01:00	02:00	03:00
Weekday Load	847.7	847.7	847.7	847.7
Weekend Load	149.6	149.6	149.6	149.6
Time of Charge	04:00	05:00	06:00	07:00
Weekend Load	423.8	423.8	169.5	169.5
Weekend Load	74.8	74.8	29.9	29.9

Table 6. Data of EPRI charging.

Time	01:00	02:00	03:00	04:00	05:00	06:00
Charge percent	10%	10%	9%	6%	4%	2%
Time	07:00	08:00	09:00	10:00	11:00	12:00
Charge percent	1%	0.5%	0.5%	1.5%	2.5%	2.5%
Time	13:00	14:00	15:00	16:00	17:00	18:00
Charge percent	2.5%	2.5%	2.5%	1%	0.5%	0.5%
Time	19:00	20:00	21:00	22:00	23:00	24:00
Charge percent	2%	4%	6%	9%	10%	10%

load is met mainly by gas and wind. Meanwhile, the gas generators will act as support generator when wind is low. The results of baseline model for 2025 are shown in **Table 7**.

4. Resultsand Analysis

The validated SEM model [13] was a fundamental starting point to carry out this analysis as a complete set of baseline transmission and generation data was already set-up in PLEXOS. As the EV load demand and charging behaviors are highly unpredictable, different scenarios of EV charging, a range for load demands and charging profiles were examined in the 2025 SEM model.

4.1. Off-Peak Charging in Model

Off-peak charging means to allocate all the charging power during off-peak load time periods. It can be clearly seen from **Figure 1** and **Table 6** that the off-peak EV charging scenario makes a huge load on the baseline system during the night from 00:00. The profiles of dispatch change due to the off-peak EV charging are illustrated in **Figure 2**. Obviously, it could be found from **Figure 2** that gas is the predominant energy to meet the EV load. However, the additional demand from EV is also powered by other generators.

The profiles of off-peak load and base line load at 8th June 2025 were simulated using our model. It could be found that the maximum power appears during the day time (**Figure 3**). The annual cost to all EV and average price paid by purchaser are calculated as £61,568 and £195 respectively. EV renewable load of wind variability during off-peak is 503.87 GWh and EV Renewable Load is 1259.67 GWh. EV Renewable Load (kTOE) is 108.33 GWh. Thus the contribution to 10% renewable energy target (%) of wind variability during off-peak is estimated to be 2.07%. All outputs from the simulations in PLEXOS for each charging scenario, such as generation cost, system marginal price (SMP), emissions, cost of EV and so on, are reported in **Tables 8-10**, respectively.

4.2. Annual Power System with EV Load

Annual power system characteristics for each scenario are reported in **Table 11**. It is found that the ability of EVs to reduce GHGs emissions is highly relative to the multitudinous factors of generators in a generation portfolio, the type of the charging portfolio and the time of day when charging.

Table 7. Power system characteristics of 2025 SEM.

Total Generation Cost (£,000)	1693576.61
Load-Weighted Average SMP (£/MWh)	55.56
Wind Curtailment Factor (%)	0.26
CO_2 Emissions(kt)	15,851
NO_x Emissions (kt)	50.12
SO_x Emissions (kt)	39.38

Table 8. Power system characteristics of the 2025 sem with off-peak charging.

Charges in Total Generation Cost (£)	61568.29
Load-Weighted Average SMP (£/MWh)	55.91
Wind Curtailment Factor (%)	1.5
Annual Cost to Load per EV (£)	195
Average Load-Weighted Price Paid by Purchase r(£)	52.57

Table 9. Emissions generated in 2025 sem due to off-peak charging.

Changes in Emissions	PHEV
CO_2 Emissions (kt)	510.69
NO_x Emissions (kt)	1.69
SO_x Emissions (kt)	0.76

Table 10. Target contributions due to wind variability and off-peak.

EV Renewable Load (GWh)	EV Renewable Load-2.5 Weighting (GWh)	EV Renewable Load (kTOE)
503.87	1259.67	108.33
10% renewable energy target (%)	Net Reduction in CO_2 (kt)	20% emissions target (%)
2.07	229.3	1.74

Table 11. Annual power system characteristics for each scenario.

Model	Demand (GWh)	Generation (GWh)	CO_2 (kt)	NO_x (kt)	SO_x (kt)	SMP (£/MWh)	Total generation costs (£)	10% renewable target (%)	Net reduction in CO_2 (kt)	20% emissions target (%)
Baseline	57536.4	57536.4	15,851	50.12	39.38	55.56	1.693576	-	-	-
Off-peak	58792.6	58790.3	510.69	1.69	0.76	55.91	1.755145	2.07	229.3	1.74
Peak	58857.8	58855.3	512.21	1.63	0.46	69.6	1.765830	1.74	227.75	1.68
Stochastic	58846.7	58846.8	516.26	1.65	0.45	59.02	1.759621	1.92	223.7	1.65
EPRI	58843.8	58843.6	517.28	1.66	0.49	55.96	1.760463	2.04	222.68	1.74

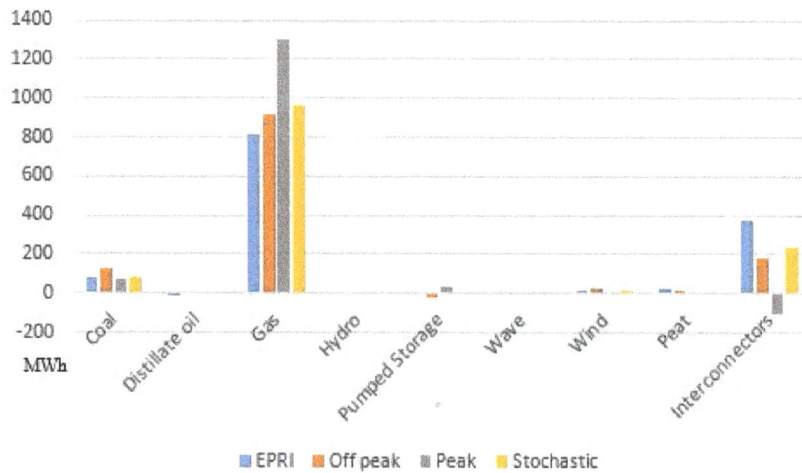

Figure 2. Dispatch changes due to EV charging.

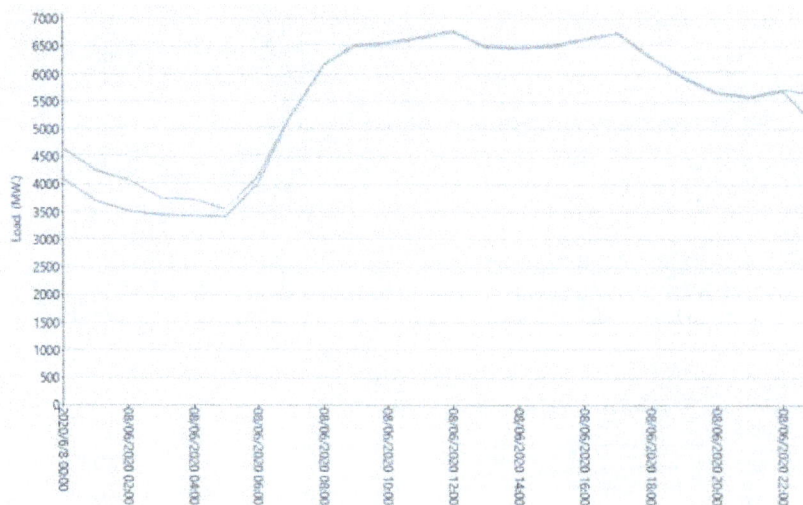

Figure 3. Profiles of off-peak load and base line load at 8th June 2025.

The EV load was predominantly charged by gas dispatch in all the charging scenarios as showed in **Figure 2**. The gas was assisted by coal, wind, pumped hydro and interconnector power depending on the size of the EV load and the period of charging. In addition, the same demand in different charging time will cause the change of totally generation cost and emissions. The SMP is reduced from peak charging to off-peak or EPRI charging. This indicates that if the EV owners charge their EV during the night time will save more for both themselves and the SEM. The total generation and generation cost also will reduce by a proper time charging.

5. Conclusion

This paper has investigated the impacts of EV charging on the power system and SEM in a future case of 2025. The SEM model and the four EV charging profiles are built in PLEXOS. The results from these all models were analyzed to outline the effects of additional EV load combined in a future power system. The present investigation confirmed that the increasing penetration of EVs could contribute to approaching the target of the EU and Ireland government in terms of emission reduction, regardless of different charging scenarios. In addition, it could also be found that the off-peak charging is the best way to charge EV load by the comparison with other three types of charging as shown in the previous research, contributing 2.07% to the target of 10% reduction of Greenhouse gas emissions by 2025.

Acknowledgements

This work was financially supported by UK EPSRC under grant EP/L001063/1 and China NSFC under grants 51361130153 and 61273040. The authors would also like to thank PLEXOS for providing software and Eirgrid SEMO and SONI for the datasets.

References

[1]   International Energy Agency (IEA) (2008) World Energy Outlook. IEA and Organization for Economic Co-Operation and Development (OECD), Paris.

[2]   Commission of the European Communities (2008) 2020 Europe's Climate Change Opportunity. COM 30 Final.

[3]   Hao, H., Wang, H. and Yi, R. (2010) Hybrid Modelling of China's Vehicle Ownership and Projection through 2050. *Energy*, **36**, 1351-1361. http://dx.doi.org/10.1016/j.energy.2010.10.055

[4]   Zivin, J., Kotchen, M. and Mansur, E. (2014) Spatial and Temporal Heterogeneity of Marginal Emissions: Implications for Electric Cars and Other Electricity-Shifting Policies. *Journal of Economic Behavior & Organization*, **107**, 248-268. http://dx.doi.org/10.1016/j.jebo.2014.03.010

[5]   Commission of the European Communities (2008) Communication from the Commission to the European Parliament, the Council, the European Economic and Social Committee and the Committee of the Regions.2020 by 2020 Europe's Climate Change Opportunity. COM (2008) 30 Final.

[6]   Gormley, J. ()2008   Government Sets Binding Target for 40% Power from Renewable Energy. Minister for Environment, Heritage and Local Government. [Press Release]

[7]   European Parliament and Council of European Union. Decision No. 406/2009/EC, 2009.

[8]   Foley, A., Gallachoir, B., Leahy, P. and McKeogh, E. (2009) Electric Vehicles and Energy Storage—A Case Study on Ireland. *IEEE Vehicle Power and Propulsion Conference*, Dearborn, 7-10 September 2009, 524-530. http://dx.doi.org/10.1109/VPPC.2009.5289805

[9]   Doucette, R.T. and McCulloch, M.D. (2011) Modeling the Prospects of Plug-In Hybrid Electric Vehicles to Reduce CO_2 Emissions. *Applied Energy*, **88**, 2315-2323. http://dx.doi.org/10.1016/j.apenergy.2011.01.045

[10] Foley, A., Daly, H. and Gallachóir, B. (2010) Quantifying the Energy & Carbon Emissions Implications of a 10% Electric Vehicles Target. In: *Proceedings of the* 2010 *International Energy Workshop.*

[11] Foley, A., Winning, J. and Gallachoir, B. (2010) State-of-the-Art in Electric Vehicle Charging Infrastructure. *IEEE Vehicle Power and Propulsion Conference* (*VPPC*), Lille, 1-3 September 2010, 1-6.

[12] PLEXOS User Manual.

[13] Foley, A., Tyther, B., Calnan, P. and Gallachóir, B. (2013) Impacts of Electric Vehicle Charging under Electricity Market Operations. *Applied Energy*, **101**, 93-102. http://dx.doi.org/10.1016/j.apenergy.2012.06.052

[14] Deane, J., Chiodi, A., Gargiulo, M., *et al.* (2012) Soft-Linking of a Power Systems Model to an Energy Systems Model. *Energy*, **42**, 303-312. http://dx.doi.org/10.1016/j.energy.2012.03.052

[15] Commission for Energy Regulation and Utility Regulator. Annual Load, Forecast, 2008.

[16] Medium and Long-Term Scenarios for Global Growth and Imbalances. http://www.oecd.org/berlin/50405107.pdf

[17] EirGrid and SONI Ltd. (2014) All-Island Generation Capacity Statement 2014-2023.

[18] DECC Fossil Fuel Price Projections.
https://www.gov.uk/government/uploads/system/uploads/attachment_data/file/65698/6658-decc-fossil-fuel-price-proje
ctions.pdf

[19] Electric Power Research Institute, Environmental Assessment of Plug-In Hybrid Electric Vehicles, Volume 2: United States Air Quality Analysis Based on AEO-2006 Assumptions for 2030, 2007, Palo Alto, CA and National Resources Defense Council, New York.

10

Obstacles Facing Developing Countries in Power System Planning

Abdullah M. Al-Shaalan

Electrical Engineering Department College of Engineering King Saud University, Riyadh, Saudi Arabia
Email: Shaalan123@gmail.com

Abstract

The problem of power system planning, due to its complexity and dimensionality, is one of the most challenging problems facing the electric power industry in developing as well as developed countries. In planning phase, two of the most important decision-making parameters are the reliability and costs. The latter includes both system investment costs and outages costs. In this paper, these parameters are described and the interrelation between them is evaluated. Some previous approaches and developed techniques will he applied to a particular planning problem in a developing country and some aspects having a significant impact on the decision making process in the planning phase will be considered.

Keywords

Power System Planning, Developing Countries, Reliability Evaluation, System Cost, Outages Cost, Systems Interconnection, Load Uncertainty

1. Introduction

The main issue regarding power system planning in developing countries is to establish basic principles and criteria to serve as a framework within which the process of planning may proceed. The framework of power system planning should be flexible, with the broad objectives of finding a plan (or plans) which guarantee a desired degree of continuous and least cost service. Good service or, in other words, acceptable reliability level of a power system usually requires the addition of more generating capacity to meet the ever increasing electrical demand [1] [2]. However, in many fast developing countries with vast, sparsely populated areas reliability-cost tradeoffs exist to satisfy the fast load growth either by investment in additional generating capacity for isolated systems or by building transmission lines to interconnect these systems in such a way as to transfer power between their load centers in case of emergencies and power shortages. Therefore, reliability and cost constraints are major considerations in power system planning process [3].

2. Reliability Aspects in Power System Planning

Reliability is one of the most important criteria which must be taken into consideration during all phases of power system planning process. Reliability criterion is required by the system planers and authorized manage-

ments in the utilities to establisharget reliability levels and to consistently analyze and compare the future reliability levels with feasible alternative expansion plans. This need has resulted in the development of comprehensive reliability evaluation in power system planning modeling and techniques [4]-[8].

One reliability index, known as the Loss of Load Expectation (LOLE), is presently the most commonly
adopted and used probabilistic criterion in power system generation expansion planning. This index computes
the expected (long term average) number of days per year on which the available generating capacity is not sufficient to supply all the period peak load levels. The first step in evaluating this index is to define models that
describe two situations of interest, namely, the capacity model and the load model. These two models are convolved together to produce the risk index known as the LOLE.

In order to derive the capacity model, each unit in the system is represented by a 2-state capacity model (*i.e.*
available full capacity or unavailable capacity). Each state is then weighted by its probability of occurrence. This
process creates the capacity model known as the "Capacity Outage Probability Table, COPT" which contains an
array of capacity states and their associated probabilities of occurrences.

The other load model needed for the LOLE index evaluation is known as the "Load Duration Curve, LDC"
which represents an arrangement of a typical period individual load levels in a descending order of magnitude
(*i.e.* starting with the peak load level and ending with the minimum load level) as the one shown in **Figure 1**.

2.1. Loss of Load Expectation (LOLE)

The two models: COPT and LDC mentioned above are to be combined together in order to yield the required reliability risk index LOLE as can be exhibited in the following equation:

$$\text{LOLE} = \sum_{i=1}^{n} t_i \cdot p(O_i) \quad (\text{days/year}) \quad (\text{outage} > \text{Reserve})$$

where

t_i : time duration of that sever outage O_i

$p(O_i)$: probability of loss of load due to the ith severe outage of size O_i

n : total number of severe outages ocurred during that period considered

2.2. Expected Load Not Served (ϵLNS)

In power system reliability evaluation, sometimes another reliability index beside the *LOLE* is needed to know the
magnitude of loads that have been lost due to severe outages (*i.e.* when the existing loads exceed the available system capacity). So, this index is known as the Expected Load Not Served (ϵLNS) and can be evaluated as in the
following equation:

$$\epsilon\text{LNS} = \sum_{i=1}^{n} (\text{LNS})_i \cdot p(O_i) \quad \text{MW}/y \quad (\text{outage} > \text{Reserve})$$

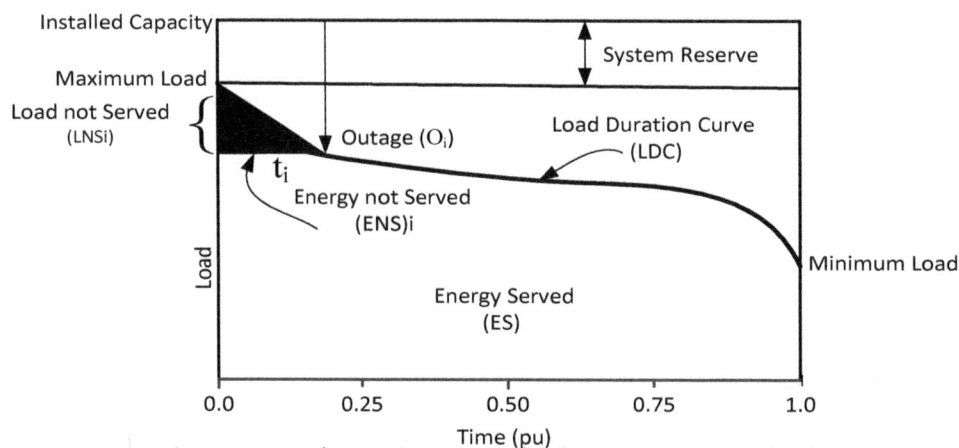

Figure 1. Load duration curve displaying various load-related variables.

2.3. Expected Energy Not Served (ϵ ENS)

Since the energy not served $(\text{ENS})_i$ caused by power outages reflects great damages and heavy losses to the entire consumers classes, so, another essential and most needed reliability index known as the Expected Energy not Served (ϵENS) can be deduced as in the following equation:

$$\epsilon\text{ENS} = \sum_{i=1}^{n}(\text{ENS})_i \cdot p(O_i) \quad \text{MWh}/y \quad (\text{outage} > \text{Reserve})$$

where

$(\text{ENS})_i = (\text{LNS})_i \cdot t_i$: *The energy not served due to severe* i^{th} *outage of size O in time* t

3. Economic Aspects in Power System Planning

There are several costs that are associated with power systems planning and can be manifested in the following sections

3.1. Fixed Cost

The fixed cost (FC) represents the cash flow at any stage of the planning horizon resulting from the costs of installing new generating units during the planning period. It depends on the current financial status of the utility, the type and size of generating units and the cost of time on money invested during the planning period. The total fixed costs (FC_T) for unit(s) being installed can be computed as:

$$\text{FC}_T = \sum_t \sum_k (\text{CAP}_k \cdot \text{CC}_k \cdot \text{NU}_k)^t$$

where

CAP_k : unit capacity added to the system of type k .
CC_k : capital cost of unit of type $k\,(\$/\text{kW})$.
NU_k : number of unit(s) added to the system of type k at each interval of time t
t : interval period of time considered in the planning horiozon, $t = 1, \cdots, T$

3.2. Variable Cost

The variable cost, (VC), represents the cost of energy served by the system. It is affected by the load variation, the type and size of generating units and the number of hours of operation. Also these costs are related to the cost of operation and maintenance (fuel, scheduled maintenance, interim spare parts, repair, staffing, wages and miscellaneous expenses) and can be evaluated as:

$$\text{VC}_T = \sum_t \sum_k (\epsilon\text{ES}_k \cdot \text{ESC}_k \cdot \text{NU}_k)^t$$

where

ϵES_k : expected energy served by unit of type k
ESC_k : energy served cost of unit of type k $(\$/\text{kWh})$
The total system costs (SC_T) for the entire expansion plan can be estimated by summing all the above individual costs at every stage of the planning period as being expressed in the following equation:

$$\text{SC}_T = \text{FC}_T + \text{VC}_T$$

3.3. Outages Cost

In power system cost-benefit analysis, the outages cost (OC) form a major part in the total system cost These costs are associated with that energy demanded but cannot be served by the system due to severe outages, and is known as the expected energy not served, (ϵENS). Outages cost is usually borne by the consumers as well as by the utility. The utility outages cost include loss of revenue, loss of goodwill, loss of future sales and increased maintenance and repair expenditure. However, the utility losses are seen to be insignificant compared with the losses incurred by its consumers when power interruptions and energy cease occur. The consumers perceive power outages and energy shortages differently. A residential consumer may suffer a great deal of anxiety and inconvenience if an outage occurs during a hot summer day or deprives him from domestic activities or causes food spoilage. For a commercial user, he will also suffer a great hardship and losses of being forced to close until power is restored. Also, an outage may cause a great damage to an industrial customer if it occurs disrupting and disabling the production processes [9] [10].

One method of evaluating the ϵENS is described in [11]. Therefore, for estimating the outages cost (OC) is to multiply the value of that ϵENS by an appropriate Outage Cost Rate (OCR), as follows:

$$OC_T = \sum_t (\epsilon ENS \cdot OCR)^t$$

where OCR: US\$/kwh and ϵENS : kWh lost.

The total cost of supplying the electric energy to the consumers is the sum of system cost that will generally increase as consumers are provided with higher reliability and customer outages cost that will, however, decrease as system reliability increases or vice versa. This total system cost (TSC) can be expressed as in the following equation:

$$TSC_T = SC_T + OC_T$$

The prominent role of outage cost estimation, as revealed in the above equation, is to assess the worth of power system reliability by comparing this cost (OC) with the size of system investment (SC) in order to arrive at the least overall system cost that will establish the most appropriate system reliability level that ensures energy continuous flow as well as the least cost of its production.

The incorporation of customer outage costs in investment models for power system expansion plans is very difficult for planners in fast developing countries. This difficulty stems principally either from the lack of system records of outage data, failure rate, frequency, duration of repair etc., or the failure to carry out customer surveys to estimate the impact and severity of such outages in terms of monetary value.

4. Models Developed for the Reliability and Cost Evaluation Utilized in This Study

To perform the assessments and analyses of this study, a computer program containing four basic models has been developed at the King Saud University. These models, shown in **Figure 2**, assess the requirements of developing power systems in order to satisfy specified reliability and economic criteria and they are briefly described as follows:

SYSDAT model: This model prepares two essential data files, namely, the capacity file and the load file. The first one is known as the Capacity Outage Probability Table (COPT) which contains all the outage capacity states with their associated probabilities of occurrence. The second one is known as the Load Duration Curve (LDC) which arranges the load levels in a typical load variation curve in a descending order of magnitude starting with the maximum load and ending with the minimum load. This preparation process starts by the SYSDA at the beginning of each year of the planning period and then is sent to the SYSREL in the next stage of the planning process in order to perform system reliability evaluation task.

SYSREL model: Receives the data files of both the COPT and the LDC. These two files are then convolved (combined together) to yield the system evaluated risk level, *i.e.*, the Loss of Load Expectation (LOLEe). This LOLEe is then compared with the risk level prescribed by the utility management (LOLEp). If The LOLEe exceeds the LOLEp, an additional capacity should be added to the system in order to maintain itsrisk level within the satisfactory and accepted level prescribed by management decision, otherwise it proceeds to the next year.

SYSENR model: estimates the energy served (ES) by each generating unit residing in the system as well as the energy not served (ENS) due to forced power outages. It adopts a priority loading order, *i.e.* the generating units are loaded according to their least operation cost. Hence, operating, first, the most efficient and economic operating units (called the base units), followed by the more costly operating units (called the intermediate units), then followed by the most costly operating units (called the peak units), and so on. This means that the least cost operating units occupy the lower levels in the LDC area, and the more expensive operating units occupy the upper levels in the LDC respectively.

SYSCOS model: computes all system pertinent costs mentioned in the scope and contained in the context of this study like: fixed costs (FC), variable cost (VC), outages cost (OC) and total system cost (TSC).

SYSCON model: evaluates system reliability levels (LOLE) for systems after being interconnected. It reveals

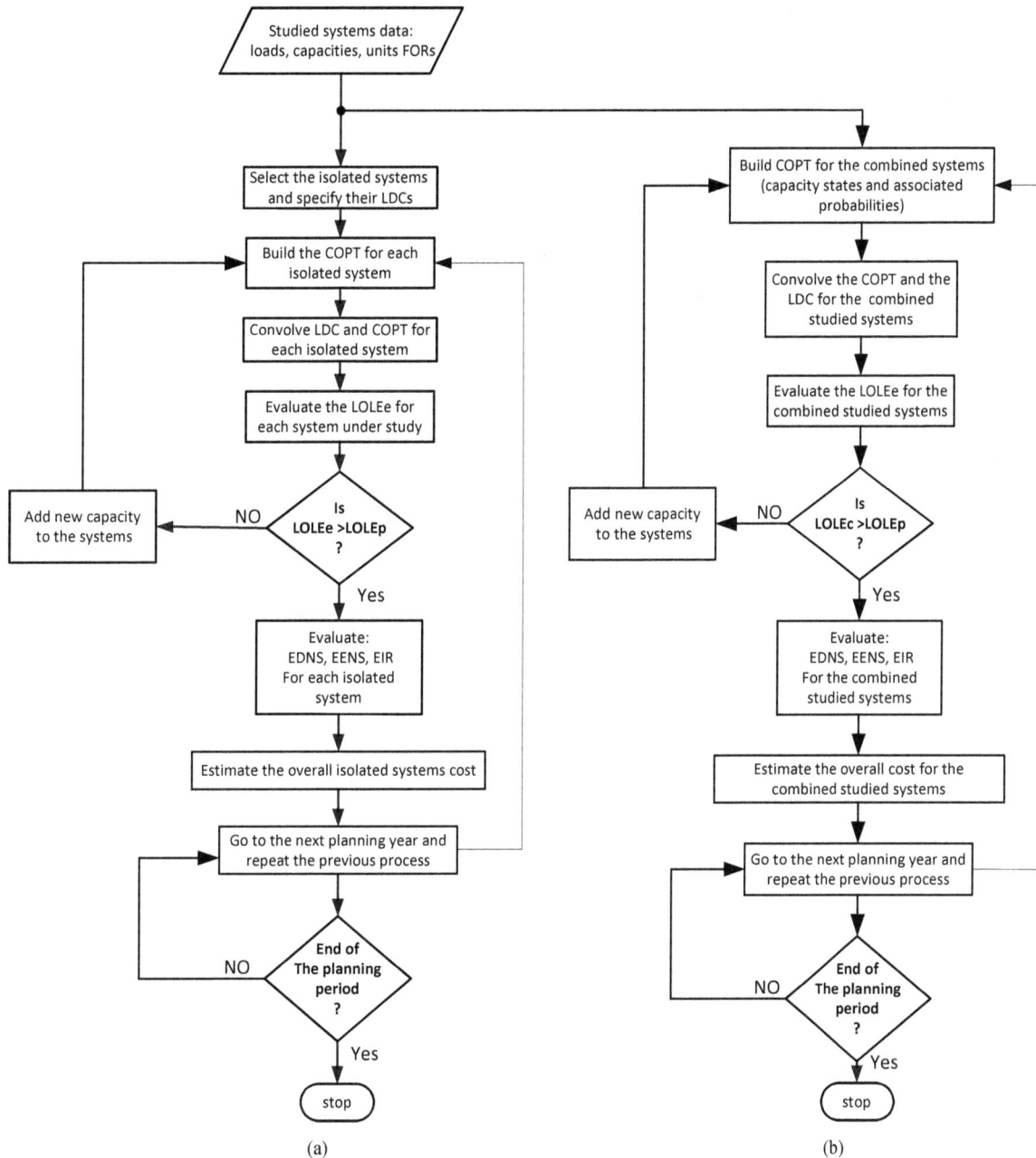

Figure 2. Flowchart for the proposed planning approach. (a) Isolated systems. (b) Interconnected systems.

the merits and advantages of system interconnection in terms of reliability level improvement and reserve capacity saving.

5. Case Study

The previous techniques have been applied to a particular case in a developing country. This case study is based on two real power systems situated in the southern part of the Kingdom of Saudi Arabia where are abbreviated in this study as systems A and B respectively. These two power systems are supposed to serve a major populated community with a potential future commercial and industrial load growth. The study considers that uncertainty is a vital aspect of power systems planning in developing countries that must be taken into consideration. Thus, the analysis procedure generally involves identifying the potential uncertain events and assigning a probability to the event occurrence. The impacts may then be probability-weighted, and a composite system impact value can be computed. This process may be repeated by examining alternative or contingency plans [12] [13].

5.1. Separate and Integrated Systems

Most power systems have grid interconnections either within the country or among neighboring countries. One objective reported in this paper is to evaluate the reliability benefits associated with the interconnection of power systems. Therefore, study is focused on reliability evaluation of systems A and B both as isolated systems and as interconnected systems. Analysis of this type enables the benefits, if any, that may accrue from integrated rather than isolated systems, to be explored as well as deciding viable generation expansion plans. Therefore, a 6-year expansion plan for systems A and B, assuming a reliability criterion (LOLE) of 0.1 days/year (frequently quoted as a practical value), is determined by implementing the methodology exhibited in **Figure 2**. The analysis represents the expansion plans for both systems as being isolated and interconnected. A summary of these expansion plans is shown in **Table 1** and plotted in **Figure 3**.

The two systems are reinforced whenever the reliability index exceeds the prescribed risk level at any year of the planning horizon. At years 2 and 4, when both systems reliability levels exceed the prescribed limit, unit (s) must be added. The results, displayed in **Table 2**, show that the number of units and the present value costs are reduced if the two systems are interconnected rather than being isolated which means saving in both units numbers and installation costs.

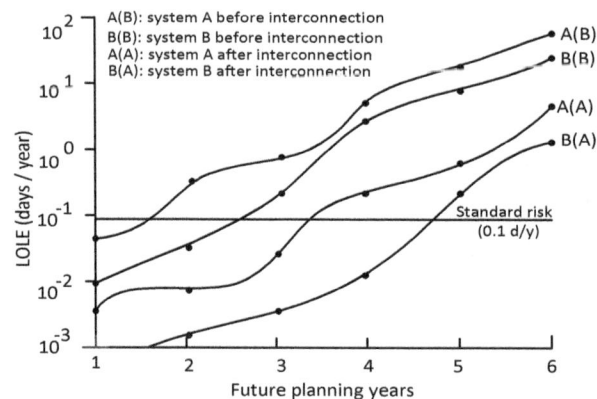

Figure 3. Variations of LOLE before and after interconnection.

Table 1. LOLE index for both systems as isolated and as interconnected.

Year	A(B)	A(A)	B(B)	B(A)
1	00.07234	0.00637	0.00926	0.00006
2	00.68861	0.00709	0.06790	0.00313
3	00.96242	0.08670	0.51488	0.00645
4	00.85371	0.13842	4.62191	0.08838
5	30.16790	0.75614	9.95545	0.21623
6	80.93185	8.29678	30.60693	0.94383

Table 2. Systems costs for isolated and interconnected systems.

System	Isolated			Interconnected		
	No. of units	Cost (MUS$)	ϵENS (MWh)	No. of units	Cost (MUS$)	ϵENS (MWh)
A	3	18.62	8.652	2	9.44	1.054
B	2	10.42	6.852	1	6.75	2.045

From the above analysis, it can be concluded that both systems will benefit from the interconnection in terms of saving in installation costs as well as reduction in size of energy interruptions. However, the next step must assess the economic and technical merits that may result from either building costly high voltage transmission lines in order to integrate isolated systems or instead, adding generating capacity to each system independently.

5.2. Uncertainty in Future Loads Growth

In some existing generation expansion planning methods it has been assumed that the forecast peak demands do not change. In fast developing countries, this assumption does not hold rigorously and there is always some degree of uncertainty in future loads growth forecasting. This uncertainty is likely to affect the system reliability levels and consequently to influence the capacity planning decision [8]. To investigate the uncertainty impact System "A" was chosen to be analyzed. The forecasted peak load was represented by a normal distribution having a standard deviation of 10% and this remained constant for the planning period. There are several important aspects associated with load uncertainty were evaluated such as system cost and outages costs using a load model having 7 discrete intervals. The effect of load uncertainty upon both system costs and outages costs are tested and the results are shown in **Table 3** and depicted in **Figure 4** which reveals that these costs (*i.e.* SC and OC) increase with increasing loads which implies reduction in the prescribed reliability level and hence requires more investment and operation costs [14].

5.3. Uncertainty in Unit Installation Date

In developing countries, a delay in unit installation date, due to probable undesirable economic conditions (e.g. lack of investment, rare resources, political havoc etc.) should be expected and taken into consideration in the planning horizon. Hence, to show the effect of more than on year's delay in unit installation dates has been investigated using system A. **Figure 5** shows the effect of delaying unit installation date for an extended number of future years and its consequences upon both system cost (SC) and outage cost (OC). It is evident from the table that installation delay has an adverse impact upon system reliability index (LOLE) as well as system costs that are directly related to it.

If more uncertainties in installation dates are assumed, results depicted by **Figure 4** show that, as unit deferring is increased, the outages cost increase rapidly but that the system cost steadily decreases. On the contrary, the timely installation has less effect on the outage costs than in the delayed case. Consequently, incentives should exist to justify decisions upon delaying or complying with the scheduled date of unit addition. One reason could be that it would be a catastrophic if unit installation is postponed for longer periods as shown in **Figure 5**.

In developing countries data collection is not an easy task and it is often difficult to establish probabilistic data for a system which did not have regular and organized collection of data for the use in probabilistic techniques. It is, therefore, important to establish systematic data collections describing all behavior aspects of power system which can then be used in reliability and economic evaluation for future planning and studies which are critically needed for power system planning in developing countries.

The effect of advancing unit installation date has been analyzed again using the same system A. The sensitivity analysis results are is shown in **Table 4** and depicted in **Figure 5**, The results indicate that advancing unit

Table 3. Data for load forecast uncertainty.

No. of St. Dev.	Load Uncertainty (%)	Load Level (MW)	Probability	System cost (MUS$)	Outages Costs (MUS$)
−3	−15	29	0.006	1	0.1
−2	−10	32	0.061	4	0.2
−1	−5	36	0.242	6	0.3
0	0	40	0.382	8	0.4
1	5	44	0.242	13	0.8
2	10	48	0.061	17	1.3
3	15	53	0.006	25	2.2

Table 4. System costs variation for timely and (delayed) installation dates.

Year	Peak Load (MW)	LOLE (days/yea)	Unit added	SC (MUS$)	OC (MUS$)	Percentage Increase (decrease)
1	435	0.02 (2.1)	0 (0)	00 (00)	2.2 (2.2)	2.0 (2.3)
2	511	0.08 (1.8)	0 (0)	00 (00)	4.4 (4.4)	5.4 (5.4)
3	571	0.03 (2.3)	1 (0)	44 (00)	3.2 (6.8)	7.2 (7.8)
4	632	0.09 (2.7)	0 (1)	00 (38)	5.3 (4.3)	10.3 (10.4)
5	712	0.05 (3.2)	1 (0)	31 (00)	4.9 (7.4)	13.7 (14.8)

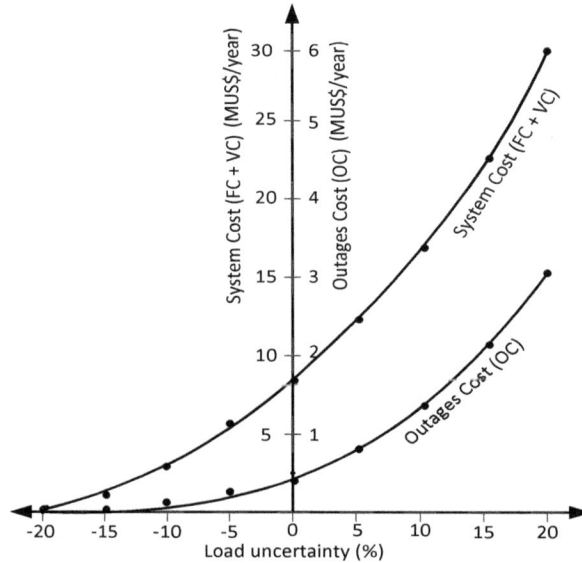

Figure 4. Effect of load uncertainty on system and outages costs.

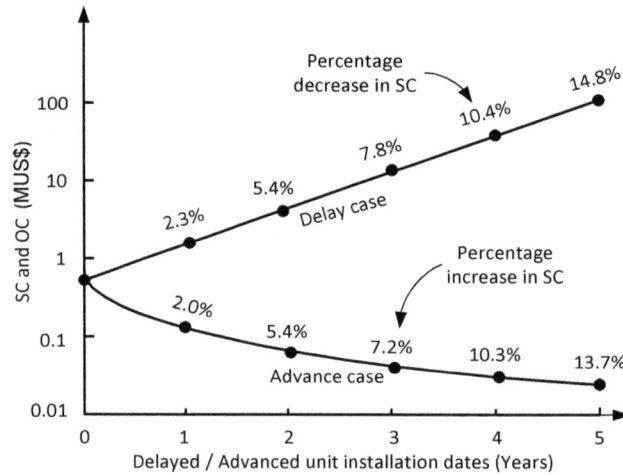

Figure 5. Effect of delay/advance in installation dates.

installation date has less effect compared with delay case on the outage cost In the advance case, outage costs are less sensitive because there is no critical need to advance the installation date. This can be argued on the basis that the system under analysis (System A) is reliable enough due to the selected reliability criterion of 0.1 days/year. However, the system cost is greater due to the earlier time investment. For convenient comparison, both delay case and advance case are plotted as function of successive future years.

6. Conclusions

In this paper, significant issues that may arise in power system planning in developing countries have been considered, analyzed and discussed. Two major constraints associated with power planning process, namely, reliability and cost have been modeled and applied to particular systems expansion planning in a developing country. The results demonstrate the benefits and merits associated with both reliability and cost of interconnecting isolated power systems into an integrated system. The uncertainly in future loads growth and unit installation time can be costly and undesirable. Therefore, their effects should be anticipated and studied in order to mitigate their effects so that possible deterioration in system reliability level as well as unnecessary additional expenditure can be averted.

In developing countries data collection is not an easy task and it is often difficult to establish probabilistic data for a system which did not have regular and organized collection of data for the use in probabilistic techniques. It is, therefore, important to establish systematic data collections describing all behavior aspects of power system which can then be used in reliability and economic evaluation for future planning and studies which are critically needed for power system planning in developing countries.

References

[1] Covarrubias, A.J. (1988) Expert Systems and Power System Planning in Developing Countries. *IEEE Conference Proceedings, Southeastcon'88*, 11-13 April 1988, 551-555.

[2] Shaalan, A.M. (2011) Essential Aspects of Power System Planning in Developing Countries. *Journal of King Saud University—Engineering Sciences*, **23**, 27-32.

[3] Schramm, G. (1990) Electric Power in Developing Countries: Status, Problems, Prospects. *Annual Review of Power*, **15**, 307-333. http://dx.doi.org/10.1146/annurev.energy.15.1.307

[4] El-Zayyat, H., *et al.* (1991) Power System Planning in Developing Countries, the Case of Jordan. *Mansoura Engineering Journal (MEJ)*, **16**.

[5] Meier, P. (1990) Power Sector Innovation in Developing Countries: Implementing Investment Planning under Capital and Environmental Constraints. *Annual Review of Power*, **15**, 277-306. http://dx.doi.org/10.1146/annurev.eg.15.110190.001425

[6] Agalgaonkar; A.P. and Kulkarni, S.V. (2006) Evaluation of Configuration Plans for DGs in Developing Countries Using Advanced Planning Techniques. *IEEE Transaction on Power Systems*, **21**, 973-981.

[7] Wilbanks, T.J. (1990) Implementing Environmentally Sound Power Sector Strategies in Developing Countries. *Annual Review of Power*, **15**, 255-276. http://dx.doi.org/10.1146/annurev.eg.15.110190.001351

[8] Sadeghzadeh, S.M. and Ansarian, M. (2006) Techno-Economic Power System Planning in Developing Countries. *7th IET International Conference on Advances in Power System Control, Operation and Management (APSCOM)*, Hong Kong, 30 October-2 November 2006, 216. http://dx.doi.org/10.1049/cp:20062083

[9] Billinton, R. and Allan, R.N. (1984) Reliability Evaluation of Power Systems. Pitman Books, London. http://dx.doi.org/10.1007/978-1-4615-7731-7

[10] Shaalan, A. M. (2012) Reliability Evaluation in Generation Expansion Planning Based on the Expected Energy Not Served. *Journal of King Saud University (Engineering Sciences)*, **1**.

[11] Shaalan, A.M. (1984) Reliability Evaluation in Long-Range Generation Expansion Planning. PhD Thesis, Victoria University of Manchester, Manchester, UK, April 1984, 51-58.

[12] Rudnick, H. and Quinteros, R. (1998) Power System Planning in the South America Electric Market Restructuring. *VI Symposium of Specialists in Electric Operational and Expansion Planning—VI SEPOPE*, Bahia, 24-29 May 1998.

[13] Weber, A., *et al.* (2013) Long-Term Power System Planning in the Context of Changing Policy Objectives. *Workgroup for Infrastructure Policy (WIP)*, University of Technology, Berlin, June 2013.

[14] Flage, R. and Aven, T. (2009) On Treatment of Uncertainty in System Planning. *Reliability Engineering & System Safety*, **94**, 884-890. http://dx.doi.org/10.1016/j.ress.2008.09.011

11

Dynamic Equivalent Method of Motor Loads for Power Systems Based on the Weighted

Hanmei Hu[1], Bo Hong[1], Ting Chen[2], Qinfeng Li[1]

[1]College of Electrical Engineering and New Energy, China Three Gorges University, Yichang, China
[2]College of Electrical Engineering, Wuhan University, Wuhan, China
Email: hongbo8966@163.com

Abstract

Dynamic equivalence can not only largely reduce the system size and the computation time but also stress the dominant features of the system [1]-[3]. This paper firstly recommends the basic concept of dynamic equivalent and the status of both domestic and abroad development in this area. The most existing equivalent methods usually only deal with static load models and neglect the dynamic characteristics of loads such as induction motors. In addition, the existing polymerization method which is based on the frequency domain algorithm of induction electric machines parameters takes a long time to equivalent for the large system, then the new method based on the weighted is proposed. Then, the basic steps for dynamic equivalence with the weighted method are introduced as follows. At first, the clustering criterion of motor loads based on time domain simulation is given. The motors with similar dynamic characteristics are classified into one group. Then, the simplication of the buses of motors in same group and network is carried out. Finally, parameters of the equivalent motor are calculated and the equivalent system is thus obtained based on the weighted. This aggregation method is applied to the simple distribution system of 4 generators. Simulation results show that the method can quickly obtain polymerization parameters of generator groups and the aggregation model retains the dynamic performance of the original model with good accuracy, the active and reactive power fitting error is smaller as well.

Keywords

Power System; Dynamic Equivalent; Induction Motors; Parameter Aggregation; The Weighted Method

1. Introduction

With the increasing growth of power grid interconnection, the scale of power system becomes larger and larger. Power systems consist of many synchronous generators and each generator includes many elements. In the study of power system dynamics, employing all of the detailed elements of generators in modeling creates a sophisticated and large system which put a heavy computational burden in simulation. Generally we only interested in one local system which is called research system (also called internal system). For external subsystem far side, we just need to keep its dynamic effects on research system unchanged in the study. There is no need to investigate its internal structure in detail, at the same time we can simplify the system by making appropriate equivalent. This area we want to simplify is the so-called external equivalent system. These are the fundamentals of dynamic equivalents.

Dynamic equivalents are used to reduce the computing effort and manifest the principal characteristics of Power systems. In general it consists of coherency method, mode method and identification method so far. The coherency method is widely used for its advantages in physical transparence, adaptation to non-linear systems and big disturbances, and it can be used directly in transient stability analysis. It is also fit for equivalent of large scale system with a great velocity and a controllability of precision.

Many research works have been carried out in the dynamic equivalence area [4]-[17]. Researchers in China also have made important contributions. The most existing equivalent methods usually only deal with static load models and neglect the dynamic characteristics of loads such as induction motors. In addition, most equivalence methods existing in the domestic adopt the frequency domain aggregation algorithm. This approach assumes that the transfer function of the generator and its control system are divided into several links which are aggregated respectively. Since the polymerization of coherent generators is complex, it takes a long time to equivalent for the large system. In order to solve this situation, in this paper, a dynamic equivalent method based on the weighed which considers motor dynamics is presented. The basic principle and procedure for aggregation are described. Clustering criterion of motors and parameter aggregation for the equivalent motor are also discussed in detail. Simulation results with a simple distribution system have validated the proposed method. During the disturbance, the power generation unit has the same speed, voltage, total mechanical power and total active power as the original group.

2. Clustering of the Motors

Most generator clustering is based on the coherency principle. It depends on whether the rotor angle of each generator can swing coherently. In fact it demands the synchronous rotors speed to be the same in a group.

Similarly, the clustering of motors may be determined according to whether each motor rotor speed ω_r is the same.

The spectral coefficient clustering analysis method is applied to motor clustering. It can be done as follows.

1) Forming the parametric index set $\{X_1, X_2\}$, where $X_1 = T_J R_r$, $X_2 = K_L$

2) Normalizing the original indexes. $\{X_1', X_2'\}$ is used to represent the parametric index set after normalization.

3) Preliminary clustering. At the beginning each motor is clustered into a group. It can be written as

$$l = 0, m = N_M, G_i^{(0)} = \{X_i'\}, i = 1, 2, \cdots, N_M,$$

where l is used to count the loops, m is the number of motor groups and N_M is the number of motor buses.

4) Calculating the parametric distance between each group. The parametric distance is calculated by

$$D_c'(i, j) = \|X_i' - X_j'\|$$

A $m \times m$ symmetric parametric distance matrix D_c' is available after this step.

5) Finding out the closest groups in the distance space.

The parametric distance D_c' and electrical distance D_e' are comprehensively considered.

6) Checking the number of motor groups m. Generally speaking, the simulation results will have satisfying accuracy when the motors are classified into two groups. So, if m is greater than 2, then return to step 4) to repeat the clustering of motors; otherwise stop.

3. The Simplication of Buses and Network

3.1. The Simplication of Motor Buses

The motor buses will be combined and simplified first after identification of the coherent motor group. Assume the bus set of the motor group which is used for combine as $\{c\}$, and the system bus set associated with $\{c\}$ as $\{b\}$, and the system bus set not related to $\{c\}$ as $\{a\}$. In the simplify, use an equivalent bus set $\{t\}$ to replace bus set $\{c\}$ and retain bus set $\{b\}$, but the associated branch of $\{c\}$ need to be converted into the associated branch of equivalent bus set $\{t\}$, keep the bus set $\{a\}$ and all of its power system the same before and after simplification. The node equation of original system is obtained as follows (the subscripts A, B, C respectively represent bus $\{a\}$, $\{b\}$, $\{c\}$):

$$\begin{bmatrix} I_A \\ I_B \\ I_C \end{bmatrix} = \begin{bmatrix} Y_{AA} & Y_{AB} & 0 \\ Y_{BA} & Y_{BB} & Y_{BC} \\ 0 & Y_{CB} & Y_{CC} \end{bmatrix} \cdot \begin{bmatrix} U_A \\ U_B \\ U_C \end{bmatrix} \tag{1}$$

the node equation for the new system after the simplification of the coherent motor bus is as follows:

$$\begin{bmatrix} I_A \\ I_B \\ I_t \end{bmatrix} = \begin{bmatrix} Y_{AA} & Y_{AB} & 0 \\ Y_{BA} & Y_{BB}^* & Y_{Bt} \\ 0 & Y_{tB} & Y_{tt} \end{bmatrix} \cdot \begin{bmatrix} U_A \\ U_B \\ U_t \end{bmatrix} \tag{2}$$

Remain Y_{AA}, Y_{AB}, Y_{BA} the same in the equivalent and calculate Y_{Bt}, Y_{tB}, Y_{tt}. And the diagonal element of Y_{BB} in the original system node equation need to correct as Y_{BB}^* accordingly. The equivalent need to satisfy the constraint conditions for $\dot{U}_A$ and $\dot{U}_B$ in the steady state. The exchanged power between different bus in $\{b\}$ and $\{c\}$ is unchanged in the steady state, so it is also called identical power transformation.

3.2. Network Simplication

The key of network simplification is the elimination of nonlinear load. The network steady trend deviation can be reduced to zero and the dynamic error can be as small as possible by displacing the nonlinear load to the remained bus equally in the process of elimination .The common methods just like REI (Radial Equivalent Independent).

4. Aggregation of Motor Parameters Based on the Weighted

The traditional frequency domain polymerization is rigorous in theory, but it also has some weaknesses as below: aggregation algorithm is complex, and it takes a long time to equivalent for the large system. The following weighted method simplifies the parameter aggregation process under the condition of guaranteeing accuracy, which can save the calculation time and advantageous to project realization.

We can get one motor group $G_M = \{1,,,j\}$ by the correlation recognition. Since the capacity of the synthesis equivalent motor is the sum of the capacity of each motor, namely

$$S_M = \sum_{\forall j \in M}^n S_j \tag{3}$$

the subscript M represents equivalent motor, then we can export the parameters for the model of equivalent motor and its control system in detail.

4.1. Basic Principle

The so-called equivalent means that the external characteristic of equivalent motor is the same or similar as the overall external characteristic of the m motors in parallel. Each motor adopts the T-end equivalent circuit, as shown in **Figure 1**, now simplify the m parallel circuits to an equivalent circuit.

4.2. The Equivalent of Inertia Time Constant

Assume that the kinetic energy of equivalent motor in synchronous speed is the sum of that of each motor. According to the definition of inertia time constant, T_J is the value that kinetic energy in synchronous speed multiply two then divide by capacity, so

$$\frac{1}{2}T_{JM}S_{NM} = \sum_{i=1}^m \frac{1}{2}T_{Ji}S_{Ni} \tag{4}$$

where T_{Ji} is inertial time constant of ith asynchronous motor; T_{JM} is the inertial time constant of the equiva-

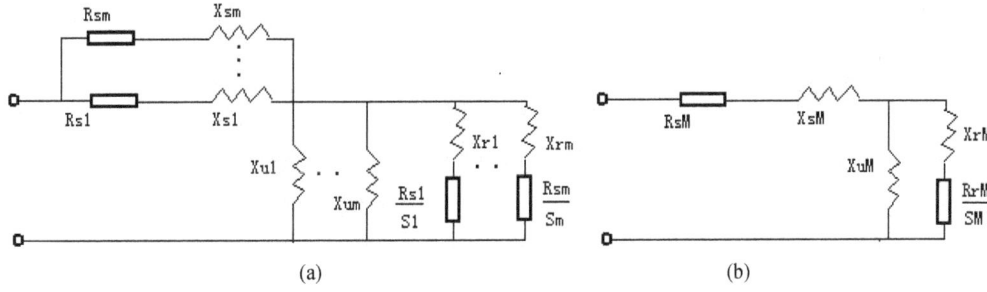

Figure 1. Equivalent T–end Circuit of Motor. (a) Equivalent Circuit before simplified (b) Equivalent Circuit after simplified.

lent motor; S_{Ni} is the rated capacity of ith asynchronous motor; S_{NM} is the capacity of the equivalent motor. Since

$$S_{NM} = \sum_{i=1}^{m} S_{Ni} \tag{5}$$

suppose

$$\rho_i = \frac{S_{Ni}}{S_{NM}} = S_{Ni} \bigg/ \sum_{i=1}^{m} S_{Ni} \tag{6}$$

with the both sides of Equation (10) divided by the half of S_{NM}, we get

$$T_{JM} = \sum_{i=1}^{m} \rho_i T_{Ji} \tag{7}$$

4.3. The Equivalent of Electrical Parameters

Equivalent excitation reactance X_{um} is the value of excitation reactance in parallel of the m motors. It should be noted that the impedance Z_i in figure are all per unit values under the total capacity, and the impedance Z_i' are per unit values under the capacity of each motor. The connection between Z_i and Z_i' is as follows:

$$Z_i = Z_i' \times \frac{S_{NM}}{S_{Ni}} = \frac{Z_i'}{\rho_i} \tag{8}$$

from what has been discussed above, we get

$$\frac{1}{X_{uM}} = \sum_{i=1}^{m} \frac{\rho_i}{X_{ui}} \tag{9}$$

in the same way, we get

$$\frac{1}{\dfrac{R_{rm}}{S_M} + jX_{IM}} = \sum_{i=1}^{m} \frac{\rho_i}{\dfrac{R_{ri}}{S_i} + jX_{li}} \tag{10}$$

where S_i, S_M is the slip of ith motor and the equivalent motor respectively; R_{ri}, R_{rm} is the rotor resistance of them; X_{li}, X_{IM} is the sum of the stator leakage reactance and the rotor leakage reactance of them .
    When $S_i = 1(i = 1, 2, \cdots\cdots, m, M)$, the same is true for Equation (14), thereby

$$R_{rM} = \frac{a}{a^2 + b^2} \quad X_{IM} = \frac{b}{a^2 + b^2} \tag{11}$$

where

$$a = \sum_{i=1}^{m} \frac{\rho_i R_{ri}}{R_{ri}^2 + X_{li}^2} \quad b = \sum_{i=1}^{m} \frac{\rho_i X_{ri}}{R_{ri}^2 + X_{li}^2} \tag{12}$$

4.4. The Equivalent of Slip

Assume that the motor runs at a constant slip S_i, set

$$a' = \sum_{i=1}^{m} \frac{\rho_i R_{ri}/s_i}{\left(R_{ri}/s_i\right) + X_{li}^2}, \quad b' = \sum_{i=1}^{m} \frac{\rho_i X_{li}}{\left(R_{ri}/s_i\right) + X_{li}^2} \tag{13}$$

we get

$$S_M = \frac{a}{a'} \times \frac{a'^2 + b'^2}{a^2 + b^2} \tag{14}$$

4.5. The Weighted Sum Method

Equivalent inertia time constant and equivalent admittance is the weighted sum of inertia constant and admittance of each motor, weights is the proportion ρ_i that each motor capacity accounts in total capacity. In order to simplify the analysis and calculation, the method that add them together after multiply weights is sometimes extended to the calculation of the equivalent slip, equivalent mechanical torque and the equivalent parameters. When the motor takes the T–end equivalent circuit as shown in figure, the Weighted Sum Method may be adopted approximatively to calculate the equivalent electrical parameters. Then we can assume that inner node K in each motor equivalent circuit is in parallel, so

$$\frac{1}{Z_M} = \sum_{i=1}^{m} \rho_i \frac{1}{Z_i} \tag{15}$$

where Z_i is the electrical branch impedance of the ith electric motor; Z_M is electrical branch impedance of the equivalent motor. For the branch of stator, $Z = R_s + jX_s$; for excitation branch, $Z = jX_\mu$ and for the branch the rotor, $Z = R_r/s + jX_r$.

5. Simulation Example

A simple power distribution network is shown in **Figure 2**, all the impedance of the transformer and line shown in the figure is per unit values. **Table 1** lists the four electric motor parameters. The active and reactive power models both use the power function model, and the power function adopt the IEEE recommended parameters. Equivalent motor parameters is shown in **Table 2**. Make the mistress line voltage drop 50%, monitor the total output active and reactive power of mistress line 1 respectively before and after polymerization. The curve is shown in **Figures 3** and **4**.

Use the fitting degree of active and reactive power absorbed in motor before and after polymerization as the basis for aggregation effect judgment. From the curve, we can conclude that the aggregation model retains the dynamic performance of the original model with good accuracy, the active and reactive power fitting error is smaller as well.

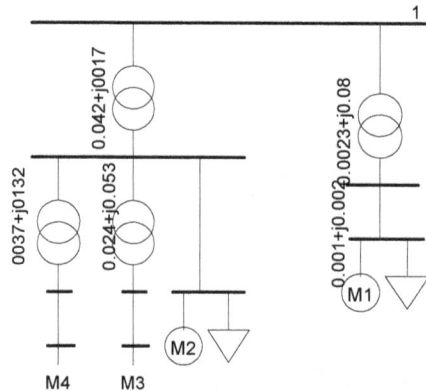

Figure 2. A sample distribution network.

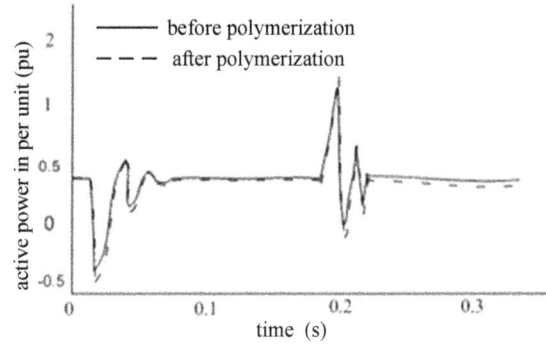

Figure 3. Active curves before and after polymerization.

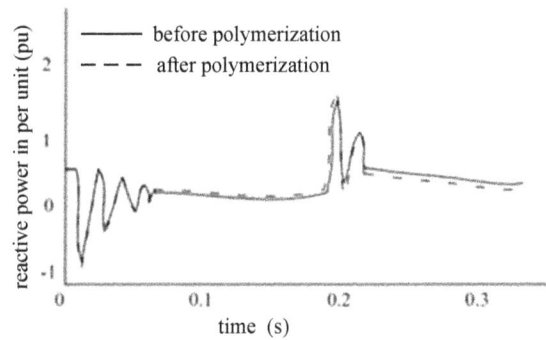

Figure 4. Reactive curves before and after polymerization.

Table 1. Motor parameters.

Motor	R_s	X_s	X_m	R_r	X_r	$T_{j/s}$	S_b/KVA	U_b/KV
1	0.0163	0.0816	2.25	0.0287	0.0836	1	597	4.16
2	0.0022	0.0759	2.62	0.0288	0.1037	0.66	3420	4.16
3	0.0235	0.1353	2.58	0.044	0.143	0.2	4269	1.1
4	0.0235	0.1353	2.58	0.044	0.143	0.2	2712	1.06

Table 2. Equivalent motor parameters.

Equivalent motor	R_s	X_s	X_m	R_r	X_r	$T_{j/s}$	S_b/KVA	U_b/KV
Parameters	0.0165	0.114	2.57	0.0384	0.1276	0.39	10998	13.8

6. Conclusion

This paper proposes the weighted method which follows the principle that the total output active and reactive power of distribution network respectively remains unchanged before and after polymerization to calculate the equivalent parameter of the distribution network. By adding the coherency identification to motors and introducing the calculation method of the weighted summation on the basis of traditional equivalence calculation, aggregation model of active and reactive power of the steady-state error is very small and it can accurately maintain oscillation mode of the original system, in addition, the deviation of oscillation amplitude is also small.

Acknowledgements

I would like to express my gratitude to all those who helped me during the writing of this thesis. I gratefully acknowledge the help from my supervisor, Ms. Hu Hanmei, who has offered me valuable suggestions in the academic studies. In the preparation of the thesis, she has spent much time reading through each draft and provided me with inspiring advice. Without her patient instruction, insightful criticism and expert guidance, the completion of this thesis would not have been possible. In addition, I deeply appreciate the contribution to this thesis made in various ways by my friends and classmates.

References

[1]   Zhou, X.X. (1997) To Develop Power System Technology Suitable to the Need in 21st Century. *Power System Technology*, **21**, 11-15.

[2]   Ni, Y.X. (2002) Dynamic Power System Theory and Analysis. Tsinghua University Press, Beijing.

[3]   Yu, Y.X. and Chen, L.Y. (1988) Power System Security and Stability. Science Press, Beijing.

[4]   Chen, L.Y. and Sun, D.F. (1989) Aggregation of Generating Unit Parameters in the System Dynamic Equivalents. *Proceedings of the CESS*, **9**, 30-39.

[5]   Min, Y. and Han, Y.D. (1991) Co-Frequency Dynamic Equivalence Approach for Calculation of Power System Frequency Dynamics. *Proceedings of the CESS*, **11**, 29-36.

[6]   Zhou, Y.H., Li, X.S., Hu, X.Y., *et al.* (1999) Dynamic Equivalents Based on the Transient Power Flow of the Connecting Lines. *Proceedings of the CSU-EPSA*, **11**, 29-33.

[7]   Xu, J.B., Xue, Y.S., Zhang, Q.P., *et al.* (2005) A Critical Review on Coherency-Based Dynamic Equivalences. *Automation of Electric Power Systems*, **29**, 92-95.

[8]   Wang, G. and Zang, B.M. (2006) External Online Dynamic Equivalents of Power System. *Power System Technology*, **30**, 21-26.

[9]   Wen, B.J., Zhang, H.B. and Zhang, B.M. (2004) Design of a Real-Time External Network Auto-Equivalence System of Subtransmission Networks in Guangdong. *Automation of Electric Power Systems*, **28**, 77-79.

[10]   Jiang, W.Y., Wu, W.C., Zhang, B.M., *et al.* (2007) Network model reconstruction in online security eEarly warning system," *Automation of Electric Power Systems*, **31**, 5-9.

[11]   Hu, J. and Yu, Y.X. (2006) A Practical Method of Parameter Aggregation for Power System Dynamic Equivalence. *Power System Technology*, **30**, 24-30.

[12]   Ju, P., Wang, W.H., Xie, H.J., *et al.* (2007) Identification Approach to Dynamic Equivalents of the Power System Interconnected with Three Areas. *Proceedings of the CESS*, **27**, 29-34.

[13]   Price, W.W., Chow, J.H. and Haqgave, A.W. (1998) Large-Scale System Testing of Power System Dynamic Equivalence Program. *IEEE Transactions on Power Systems*, **13**, 768-774. http://dx.doi.org/10.1109/59.708595

[14]   Sebastiao, E.M., Oliveira, D. and De Queiroz, J.F. (1988) Modal Dynamic Equivalent for Electric Power Systems. *IEEE Transactions on Power Systems*, **3**, 1723-1730. http://dx.doi.org/10.1109/59.192987

[15]   Joe, H.C., Galarza, R., Accari, P., *et al.* (1995) Inertial and Slow Coherency Aggregation Algorithms for Power System Dynamic Model Reduction. *IEEE Transactions on Power Systems*, **10**, 680-685. http://dx.doi.org/10.1109/59.387903

[16]   Wallace do, C.B., Reza, M.I. and Amauri, L. (2004) Robust Sparse Network Equivalent for Large Systems: Part I-Methodology. *IEEE Transactions on Power Systems*, **19**, 157-163. http://dx.doi.org/10.1109/TPWRS.2003.818603

[17]   Zhou, H.Q., Ju, P., Yang, H., *et al.* (2010) Dynamic Equivalent Method of Interconnected Power Systems with Consideration of Motor Loads. *Science China Technological Sciences*, **53**, 902-908. http://dx.doi.org/10.1007/s11431-010-0110-8

12

A 3D Modelling of Solar Cell's Electric Power under Real Operating Point

Mayoro Dieye[1], Senghane Mbodji[2], Martial Zoungrana[3], Issa Zerbo[3], Biram Dieng[2], Gregoire Sissoko[1]

[1]Laboratory of Semiconductors and Solar Energy, Department of Physics, Faculty of Science and Technology, Cheikh Anta Diop University, Dakar, Senegal

[2]Department of Physics, Alioune DIOP University of Bambey, Bambey, Senegal

[3]Laboratoire d'Energies Thermiques et Renouvelables (L.E.T.RE), Departement de Physique, U.F.R-S.E.A, Universitede Ouagadougou, Ouagadougou, Burkina Faso

Email: gsissoko@yahoo.com

Abstract

This work, based on the junction recombination velocity (Sf_u) concept, is used to study the solar cell's electric power at any real operating point. Using Sf_u and the back side recombination velocity (Sb_u) in a 3D modelling study, the continuity equation is resolved. We determined the photocurrent density, the photovoltage and the solar cell's electric power which is a calibrated function of the junction recombination velocity (Sf_u). Plots of solar cell's electric power with the junction recombination velocity give the maximum solar cell's electric power, P_m. Influence of various parameters such as grain size (g), grain boundaries recombination velocity (Sgb), wavelength (λ) and for different illumination modes on the solar cell's electric power is studied.

Keywords

Electric Power, Grain Size, Grain Boundary Recombination Velocity, Polycrystalline, Solar Cell, Junction Recombination Velocity

1. Introduction

Our work is grounded on the junction recombination velocity (Sf_u) [1] and the back side recombination velocity (Sb_u) [2] of the solar cell; subscript u refers to the illumination mode. With the junction recombination velocity concept, introduced in 1996 [1], modelling and characterization of solar cells were made possible for any operating point of the solar cell from the open circuit to short-circuit real operating points. The junction recombination velocity is related to external load, to their time of use and the season, the month and day of use.

Thus, one-dimensional (1D) studies which used the steady and transient states of the solar cell have determined with great precision, the lifetime of excess minority carriers, their diffusion length, the effective recombination velocity at the backside of the solar cell, the intrinsic junction recombination velocity ($Sf_{0,u}$), the characteristic I-V and shunt and series resistances [3] [4]. For studying the influence of electromagnetic waves produced by an amplitude modulation radio antenna on the electric power delivered by a silicon solar cell, a 1D study is used by [5]. In these works [3]-[5], authors proved that the intensity of the electromagnetic field depended on the distance between the solar cell and the amplitude modulation radio antenna. Taking into account the wavelength of the monochromatic radiation, they [5] also determined the maximum electric power and the corresponding operating point of the solar cell according to distance or electromagnetic field intensity.

However for polycrystalline silicon solar cells that provide the best efficiencies [6] but are made by small grains with various geometrical shape, it is necessary to make a 3D study to clearly identify the effect of grain size (g) and grain boundaries recombination velocity (Sgb). The solar cell's extension region width of the junction which could be considered as a plane capacitor with two identical plane electrodes separated by a thickness (d) is a function of these two parameters [7]. For high grain boundary recombination velocity (Sgb), the thickness obtained under open-circuit condition reaches to those obtained in short-circuit condition and the electron doesn't cross the junction [7]. But for high grain size, there is an important gap between thicknesses obtained respectively under open-circuit and short-circuit conditions and corresponding to the best solar cell [7].

The efficiency of the solar cell is calculated as the ratio between the maximum power, P_m, generated by the solar cell and the power of the incident light's flux, P_{in}; subscript m refers here to the maximum power point in the module's I-V curve [8].

Calculation of the solar cell's maximum electric power is then fundamental for photovoltaic devices characterization. That is why, the maximum powerpoint tracking (MPPT) control [9] [10] is developed and its role is to follow the maximum power point (MPP) of the photovoltaic module [11].

In this paper, we used the junction recombination velocity (Sf_u) to determine the generated power of the solar for any operating point. Within the first section, the basic theory is presented while the junction recombination velocity's role and the results related to the influence of grain size (g), grain boundary recombination velocity (Sgb), the wavelength (λ) and the illumination modes are presented in the second part of this paper.

2. Theoretical Analysis

A bifacial solar cell is a device which generates electricity directly from visible light. When light quanta are absorbed, electron hole pairs are generated as it can be seen in **Figure 1(a)**.

An n^+p-p^+ poly crystalline solar cell, made of many small individual grains, is considered.

Taking into account of the physical process simulation, the 2D representation of the solar cell is illustrated in **Figure 1(a)** and in **Figure 1(b)**, the fibrously oriented columnar grain is considered.

The following three illumination modes are considered: front illumination, rear side illumination and simultaneous front and back side illumination. Hence, the electron-hole pairs generation rate $G_u(z)$, related to each illumination mode is expressed as [12]:

$$G_u(z) = \alpha(\lambda) \cdot I_0 \cdot (1 - R(\lambda)) \cdot (\varepsilon \cdot \exp(-\alpha \cdot z) + \gamma \cdot \exp(-\alpha \cdot (H - z))). \tag{1}$$

Table 1 illustrates the values of ε and γ.

Coefficient $\alpha(\lambda)$ denotes the absorption of the monochromatic illumination [13]. I_0 is the incident photon flux and $R(\lambda)$ is the reflection at the wavelength, λ [13].

At the junction, N$^+$-P interface (z = 0)), Sf_u quantifies how the excess carriers flow through the junction in actual operating conditions and then Sf_u characterizes how electrons cross to the junction [1] [3]-[5].

At the back side of the solar cell, (Sb_u) is used to translate the losses in this zone. It quantifies hence, the rate at which excess minority carriers are lost at the back side of the cell [1]-[7].

2.1. Excess Minority Carriers Density

The solar cell's emitter is considered as a dead zone. So, the excess minority carriers density is determined taking account into only the contribution of the solar cell's base.

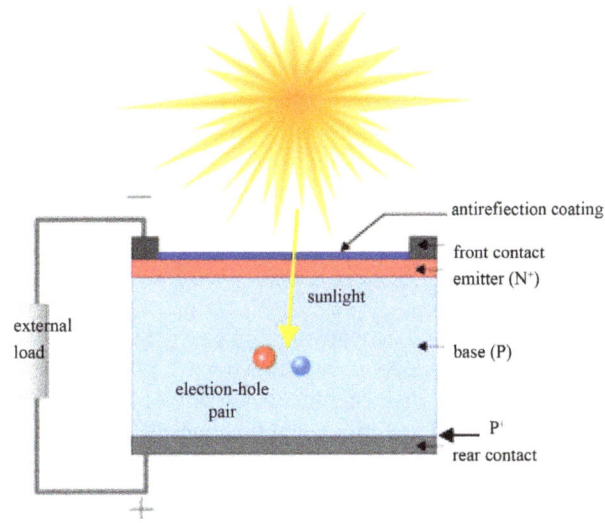

(a)

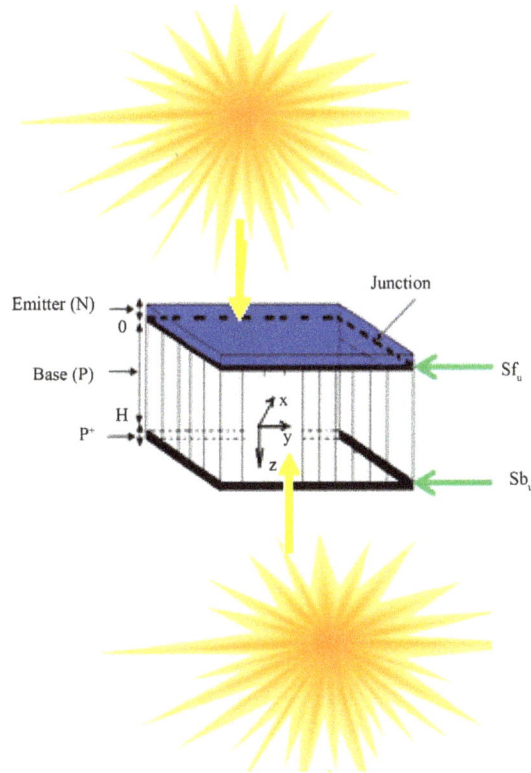

(b)

Figure 1. (a) 2D representation of the monofacialsolar cell [11]; (b) 3D representation of the solar cell $H = 130$ μm, $D = 26$ cm^2·s^{-1}.

Table 1. Range of ε and γ.

Illumination mode	ε	γ
Front side illumination ($u = fr$)	1	0
Rear side illumination ($u = r$)	0	1
Simultaneous illumination ($u = s$)	1	1

The excess minority carriers density is derived from the continuity equation [7]:

$$D \cdot \left(\frac{\partial^2 \delta_u (x,y,z)}{\partial^2 x} + \frac{\partial^2 \delta_u (x,y,z)}{\partial^2 y} + \frac{\partial^2 \delta_u (x,y,z)}{\partial^2 z} \right) - \frac{\delta_u (x,y,z)}{\tau} = -G_u (z) \qquad (2)$$

where D is excess minority carriers diffusion constant while τ is the lifetime of the excess minority carriers in the base of the solar cell.

The general solution of this equation is:

$$\delta_u (x,y,z) = \sum_k^{\infty} \sum_j^{\infty} Z_{kj} (z) \cdot \cos (x \cdot c_k) \cdot \cos (y \cdot c_j). \qquad (3)$$

The factors c_k and c_j are eigen values and depend on grain size (g) and grain boundary recombination velocity (Sgb) only. Parameter $Z_{kj}(z)$ express the z-dependence of $\delta_u (x,y,z)$. k and j vary form 1 to 30.

Inserting the Equation (3) into (2) and replacing the expression of generation by its value and taking into account of the fact that $\cos(c_k x)$ and $\cos(c_j x)$ are orthogonal functions, we obtain:

$$Z_{kj,u}''(z) - \frac{Z_{kj,u}(z)}{L_{kj}^2} = - \frac{16 \sin \left(\dfrac{g \cdot c_k}{2} \right) \cdot \sin \left(\dfrac{g \cdot c_j}{2} \right)}{D \cdot c_k \cdot c_j \cdot F_{kj}} \cdot G_u (z) \qquad (4)$$

where $L_{kj} = \left(\dfrac{1}{L^2} + c_k^2 + c_j^2 \right)^{-\frac{1}{2}}$ and $F_{kj} = \left(g + \dfrac{\sin (c_k \cdot g)}{c_k} \right) \cdot \left(g + \dfrac{\sin (c_j \cdot g)}{c_j} \right)$.

The solution of Equation (4) named $Z_{kj}(z)$ can be written as follows:

$$Z_{kj,u}(z) = A_{kj,u} \cdot ch \left(\frac{z}{L_{kj}} \right) + B_{kj,u} \left(\frac{z}{L_{kj}} \right) - \frac{16 \cdot L_{kj}^2 \cdot \sin \left(\dfrac{g \cdot c_k}{2} \right) \cdot \sin \left(\dfrac{g \cdot c_j}{2} \right)}{D \cdot c_k \cdot c_j \cdot F_{kj}} \cdot G_u (z). \qquad (5)$$

The constants $A_{kj,u}$ and $B_{kj,u}$ in Equation (5) were determined using the boundary conditions at two interfaces [14]; one interface at (a) the N$^+$-P boundary $z = 0$ [14]:

$$D \cdot \left. \frac{\partial \delta_u (x,y,z)}{\partial z} \right|_{z=0} = Sf_u \cdot \delta_u (x,y,z=0) \qquad (6)$$

and (b) at the back side of the bifacial solar cell, $z = H$ [2] [14]:

$$D \cdot \left. \frac{\partial Z_{kj}(z)}{\partial z} \right|_{z=H} = -Sb_u \cdot Z_{kj,u}(H). \qquad (7)$$

Using boundary conditions at the contact of two grains respectively in the x-direction at $x = \pm g/2$, and y-direction at $y = \pm g/2$, transcendal Equations (8) and (9) are obtained:

$$\tan (c_k \cdot g/2) = \frac{Sgb}{c_k \cdot D} \qquad (8)$$

and

$$\tan (c_j \cdot g/2) = \frac{Sgb}{c_j \cdot D}. \qquad (9)$$

2.2. Photocurrent Density

The photocurrent density can be calculated by the following equation [2] [6]:

$$I \left(Sgb, g, \lambda, Sf_u, Sb_u \right) = q \cdot D \cdot \left. \frac{\partial \delta_u (z, Sgb, g, \lambda, Sf_u, Sb_u)}{\partial z} \right|_{z=0} . \qquad (10)$$

2.3. Photo Voltage

Using the Boltzmann's relation, the photo voltage V_{ph} can be expressed as [2] [6]:

$$V_{ph}\left(Sgb, g, \lambda, Sf_u, Sb_u\right) = V_T \cdot Ln\left(\frac{\delta_u\left(z = 0, Sgb, g, \lambda, Sf_u, Sb_u\right)}{m_0} + 1\right). \tag{11}$$

Here, V_T is the thermal voltage, $m_0 = \dfrac{n_i^2}{Nb}$ with Nb the base doping density and n_i the intrinsic carriers density.

2.4. Solar Cell's Electric Power

The power generated by the cell is given by [6]:

$$P\left(Sgb, g, \lambda, Sf_u, Sb_u\right) = V_{ph}\left(Sgb, g, \lambda, Sf_u, Sb_u\right) \cdot I_{ph}\left(Sgb, g, \lambda, Sf_u, Sb_u\right). \tag{12}$$

The solar cell's generated power depends on Sf_u and then is function of the solar cell's real operating point varying from the short-circuit operating point to the open one.

3. Results and Discussions

In **Figures 2-5** we show curves of solar cell's electric power versus junction recombination velocity which varies from 10^0 to 10^{12} cm/s when the grain recombination velocity (Sgb), the grain size (g), the wavelength (λ) and the illumination mode varies, respectively. In **Figures 2-5**, the solar cell is illuminated by its front side.

We noted in each plot, as already shown by [5] in 1D study, that solar cell's power tends to zero, when $Sf_u < 10^2$ m/s and $Sf_u > 10^{10}$ m/s corresponding to the open-circuit operating and short-circuit operating points, respectively. The open circuit operating point is characterized by the open circuit photovoltage V_{oc} and where the photocurrent is null. The short-circuit operating point is characterised by the short-circuit photocurrent I_{sc}. In our previous studies [15], we remarked that poly crystalline solar cells tend to produce high open circuit photo voltage (V_{oc}) and short-circuit photocurrent (I_{sc}) as the grain size (g) increases, conversely, V_{oc} and I_{sc} decrease with the increase of grain boundary recombination velocity (Sgb) [15].

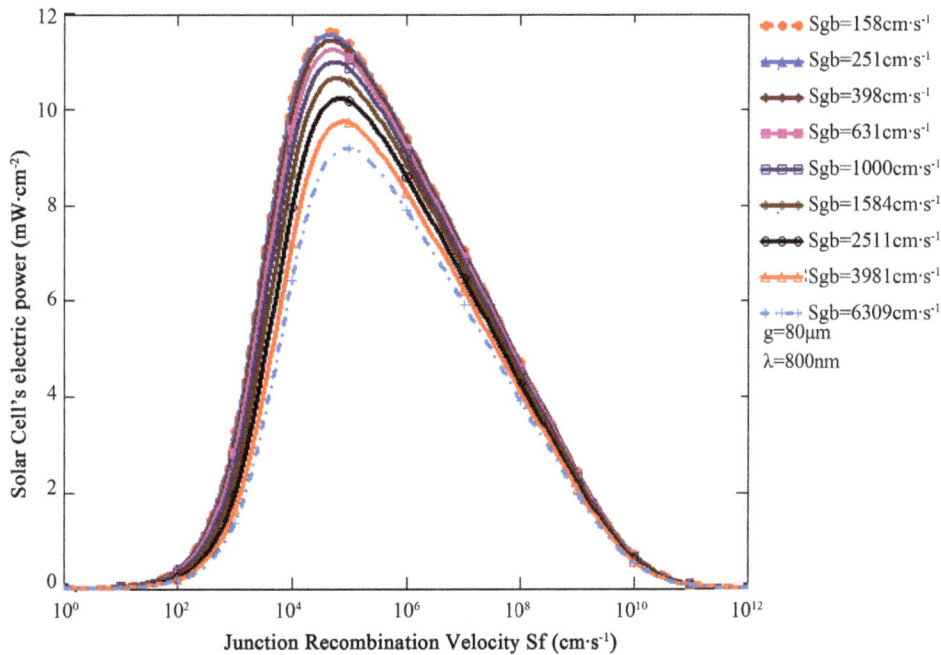

Figure 2. Solar cell's electric power versus junction recombination velocity for various grain boundary recombination velocity: $H = 130$ μm and $D = 26$ cm^2·s^{-1}.

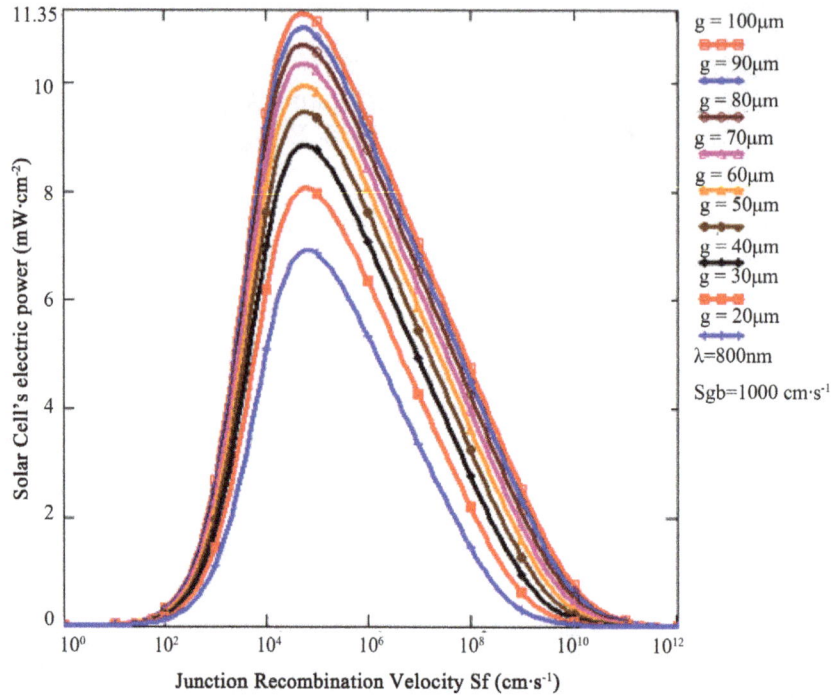

Figure 3. Solar cell's electric power versus junction recombination velocity for various grain size (*g*): $H = 130$ μm and $D = 26$ cm^2·s^{-1}.

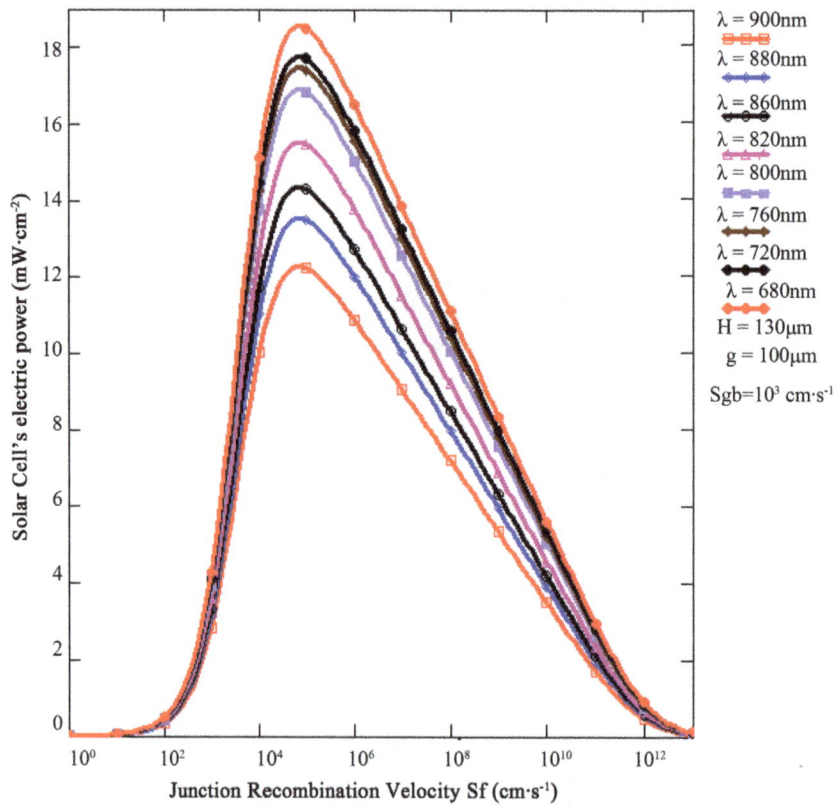

Figure 4. Solar cell's electric power versus junction recombination velocity for different incident light wavelength (*λ*): $H = 130$ μm and $D = 26$ cm^2·s^{-1}.

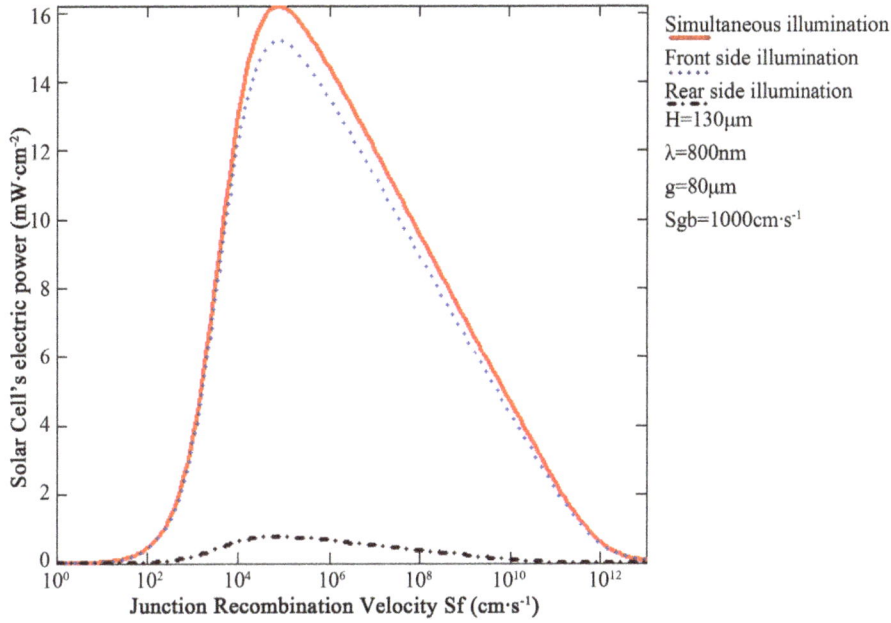

Figure 5. Solar cell's electric power versus junction recombination velocity ($u = fr, r, s$).

We also remarked that the solar cell's electric power increases as the junction recombination velocity is high and goes through a maximum, named maximum power point, P_m. There is, for each P_m, a related junction recombination velocity $Sf_{u,m}$ corresponding to the real operating point of the solar cell.

P_m which is equal to the product of maximum power point photocurrent (I_m) and photo voltage (V_m), corresponds to the "knee" of the I-V curve [15] and is an optimum operating point such that solar cell delivers the maximum possible power to the external load. It is well known that, the maximum power point changes with atmospheric conditions and is usually determined by using the maximal power point tracking (MPPT) control technic as shown by [8]-[10].

Figure 2 and **Figure 3** show respectively that the solar cell's maximum electric power P_m, increases for high values of grain size (g) and decreases for high values of grain boundary recombination velocity (Sgb). It means that, increasing the grain size leads to fewer recombination in the bulk of the solar cell and hence high electrons' flow rate crossing the junction, corresponding to an increase of the photocurrent and the efficiency of the solar cell as it can be seen in [7]. The increasing of grain boundary recombination velocity (Sgb) give high recombination in the bulk of the solar cell and low efficiency [12] [15].

We also noted, for all plots, that the solar cell's electric power decreases from the junction recombination velocity ($Sf_{u,m}$) corresponding to the solar cell's maximum electric power (P_m).

The effects of grain size (g) and grain boundary recombination velocity (Sgb) upon the junction recombination velocity ($Sf_{u,m}$) corresponding to the solar cell's maximum electricpower (P_m) is shown in **Figure 2** and **Figure 3**. The low values of $Sf_{u,m}$ are obtained with high grain size (g) and low grain boundary recombination velocity (Sgb), respectively.

We deduced that, the increase of the grain size (g) leads to the decrease of the grain boundary recombination velocity (Sgb). We can also conclude that, junction recombination velocity is, effectively, the sum of two terms as applied and demonstrated in some works [4] [5]:
- $Sf_{0,u}$, the intrinsic junction recombination velocity imposed by the shunt resistance [4] [5];
- Sf_j which is related to current flow passed through the junction is imposed by the external load [4] [5].

For a fixed external load, corresponding to a specific value of Sf_j, $Sf_{0,u}$ increases with the grain boundary recombination (Sgb). This leads to the lower shunt resistance (R_{sh}) and to the initiating short-circuit condition quickly reached [16]. For the same conditions, when the grain size (g) increases, $Sf_{0,u}$ and the shunt resistance decreases and increases, respectively. This situation corresponds to less recombination and the real operating point of the solar cell roll away of the initiating operating short-circuit condition studied in [16].

Evolution of the solar cell's electric power in relation to the junction recombination velocity when the wavelength is ranging from 680 nm to 900 nm, corresponding at the domain of high wavelength is plotted on **Figure 4**. The solar cell is illuminated on its front side.

It is shown, in **Figure 4**, that the solar cell's electric power, for any operating point, decreases with the increase of wavelength (λ) corresponding to low internal quantum efficiency (IQE) and small diffusion lengths of the solar cell [17]. Effect of this range of wavelength is explained using the thickness of emitter-base junction determination in a bifacial polycrystalline solar cell under real operating condition technic developed by [7]. This work [7] shows that the extension region's width in open circuit increases with wavelength due to the energy of the incoming photons. Hence, excess minority carriers and photocurrent density would decrease for high wavelength. Because, at high wavelength, corresponding to low energy, absorbed photons are low and then the maximum solar cell's electric power and efficiency would decrease as shown in [10].

Effect of the illumination mode on the solar cell's electric power is plotted in **Figure 5**. The n^+-p-p^+ bifacial solar cell has the advantage of receiving a light by its rear side. The bifacial solar cell thus receives a simultaneous illumination due the albedo [18]-[20]. As expected and demonstrated in [7] [14] the simultaneous illumination mode gives the highest solar cell's electric power. It is followed by the solar cell's electric power obtained by the front side illumination mode. These two solar cell's electric powers tend to merge when $Sf_u < 10^4$ m/s. The back side illumination mode's contribution can be neglected when compared to solar cell's electric power that are obtained by front side and both front and back sides. This is consistent with results found by [21] who studies a I-V characteristics curve for bifacial silicon solar cell under magnetic field. Authors demonstrated that best solar cell efficiency is obtained with both front and back sides illumination mode.

4. Conclusions

Using the junction recombination velocity concept permits us to determine solar cell's electric power for any operating point of the solar cell contrarily to others studies which used the maximal power point tracking (MPPT) control technic characterized by one operating point corresponding to the maximum output power delivered by solar cell.

It is shown that, for any real operating point, the solar cell's electric power increases with grain size (g) corresponding then to the best solar cell. It decreases with high gain boundary recombination velocity (Sgb) and with high wavelength (λ).

We noted that using an n^+-p-p^+ solar cell which could be illuminated by both front and back sides of the solar cell simultaneously had the advantage of giving a high electric power.

References

[1] Sissoko, G., Museruka, C., Correa, A., Gaye, I. and Ndiaye, A.L. (1996) Light Spectral Effect on Recombination Parameters of Silicon Solar Cell. *Proceedings of World Renewable Energy Congress.*

[2] Dugas, J. (1994) 3D Modelling of a Reverse Cell Made with Improved Multicrystalline Silicon Wafer. *Solar Energy Materials & Solar Cells*, **32**, 71-88. http://dx.doi.org/10.1016/0927-0248(94)90257-7

[3] Barro, F.I., Mbodji, S., Ndiaye, M., Maiga, A.S. and Sissoko, G. (2008) Bulk and Surface Recombination Parameters Measurement of Silicon Solar Cell under Constant White Bias Light. *Journal des Sciences*, **8**, 37-41.

[4] Nzonzolo, Lilonga-Boyenga, D. and Sissoko, G. (2014) Illumination Level Effects on Macroscopic Parameters of a Bifacial Solar Cell. *Energy and Power Engineering*, **6**, 25-36. http://dx.doi.org/10.4236/epe.2014.63004

[5] Zerbo, I., Zoungrana, M., Ouedraogo, A., Korgo, B., Zouma, B. and Bathiebo, D.J. (2014) Influence of Electromagnetic Waves Produced by an Amplitude Modulation Radio Antenna on the Electric Power Delivered by a Silicon Solar Cell. *Global Journal of Pure and Applied Sciences*, **20**, 139-148. http://dx.doi.org/10.4314/gjpas.v20i2.9

[6] Madougou, S., Kaka, M. and Sissoko, G. (2010) Silicon Solar Cells: Recombination and Electrical Parameters. In: Rugescu, R.D., Ed., *Solar Energy*, InTech, Croatia.

[7] Mbodji, S., Mbow, B., Barro, F.I. and Sissoko, G. (2011) A 3D Model for Thickness and Diffusion Capacitance of Emitter-Base Junction Determination in a Bifacial Polycrystalline Solar Cell under Real Operating Condition. *Turkish Journal of Physics*, **15**, 281-291.

[8] Skoplaki, E. and Palyvos, J.A. (2009) On the Temperature Dependence of Photovoltaic Module Electrical Performance: A Review of Efficiency/Power Correlations. *Solar Energy*, **83**, 614-624. http://dx.doi.org/10.1016/j.solener.2008.10.008

[9] Hamrouni, N., Jraidi, M. and Chérif, A. (2008) Solar Radiation and Ambient Temperature Effects on the Performances

of a PV Pumping System. *Revue des Energies Renouvelables*, **11**, 95-106.

[10] Dinçer, F. and Meral, M.E. (2010) Critical Factors That Affecting Efficiency of Solar Cells. *Journal of Smart Grid and Renewable Energy*, **1**, 47-50. http://dx.doi.org/10.4236/sgre.2010.11007

[11] Ndoye, S., Ndiaye, M., Diao, A., Dione, M.M., Diarisso, D., Bama, A.O.N., Ly, I., Sow, G., Maiga, A.S., Foulani, A., Barro, F.I. and Sissoko, G. (2010) Modeling and Simuling the Powering System of a Base Transmitter Station with a Standalone Photovoltaic Generator. *Proceedings of 25th European Photovoltaic Solar Energy Conference and Exhibition*, 5208-5211.

[12] Sissoko, G., Correa, A., Nanema, E., Diarra, M.N., Ndiaye, A.L. and Adj, M. (1998) Recombination Parameters Determination in a Double Sided Back-Surface Field Silicon Solar Cell. *Proceedings of the World Rrenewable Energy Congress*, **3**, 1856-1859.

[13] Green, M.A. and Keevers, M. (1995) Optical Properties of Intrinsic Silicon at 300K. *Progress in Photovoltaics*, **3**, 189-192. http://dx.doi.org/10.1002/pip.4670030303

[14] Diallo, H.L., Maiga, A.S., Wereme, A. and Sissoko, G. (2008) New Approach of both Junction and Back Surface Recombination Velocities in a 3D Modelling Study of a Polycrystalline Silicon Solar Cell. *The European Physical Journal Applied Physics*, **42**, 193-211. http://dx.doi.org/10.1051/epjap:2008085

[15] Mbodji, S., Ly, I., Diallo, H.L., Dione, M.M., Diasse, O. and Sissoko, G. (2012) Modeling Study of N^+/P Solar Cell Resistances from Single I-V Characteristic Curve Considering the Junction Recombination Velocity (Sf). *Research Journal of Applied Sciences, Engineering and Technology*, **4**, 1-7.

[16] Ly, I., Ndiaye, M., Wade, M., Thiam, N., Gueye, S. and Sissoko, G. (2013) Concept of Recombination Velocity Sfcc at the Junction of a Bifacial Silicon Solar Cell, in Steady State, Initiating the Short-Circuit Condition. *Research Journal of Applied Sciences, Engineering and Technology*, **5**, 203-208.

[17] Madougou, S., Made, F., Boukary, M.S. and Sissoko, G. (2007) Recombination Parameters Determination by Using Internal Quantum Efficiency Data of Bifacial Silicon Solar Cells. *Advanced Materials Research*, **18-19**, 313-324. http://dx.doi.org/10.4028/www.scientific.net/AMR.18-19.313

[18] Cuevas, A., Luque, A., Eguren, J. and Del Alamo, J. (1982) 50 Per Cent More Output Power from an Albedo Collecting Flat Panel Using Bifacial Solar Cells. *Solar Energy*, **29**, 419-420. http://dx.doi.org/10.1016/0038-092X(82)90078-0

[19] Gophen, M. (2008) Land-Use, Albedo and Air Temperature Changes in the Hula Valley (Israel) during 1946-2008. *Open Journal of Modern Hydrology*, **4**, 101-111. http://dx.doi.org/10.4236/ojmh.2014.44010

[20] Bird, R.E. and Riordan, C. (1985) Simple Solar Spectral Model for Direct and Diffuse Irradiance on Horizontal and Tilted Planes at the Earth's Surface for Cloudless Atmospheres. *Journal of Climate and Applied Meteorology*, **25**, 87-97. http://dx.doi.org/10.1175/1520-0450(1986)025<0087:SSSMFD>2.0.CO;2

[21] Madougou, S., Made, F., Boukary, M.S. and Sissoko, G. (2007) I-V Characteristics for Bifacial Silicon Solar Cell Studied under a Magnetic Field. *Advanced Materials Research*, **18-19**, 303-312. http://dx.doi.org/10.4028/www.scientific.net/AMR.18-19.303

13

Electric Vehicles Analysis inside Electric Mobility Looking for Energy Efficient and Sustainable Metropolis

Miguel Edgar Morales Udaeta*, Carolina Attas Chaud, André Luiz Veiga Gimenes, Luiz Claudio Ribeiro Galvao

Energy Group of the Electrical Power and Automation Engineering, Department of the Polytechnic, University of São Paulo—GEPEA/EPUSP, São Paulo, Brazil
Email: *udaeta@pea.usp.br

Abstract

This paper aims to study and evaluate electric mobility over time, focusing on the development of the electric car. Methodologically, in order to accomplish this intent, the characterization of the electric vehicle (EV) is made based on the variables which determine its performance, such as: assessment of speeds, distance traveled, analysis of facts related to the energy source (electrochemical accumulators) and analysis of the determining system of electric mobility (the electric engine as a function of power (W) and voltage (V)). This way, to demonstrate the effects of time, this process will be analyzed from the beginning of the 20th century (1930s) to the present (the first decade of the 21st century), methodologically structured in 4 cycles that show the performance of the EV. The results show the existence of vulnerabilities and of electric mobility potential, as well as the nuances of the development of the electric vehicle along the years and along the transformations in what is considered state-of-the-art. Thus, in the case of batteries, it is evident that the lithium-ion type used nowadays reveals better results due to its higher specific efficient energy, which maximizes energy autonomy to 200 km. In the beginning, the insertion of the electric vehicle was commercially harmed by the fundamental limitations of batteries as a power source. Conclusively, on certain occasions there have been improvements in the aerodynamics, engines, weight and size of the batteries, demonstrating the maturity of EVs.

Keywords

Electric Vehicle, EV Development, Electric Mobility, Energy Analysis, Metropolis, Sustainability

---

*Corresponding author.

1. Introduction

Urban mobility in the 21st century comes with a series of constraints—ranging from problems in availability of essential resources to difficulties in the disposal of waste by average speed in the transportation of goods and people—all of which leading to considerations about the future role of the automobile in the cities. This is an opportunity to think and implement fundamental changes in concepts and models of the last century's standardized transport. Even though cars, trucks and other mobility machines have been synonymous with urbanization development, nowadays these machines are partially responsible for the collapse of metropolises [1]. From numerous negative factors relating to urban mobility in the full sense, the biggest target for the minimization of effects is local pollution, simply because humans need well-being. In this sense, electric mobility seems as a part of a solution known to the cities since—on the contrary of what it may seem—the first electric vehicles have emerged in the turn of the twentieth century, preceding the invention of the combustion engine gas (by Daimler Benz in 1885's Germany). In fact, before the internal combustion engine imposed itself as synonymous with mobility, the history of western socioeconomic development demonstrates that the electric vehicles industry was thriving, with electric buses lines, for example, which earned space on the streets of London around 1886. This has occurred because of the supremacy of the research conducted in France by Gustave Trouvé in 1881, which has made it possible to recharge batteries.

Regular use of electric vehicles has endured a long time, the proof of which being the remarkable advances in this area, such as the construction of the electric car that reached the incredible speed, for those days, of 100 km/h (by the Belgian Camille Jenatzy in 1899). The same way, in 1918, the electric buses line between Praça Maua and the then existing Monroe Palace in the city of Rio de Janeiro was inaugurated, known as the comfortable bus battery—powered electric traction—with solid rubber wheels, no noise, no vibration and without smoke or the hassle of gasoline (Peres, 2003).

The entrance of the Ford T in the market in 1909, along with its subsequent improvements such as the electric start, meant at the time the fall of the electric car—even with the advances recharge time and autonomy. Thus, like T Ford, urban mobility initiated its way to the current days. More than that, in the consumer society the automobile has become an indispensable asset. The following well-known step, in the case of the internal combustion engine, was the investment in oil refinery fields and their availability primarily in the U.S., which has determined the necessary support fuel. It is important to recall that the diesel engine [2] had been invented in the late 19th century but the use of fossil fuels like diesel in general transportation is not well seen due to the serious pollution and GHG emissions that result from it.

In this sense, environmental initiatives have favored in recent times the technological evolution of the electric power train, such as the Kyoto protocol, which especially concerns the burning of petroleum in powered internal combustion engine vehicles. Additionally, these vehicles in large population centers are a significant source of contaminating emissions, contributing often in 100% with the emission of pollutants into the atmosphere [3] [4]. Therefore, electrical and urban mobility are inserted in a new context due to the increase in the availability and use of electric vehicles in the world.

It is interesting to evidence that the relative emission of urban mobility—fueled by petroleum derivatives—while added to the combustion products of local industries becomes lethal when the case involves temperature inversions that prevent the dispersion of pollutants [5].

Therefore, to foster the construction of an eco-economy towards a sustainable socioeconomic development that includes mobility in large cities seems irreversible in the 21st century. This means we live in a world where energy comes from clean sources such as wind turbines and not from coal mines, where recycling industries replace mining industries and where cities are planned for people and not for cars. A sustainable economy includes the welfare of future generations [1].

As a premise, the electric vehicle proves itself to be conducive to environmental issues of urban mobility, since emissions are significantly reduced, which should also include a decrease in noise. Since recharging batteries (neuralgic point of the inclusion of electric mobility) incorporates new concepts into the commercial power grid, the inclusion of the electric vehicle should be planned within the power generation mix of each country. In any case, with respect to Brazil—given that the electric energy is hydroelectric—electric mobility is a significant option for sustainable urban mobility in megacities such as São Paulo [2].

Relevant Aspects of the Electric Vehicles

At first, an electric vehicle in its basic meaning can be understood as being an automobile with an electric motor connected to the front wheels through a gearbox with one or two speeds, but there are several other possible variations of the propulsion system architectures. Thus, for example, a significant variation of this techonology is the use of four small motors for each wheel, in exchange for only one drive motor, as originally designed in the beginning of the electric car [6].

Moreover, in the context of this work, electrical mobility aims at the driving of people, objects or a specific load. However, the electric vehicle in this case, independently from other technological variants, is generally understood as a system whose only power source is the battery charged with the task of activating one or more of the automobile's electric motors. Even because, in this specific case of electric mobility, it is generally assumed that the availability of power supply determines the type of vehicle so that, for example, a subway or trolley follows a predetermined route by rail or electrical distribution network [7].

Given the diversity—in the modern world—of electric mobility among existing systems for electric vehicles, this work focuses on cars and utility vehicles for loads. In this sense, the study considers electric vehicles throughout time, in an analysis divided by the phases of its evolution, considering also the technology employed and socioeconomic importance of the period.

2. Technical Evolutionary Context of Electric Vehicles

In essence, the electric mobility's technology is based on the joint implementation of energy accumulators and of the electric motor as drive system. Two researchers stand out in this context: Alessandro Volta for the precursor of the battery in 1800 ("Battery Back") and Michael Faraday, who developed the homopolar motor in 1821.

In this sense, the predominance of the electromagnetic induction since 1831 has led to the consolidation of the electrical and electronics technologies, including engines and electrical generators. From this assumption, all findings related to the operation of an electric motor have supported the conjunction of the battery and of the electric motor connected to the wheels of a light vehicle. The first electric vehicles appear in the 1830s using non-rechargeable batteries.

In 1859 Gaston Plante introduced the rechargeable lead-acid, which is the technology currently used in acquiescence in most applications requiring energy storage. Following that, Aphonse Camille Faure improved the ability of such batteries, leading to industrial scale production, in such a way that in 1881 the autonomous electric vehicles proliferated the cities' streets. That way, it is important to mention that in the late 20th century, with the rechargeable batteries, electric cars entered the commercial market [8]. Thus, it can be said that the predominance of the accumulation of electrical energy and its conversion to mechanical energy has enabled a new, quiet and clean method for urban mobility.

As for the accumulators, other batteries have been developed—for example, the iron-zinc battery. Even Thomas Edison, interested in the potential of electric vehicles in the early 1900s, developed the nickel-iron battery, with a storage capacity 40% greater than that of the battery lead, but with a much higher cost of production. Thus, the 19th century saw the development of batteries such as the nickel-zinc and the zinc-air [9].

Thus, electrical mobility starts when the first electric vehicle (tricycle) to use lead-acid (developed by Planté) as a source of energy was demonstrated in France by Mr. Trouvé in 1881. In this period, other electric tricycles with lead batteries were also in the U.S. and UK. In this context, it is worth remembering that only in 1885 did the German Karl Benz demonstrate the first combustion powered vehicle: the Pantentmotorwagen [10].

Moreover, in 1837, Robert Davison Aberdeen built in England the first electric carriage, powered by a rustic iron-zinc battery and driven by an electric motor, but containing all the basic elements used in modern electric vehicles. In France, experiments were performed by Charles Jeantaud Raffard while, at the same time, Werner Siemens, in Germany, perfected the electric motor. Even though the mobility steam prevailed at the time—especially in the area of public transport—the electric vehicle showed itself ideal for urban traffic, not having made noises and having had a drive system that did not pollute the environment [7].

Anyway, since its inception electric mobility has gained significance, seeing that the electric motor, by not involving any combustion, was free of soot and grease, therefore being very clean. More than that, the electric vehicle with graduations for three, four and up to nine speeds was not required to carry a paraphernalia command, which is a fundamental characteristic of petrol and steam cars [11].

3. First Cycle of Electric Vehicles (1837-1912)

Schiffer [12] points out that the electric car powered by rechargeable batteries seemed to have a great future about a century ago. Twenty-eight percent of the 4192 cars produced in the U.S. in 1900 were electric. Some of the most prestigious inventors, including Thomas Edison, promoted electric cars or the participation in its development. And the first industries to produce cars in series were manufacturing electric cars. In the early twentieth century, electric, steam and gasoline cars competed more or less on equal terms. Many analysts at the time believed that each type of car would find its own "performance space" and that they would coexist indefinitely. However, by the late 20s the electric car was a product in decline when considering the commercial sphere. The gasoline-powered car had conquered all spaces with its impressive speed, performance and finish. "Tripping A spectacular", from "The electric automobile in America". **Table 1** presents schematically an analysis of the performance achieved each year by the respective electric vehicles, parameterized in speed, range, battery voltage and power, in the golden period related to the 1st cycle, which it set between 1837 and 1912.

Occasionally, around 1905 the gasoline-powered vehicle began to stand out when compared to the electric vehicles in question. The autonomy of the 63 miles (about 100 km) reached by vehicle combustion more than doubled the range of an electric car (30 miles or approximately 50 km), as seen in the table. The initial investment and the operating cost of electric cars were higher than the gasoline-powered. The available figures indicate that the 1900s petrol cars cost between $1000.00 and $2000.00, while an electric car cost U.S. $1250.00 to U.S. $3500.00. The cost of operating a gasoline car was $0.01/mile while for an electric car it was $0.02 to 0.03/mile. In 1901 big oil fields were discovered in Texas, which drove down petrol's costs in a way that it gained a sustainable competitive advantage.

Still, according to Larminie and Lowry [8], early in the development of electric vehicles, an internal combustion engine that flipped a generator was used together with one or more electric motors, called hybrid vehicle.

According to Schiffer [12], electricity best fulfills the requirements of a traction system than steam engines or even internal combustion engines.

However, in 1912, the fleet of electric cars in the United States reached its peak of 30,000 units and the amount of petrol cars was already thirty times larger, to the verge of 900,000 units [13]. Given this background, in the same period, between 1900 and 1912, there were initiatives in the pursuit of improving distance and performance of electric vehicles through the adoption of the hybrid configuration.

Based on the assumptions presented above, the trajectory of electric cars has continued in eversion. Among the main factors that have since contributed to the decline of electric cars [14], the following can be cited:

- The competitive advantage achieved by the development of the production system in series, applied by Henry Ford, has allowed an increase in the manufacture of combustion-powered cars, as shown in **Table 2**.
- Elimination of the crank used to drive vehicles powered combustion. Invention of the electric starting, in 1912 [15];
- In the mid-1920s, the highways in the United States interconnected several cities, which demanded vehicles capable of traveling long distances [14];

Table 1. Evolution of electric vehicles from 1837 to 1912.

Year	Speed	Autonomy	Battery	Motor (Power/Voltage Type)		
1837	6 km/h	2 km	Zinc-acid	5 KW	-	-
1881	15 km/h	40 km/h	Lead	0.37 KW	20 V	MCC
1890	14 km/h	23 km	24 cells	3 KW	58 V	-
1897	15 km/h	48 km	Lead	2.6 KW	-	-
1899	105.8 km/h	-	Lead	2×50 KW	200 V	MCC
1900	58 km/h	-	Lead	2×5.15 KW	80 V	-
1902	21 km/h	64 km	Ni-Fe	1 KW	40 V	-
1908	16 km/h	-	40 cells	-	40 V	-
1911	37 km/h	60 km	Ni-Fe	6 KW	84 V	-

Source: own compilation.

Table 2. Number of vehicles built in USA in the beginning of the century.

Year	Electric	Gasoline
1899	1575	936
1904	1425	18,699
1909	3826	120,393
1914	4669	564,385
1919	2498	1,649,127
1924	391	3,185,490
1929	757	4,454,421
1933	-	1,560,599

Source: own adaptation from [7].

- Oil discoveries in Texas have reduced the price of gasoline, making it an attractive fuel for the transportation sector [16];
- Development of distillation techniques in continuous and consequent cheapness of petroleum products resulting in the expansion of the technological development of the automobile industry headed for gasoline-powered vehicles [7].

4. Second Cycle of Electric Vehicles (1912-1973)

After the discovery of oil fields, gasoline vehicles and, later, diesel quickly reached levels of performance that resulted in greater speed, greater acceleration and lower weight compared to electric vehicles. The oil industry had developed into such a supreme point that virtually all of its derivatives started to present cost advantage because of its growing consumption.

Sales were minimal even in England, where Brougham (by Partridge Wilson) was marketed, powered by a battery of 60 V and 34 Ah, which enabled a speed of 51.5 km/h and a radius of 97 km per battery charge [7].

Based on the assumptions presented above, having the 2nd cycle started around 1909, the production of electric cars dropped considerably, reaching about 4.4% of the number of combustion-powered cars. In 1913, as aforementioned, Ford began producing gasoline cars in series in the first industrial assembly line in Highland Park plant. In 1912 with the advent of the electric starter motor vehicles, the explosion made these cars even more attractive.

Obviously, by 1912 there was renewed enthusiasm for the electric car with the emergence of a few technical developments. Thomas Edison had perfected and carried out the first tests using the nickel-iron battery, which had a 35% increase in storage capacity between 1910 and 1925. The lifespan of these batteries has also increased whereas the maintenance costs decreased.

However, this resurgence was most striking for small delivery trucks in companies that owned fleets of around 60 vehicles and that could have their own central recharge batteries.

The advent of the First World War in 1914 caused an increase in oil prices and also in optimism about electric cars. But despite the commercial and marketing efforts, the number of electric trucks fell from 10% in 1913 to only 3% - 4% in 1925 according to Schiffer [12].

However, the years between 1920 and 1970 were a time of steady decline in electric cars. On a global scale, the depression of the 1930s, followed by World War II, harmed a possible resurgence of electric vehicles and new experiments with alternative fuel vehicles. At that time, few studies and scientific research have been developed for electric vehicles. Even in the postwar period of economic prosperity, the electric vehicle projects remained stored in a context of small concern for energy security, due to the existence of abundant and cheap fuel and vehicles with internal combustion engine (MCI), the largest and fastest [15].

In the United States, a meager revival of electric vehicles happened in the 60s. Car technology was basic, with DC motors with brushes. However, in terms of energy accumulators there was some variety beyond the normal lead-acid: the lead-cobalt and the nickel-cadmium.

Specifically, the electric traction technology began again to be shyly explored, its development returning only from the 60s on, when the electric vehicle was seen as a way to overcome the environmental problems caused by emissions from the combustion powered vehicles. It is noteworthy that most of the electric vehicles produced in the 60s emanated from the conversion of conventional vehicles.

We observe, however, that from the 1970s on, with the emergence and worsening of the oil crisis, discussions on environmental issues in urban centers have become a worrying factor for government leaders. The electric vehicle has been considered as an energy alternative, mainly in countries with a lot of hydroelectric generation or coal-based thermal power. In this period, there were several initiatives aiming to insert them back into the market, but neither the pure electric cars nor the hybrids were able to compete in the market with conventional cars, which had a sustainable competitive advantage [9].

Thus, the performance achieved each year by the respective electric vehicles parameterized in speed, range, battery voltage and power on the 2nd cycle (between 1912 and 1973) are summarized in **Table 3**.

Because of its importance and relevance to this study, as a kind of complementary measure, the table shows one of the striking points in time due to short range (50 to 100 km) and low average speed (on average, 50 to 100 km/h). Somehow, these factors affected the introduction of electric vehicles on a large scale. Moreover, the delay on the recharging of the battery (about 8 hours) and the lack of infrastructure in use service became a concomitant factor.

Meanwhile, according to Valle Real and Balassiano [17], there are basically two paths to be taken. The first would make them more efficient vehicles from the point of view of energy consumption (as well as of the amount of emissions); and the second, by means of restrictions and adoption of specific rates, would make the user reduce the utilization of motor vehicles, particularly the car.

5. Third Cycle of Electric Vehicles (1973-1996)

Apart from being cleaner (depending on the type of energy used), electric cars manufactured and converted in the late 60s used certain conservation techniques in an attempt to increase their autonomy and maximum speed. The priority was in guaranteeing that these vehicles reached the level of performance offered by combustion powered cars, whose development had been significant throughout the century.

Interest in electric vehicles had increased considerably until the late 80s, when the problem of air pollution from large cities began to be discussed more often [6]. With respect to Brazil, for example, the fundamental importance of engineer John Augustus Conrado do Amaral Gurgel (1926-2009) cannot be forgotten, since he pro-

Table 3. Evolution of electric vehicles from 1912 to 1973.

Year	Speed	Autonomy	Battery	Motor (Power/Voltage Type)		
1915	50 km/h	161 km	Lead	32 KW	76 V	-
1916	42 km/h	60 km	Lead	1.8 KW	80 V	-
1917	40 km/h	340 km	Ni-Fe	-	-	-
1941	60 km/h	50 km	Lead	-	96 V	-
1947	75 km/ h	65 km	Lead	-	36 V	-
1960	70 km/h	160 km	Lead	2 × 6 KW	48 V	-
1961	50 km/h	55 km	Lead	2 × 6 KW	48 V	MCC
1966	100 km/h	210 km	Pb-Co	57 KW	120V	MCC
1967	60 km/h	60 km	Lead	3.7 KW	48 V	-
1968	85 km/h	190 km	Ni-Cd	-	-	MCC
1972	60 km/h	140 km	Lead	32 KW	144 V	-
1973	85 km/h	60 km	Lead	3 × 2.5 KW	48 V	-

Source: own compilation.

duced the first Brazilian electric car in 1974, with a range of 60 km—called Itaipu.

Generically speaking for this period, it is noteworthy that by the year 1973 the crisis associated with the embargo imposed by the OPEC (Organization of Petroleum Exporting Countries) brought new prospects for electric cars. However, the United States depended significantly on oil from the Arab countries and the U.S. Congress at the time was determined to reduce this dependence. There was also an economic motivation given by the U.S. trade balance. Environmental issues were not actually critical, therefore, it was not generally considered that only the use of electric cars would improve air quality, according to Schiffer [12].

Objectively, as stated before about the 80s, the government interests turned to the advantages derived from powered electric vehicle propulsion vehicles, mainly because of environmental issues. Thus, it is observed that government policy measures were introduced worldwide in the pursuit of reducing urban vehicle emissions. The prime example was the California Air Resources Board's (CARB) that implemented, in 1990, the first regulatory standards for zero-emission vehicles in California.

A number of modern vehicles was introduced by automakers between the 1980s and 1990. Along with the selling of the top vehicles, such as the General Motors EV-1, the Toyota RAV4-EV and the Ford Ranger EV, several studies about the cost of batteries have been developed in order to assess the commercial prospects of these vehicles [12]. Although they were more efficient than conventional cars, this advantage had little value at the time when the oil price was the lowest in history [18].

Based on the performances achieved each year by its respective electric vehicles, parameterized in speed, range, battery voltage and power and in comparison to previous cycles, **Table 4** shows the variables from the 3rd cycle, between 1973 and 1996.

6. Fourth Cycle of Electric Vehicles (1997 to the Present)

Within the different stages of live cycles, currently the hybrid, the electric and the plug-in vehicles emerge

Table 4. Evolution of electric vehicles from 1973 to 1996.

Year	Speed	Autonomy	Battery	Motor (Power/Voltage Type)		
1974	60 km/h	70 km	Lead	4.4 KW	48 V	MCC
1976	60 km/h	90 km	Lead	8.8 KW	72 V	MCC
1977	105 km/h	100 km	Lead	17 KW	84 V	MCC
1978	120 km/h	160 km	Lead	24 KW	36 V	-
1980	105 km/h	115 km	Lead	18 KW	96 V	MCC
1981	81 km/h	80 km	Lead	15 KW	102 V	-
1983	80 km/h	110 km	Lead	10 KW	72 V	-
1984	50 km/h	115 km	Lead	24 KW	84 V	-
1987	100 km/h	80 km	Na-S	17 KW	200 V	MCC
1989	90 km/h	100 km	Lead	18 KW	96 V	MCC
1990	105 km/h	150 km	Lead	60 KW	320 V	MI
1991	120 km/h	276 km	Ni-Cd	30 KW	200 V	MCC
1992	110 km/h	170 km	Na-NiCl	62 KW	120 V	MCC
1993	120 km/h	90 km	ZnBr2	12 KW	168 V	MI
1994	85 km/h	100 km	NiCd	30 KW	240 V	MI
1995	90 km/h	160 km	Lead	2 × 18 KW	72 V	MCC
1996	100 km /h	90 km	Li-ion	30 KW	216 V	MI

Source: own compilation.

as instruments to solve flagship-oriented issues such as energy security and climate impact. According to Anfavea [19], the Brazilian Chain of Automotive Supplies brings together a spectrum of ethnic diversity in the nationalities of their manufacturers, so that automakers are gathered here from no less than nine different countries: Germany, Brazil, South Korea, United States France, India, Italy, Japan and Sweden. This ethnic diversity has no record in any other major producer of vehicles on the planet [19].

Most of these automakers that help compose the chain of the automotive segment conduct research with universities in order to develop integrated models of hybrid and electric cars format. The first major step in this recent movement arose in 1997, when Toyota, the Japanese automaker, launched the Prius in Japan—a hybrid four-door sedan—followed by Honda, the first to launch a hybrid in the U.S. market—the Insight in 1998 [20].

Since the launch of the Toyota Prius in 1997, 1.9 million HEVs (Hybrid Electric Vehicles) vehicles and 60.0 thousand PHEVs (Plug-in Hybrid Electric Vehicles) and BEVs (Battery Electric Vehicles) vehicles have been sold in the North-American market [15]. This fact can be attributed in large part to the encouragement of the U.S. government to manufacturers and consumers of hybrid and electric vehicles. Worldwide, over the last decade, many HEVs, PHEVs and BEVs have been sold, totaling more than 2.5 million vehicles. In early 2011, the penetration of these technologies in the market is of 2% in the U.S. and 9% were sold in Japan [21]. A descriptive approach to the start of the 4th cycle will be presented below.

Although it is necessary to prioritize the previously mentioned variables that portray each historical cycle, at first, it is noticed that the electric motor is the ideal drive propulsion. Due to the strong competition and the growing consumer demand, companies in the area of transport and logistics attempt to reduce operating costs, while seeking to improve services. The electric vehicle meets those needs, since it has compatible attributes such as: it is quiet, it is highly efficient, it has excellent torque characteristics × speed and it does not pollute. However, the Mercedes-Benz CL600 has a 367 hp of 12-cylinder engine and can accelerate its 2380 kg 0 - 60 miles/h (about 100 km/h) in 6.3 s. To increase efficiency there is a mechanism that disables 6 cylinders when there is no need for high torque, according to Dettmer [22]. **Table 5** shows the variables portrayed in the 4th cycle, starting in 1997 until today.

With the new developments in batteries, electric vehicles now have their storage capacity between 20 and 60 kWh, allowing its interconnection with the electric distribution network through the consumption of energy. Plus, in the very near future, this will provide energy according to the needs of network functionality through the Vehicle Connected to the Network (VLR), according to Kempton and Tomic [23].

Figure 1 shows a partial configuration of the components of an electric vehicle that properly uses the electricity from the public distribution system to recharge the battery installed in the vehicle (battery bank). The re-

Table 5. Evolution of EVs between 1996 and the present.

Year	Speed	Autonomy	Battery	Motor (Power/Voltage Type)		
1997	100 km/h	100 km	NiMH	18.5 KW	288 V	MS
1998	120 km/h	185 km	NiMH	84 KW	-	MI
1999	100 km	115 km	Li-ion	24 KW	300 V	MIP
2000	130 km/h	200 km	Li-ion	65 KW	345 V	MIP
2002	120 km/h	80 km	NiCd	44 KW	180 V	MIP
2007	150 km/h	290 km	Li-ion	150 KW	355 V	MI
2008	110 km/h	100 km	Li-ion	2 × 15 KW	144 V	MIP
2009	150 km/h	210 km	Li-ion	150 KW	380 V	MI
2010	140 km/h	300 km	Li-ion	200 KW	380 V	MIP
2011	145 km/h	270 km	Li-ion	47 KW	360 V	MI
2012	120 km/h	120 km	Li-ion	55 KW	300 V	MIP

Source: own compilation.

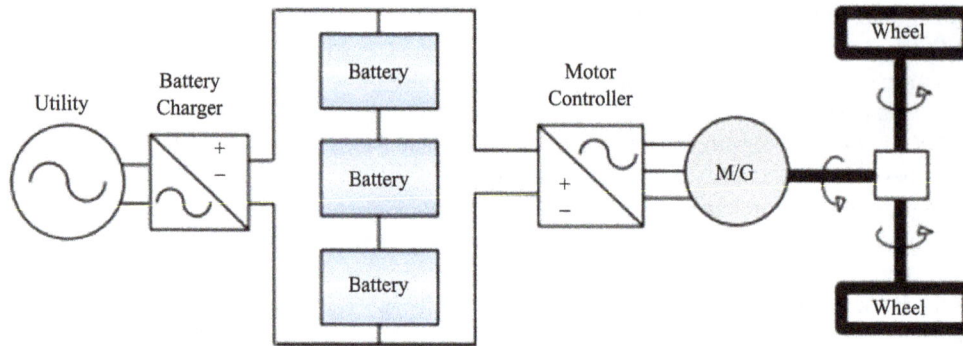

Figure 1. Evolution of batteries.

ceived energy is stored in the battery, in electrochemical format. This stored energy is converted into electrical energy that is transported to the Electric Motor (M/G), which will make its conversion into mechanical energy and thus provide the movement of the vehicle without generating emissions or noise. Also, if a Regenerative Breaking System (TR) is implemented in the electric vehicle, it is also possible to store the energy produced when breaking or slowing down, by converting the kinetic energy into electrical energy through the M/G, which will be stored in the battery [24].

7. Potentialities and Limitations Highlighted in the LV Development

The importance of batteries is irrefutable in an electric vehicle. That way, the main characteristics of an accumulator of energy are the specific energy, specific power and lifetime. Specific energy is the energy amount stored by the battery per unit weight; the specific power is the supplied power per unit mass; and the lifetime is the number of charge/discharge that may be required.

In **Graph 1**, it is possible to check, on a timeline, the progress in size and weight of the various types of batteries. The batteries of the Pb-SO4, Ni-Cd, Ni-MH type are stagnant in terms of specific energy and density.

Thus, it can be defined as energy density the ratio between the maximum amount of energy that can be stored securely in an energy storer element body and the volume of that body. The higher the energy density—which can be measured in Wh/l (watt/h per liter) or MJ/l (megajoules per liter)—the more energy can be stored or transported in one body with the same volume. As for the specific energy, it also relates to the maximum amount of energy that can be stored, but with the mass of the containing body element. Specific energy can be quantified in Wh/kg (watt-hours per kilogram).

Peças Lopes, Soares, Almeida and Moreira da Silva [11] point out that, with the new developments of batteries, vehicles powered by electric traction have the storage capacity of between 20 and 60 kWh, allowing its interconnection with the power grid distribution through the energy consumption. In the very near future, this will provide power according to the needs of the network through the functionality of a Vehicle Connected to the Network (VLR).

On the face of it, the technical requirements demanded by the energy accumulators are different for each type of vehicle. Electric vehicles require batteries with higher energy densities, which limits them due to the masses and volumes associated, contributing to the low range of these types of vehicles. Since in pure electric vehicles batteries are the only energy source, these end up suffering deeper discharges, demanding more robust batteries with long life and the acceptance of a high number of charge and recharge cycles. However, increasing the autonomy of electric vehicles requires larger batteries, greatly increasing the vehicle mass. Inversely, reducing the range of electric vehicles enables higher effective energy efficiency. In **Graphs 2-5** we can see the evolution of the range and speed before 4 cycles, taking into account previously shown factors that were considered essential for the development of these vehicles.

It is worth noting that the lead-acid (Pb) batteries are the best-known being used in cycles 1, 2 and 3, having the largest implementation in 2012. These batteries are the most inexpensive and they require little maintenance; however, they have limited power and specific energy, 40 Wh/kg and 350 W/kg respectively. The average lifespan of these batteries is one of its limitations, being it about 500 charge/discharge cycles.

Faia [25] discusses that the type of battery which has appeared most promising in recent years has been the

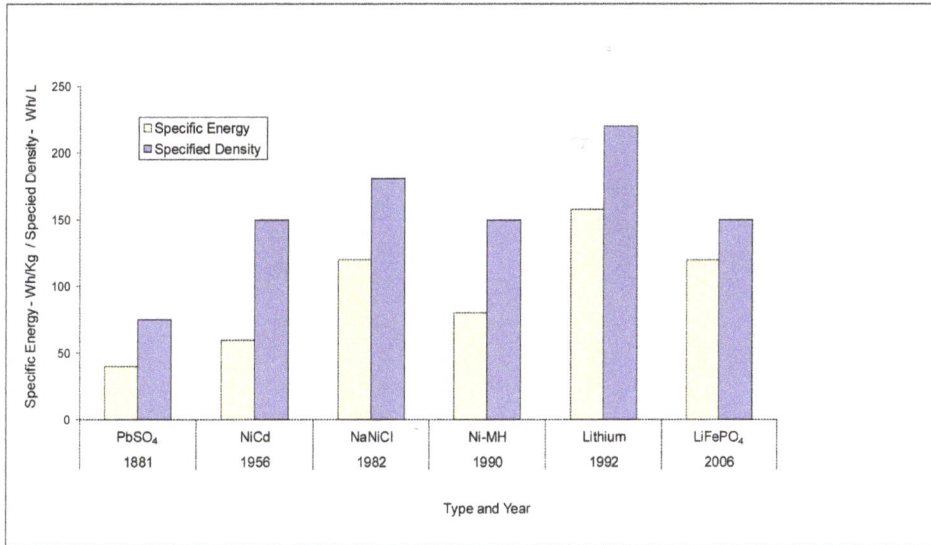

Graph 1. Evolution of batteries.

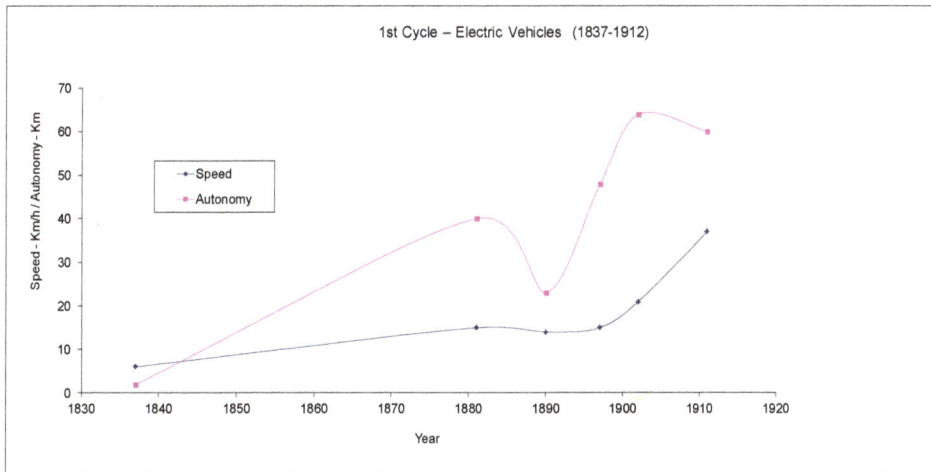

Graph 2. 1st cycle—electric vehicles (1837-1912).

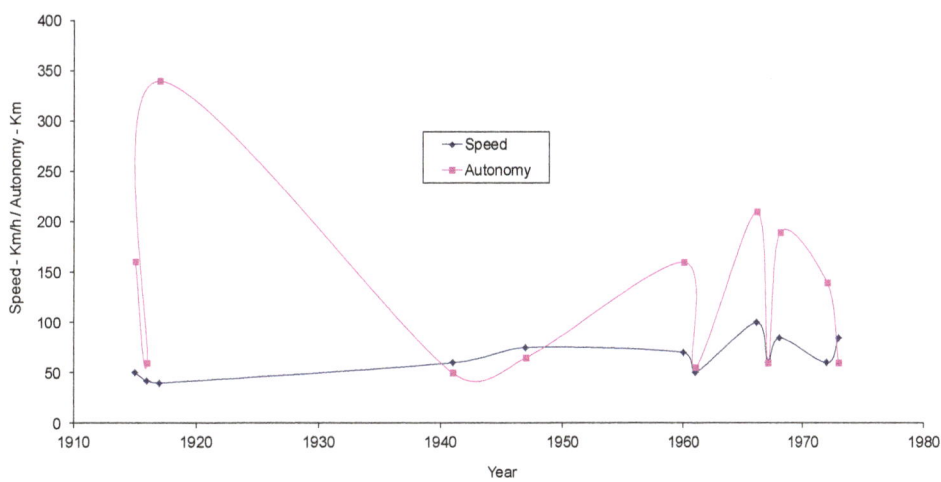

Graph 3. 2nd cycle—electric vehicles (1912-1973).

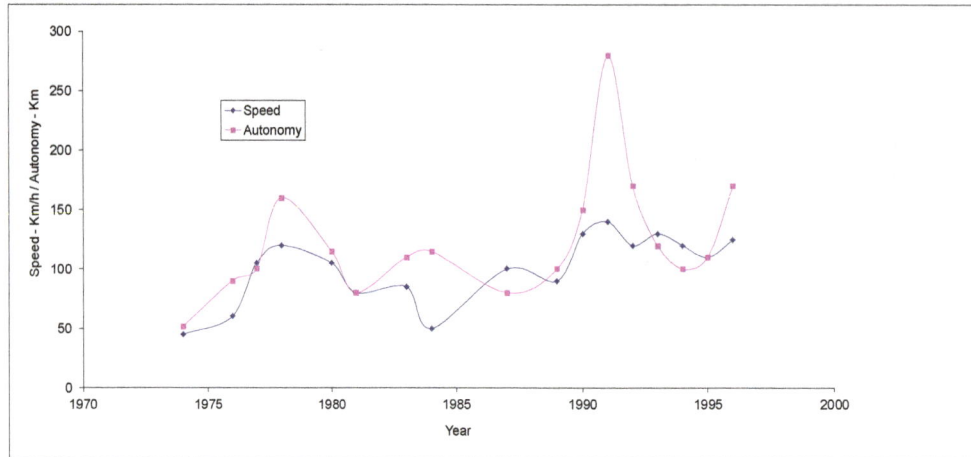

Graph 4. 3rd cycle—electric vehicles (1973-1996).

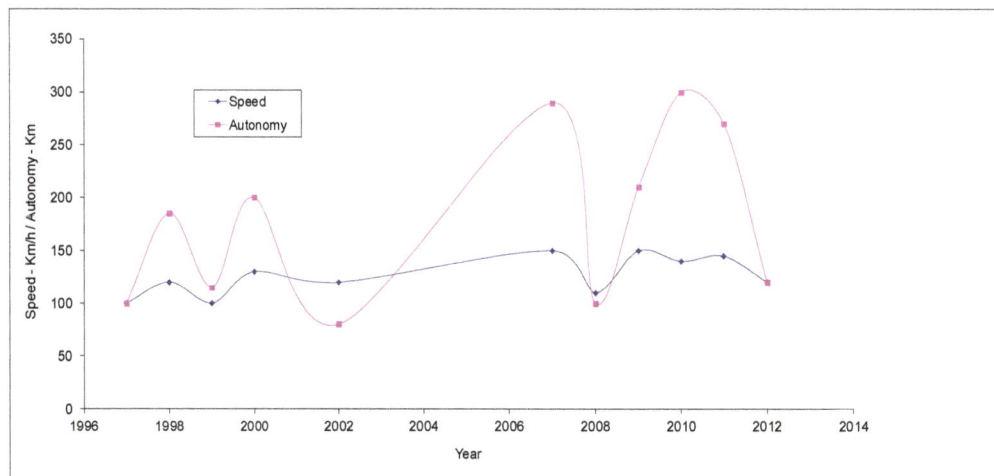

Graph 5. 4th cycle—electric vehicles (1996 to present).

lithium (Li-ion) ion. This battery, used intrinsically in the 4th cycle, has a specific energy of over 150 Wh/kg and specific powers that can go up to 2000 W/kg. Its lifespan is about 1200 charge/discharge cycles. The disadvantages of this type derive from the fact that batteries require a precise charging system due to its low tolerance for peak power and are still relatively expensive for pure electric vehicles.

A solution to the inflexibility of the batteries' autonomy has been presented by Andersen *et al.* [26], being implemented in Israel. The core of this proposal is the separation of the properties of the car and of the battery. A company would be responsible for assuming the risk of ownership of the batteries, which would be leased. The consumer would be charged for the energy they consume over the kilometers traveled. This is analogous to the collection of minutes used by mobile phone operators. Consumers would have a number of "packages" available, which will depend on the battery usage profile.

In this aspect, the main types of motors used in electric vehicles are: Continuous Current Motor (MCC), or Asynchronous Induction Motor (IM), Permanent Magnet Synchronous Motor (IPM) and Switched Reluctance Motor (MRC). In **Figure 2** one can see the qualitative assessment of various engine characteristics, with the objective of identifying the technologies that would have the best performance in the electric car applications according to Zeraoulia, Benbouzid and Diallo [27].

It is possible to compare the total ratings of the characteristics of the different propulsion systems listed in **Figure 2**, allowing us to infer that the MI OE MIP are the most suitable for electric vehicles. Thus, from the 4th cycle to the IPM motor cycle, this type of vehicle has currently become more solicited because of its gains in density, its efficiency and its cost.

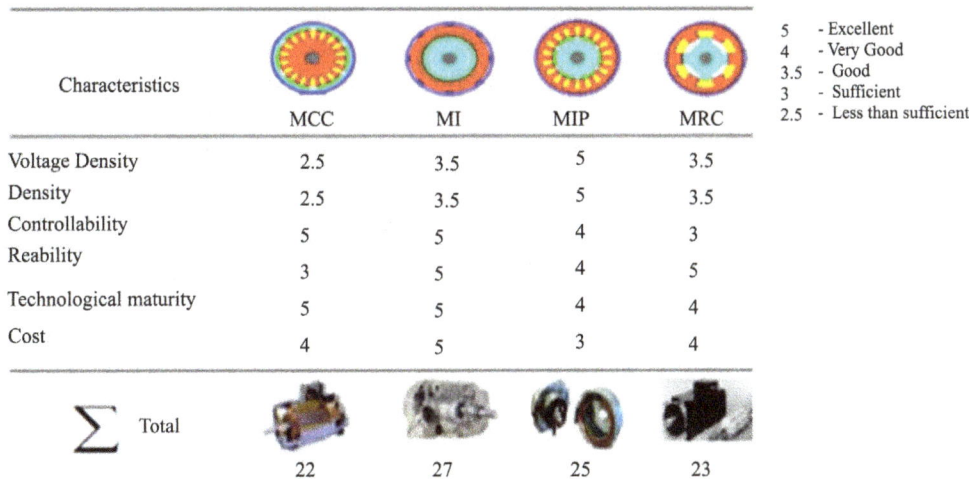

Characteristics	MCC	MI	MIP	MRC
Voltage Density	2.5	3.5	5	3.5
Density	2.5	3.5	5	3.5
Controllability	5	5	4	3
Reability	3	5	4	5
Technological maturity	5	5	4	4
Cost	4	5	3	4
∑ Total	22	27	25	23

5   - Excellent
4   - Very Good
3.5 - Good
3   - Sufficient
2.5 - Less than sufficient

Figure 2. Matters of electric motor selection drive for HEV propulsion systems: a comparative study.

8. Conclusions and Final Thoughts

The analysis in the context of this study points to the vulnerabilities and potentialities that derive from the introduction of electric vehicles (EV) in a timeline. In essence, the lithium used properly in the fourth cycle ion shows better results due to its higher specific energy and high energy efficiency maximizing autonomy of 200 km. Moreover, there have been improvements in aerodynamics, engines, and the weight and size of batteries, demonstrating the technical and marketing maturity of the EVs. That is why studies about the logistics of such vehicles in terms of autonomy become intrinsically relevant.

Accordingly, for the first two cycles, specifically in the context of a car, all variables described by means of **Tables 1-3** show that the inclusion of these commercially electric vehicles is hindered by the fundamental limitations of batteries as a power source.

Also in relation to energy accumulators, known as batteries, it is valid to emphasize the significant time (6 - 12 pm) they take to be recharged. In contrast, a gasoline tank takes about 2 - 3 minutes to be completed, which means a power flow of an order of 20 - 30 MW during its supply in a gas station—an issue addressed by Hermance and Sasaki [28].

Undoubtedly, gasoline has become an ideal fuel, but it also has its adversity such as very low efficiency and, in low rpm conditions, the available torque is low. Torque is what determines the acceleration capability; for a conventional car, the combination of the gearbox and the oversizing of the combustion engine—which carries an even greater inefficiency—is what defines this acceleration.

Once again, the main changes that took place, especially in the 3rd and 4th historical cycles of electric vehicles, should be emphasized. They range from the break-up of new technologies to the development of new motors, power converters, battery chargers and energy accumulators.

It is evident that the specific properties of electric vehicles vary according to the battery size and type of vehicle. The same way electric cars need higher specific energy per volume, hybrid vehicles in return require batteries that provide maximum power in the smallest possible size [25].

In a simplified way, after the energy crisis of 1970 and 1980, and with the significant increase in emissions of Greenhouse Gases (GHG), the revival of the interest in electric vehicles based on energy sources alternative to oil arose, as is corroborated by Tomic and Kempton [23].

In this sense, the reason why the production of electric cars remains static is related to some indicators such as high production costs—which translate into low penetration—, lack of logistics networks for the vehicles to refuel and unsatisfactory autonomy. This is undoubtedly a worrying factor for the countries, manufacturers and end customers throughout the automotive supply chain.

To firm the findings and considerations of this work, it is important to highlight the studies done by Hall, Reis and Junior [29]-[31], which have proved that the lithium ion is expected to occupy a prominent place in electric vehicles in the present days. This is due to its higher specific energy in terms of volume and mass, its high energy efficiency (near 100%), its long life cycle (approximately 3000 cycles with a discharge depth of 80%), its low rate of self-discharge, and its lack of memory effect, which decreases battery capacity when it is recharged. Mass production and development of nanostructured materials provide considerable scope for cost savings. We observed their use in large scale from the year of 1996, during the fourth and last cycle analyzed.

Based on the factors put in evidence throughout this work, we point out that the disadvantages of electric vehicles are associated with deficiencies of electrochemical energy storage. Compared to conventional fuels, the battery of electric cars has low specific energy in terms of volume and mass and low rate of refueling/recharging. These questions are well evaluated by Bradley e Frank [32].

Additional considerations need to be brought up with respect to another aspect that should receive attention in EVs. One of them is associated with the use of electricity by the transportation sector. The introduction in scale of EVs comes with an increase in electricity demand and the possible need for increased capacity to generate electricity. More than that, the penetration of electric vehicles in the electrical system can cause overload on transformers and distribution lines, as well as—mainly depending on the timing and form of clearance—overshoot in the electrical system. These aspects of the electrical system are also put in evidence by Kiviluoma and Meibom, Hadley and Tsvetkova, Green II et al., Lin et al. [33]-[36], among others.

Finally, it is concluded from the results, based on the four cycles presented here, that the inverters have improved over the years. We highlight that in old electric vehicles' DC, there was the need for a transmission to enable reverse driving (rear gear), but this is no longer needed.

References

[1]   Brown, L.R. (2003) Eco-Economy: Building and Economy to Earth. Publicado no Brasil pela Universidade Livre da Mata Atlântica (UMA), Primeira Edição, 368 p.

[2]   Pecorelli Peres, L.A. (2003) Electric Vehicles: Environmental and Energy Benefits. Rio de Janeiro and Cultural Research Noel Rosa.

[3]   Ministry of the Environment (MMA) (1999) Inspection Program Evaluation and Maintenance of Vehicles in Use in Rio de Janeiro. Document Prepared by the partnership LIMA (Interdisciplinary Laboratory Ministry of the Environment (MMA). COPPE/UFRJ as Part of the Project "Air Quality Management in Major Metropolises Brasileiras" under the Third Amendment to the Agreement MMA/Foundation COPPETEC 1999-CV-000054.

[4]   Campi, T.M., Rutkowski, E. and Lima Jr., O.F. (2004) Sustainability of Transportation Techniques. UNICAMP, Campinas.

[5]   Braga, A., Pereira, L.A.A. and Saldiva, P.H.N. (2002) Pollution and Its Effects on Human Health. *Seminar on Sustainability in Energy Generation and Use*, UNICAMP, Campinas, 20 p. http://www.bibliotecadigital.unicamp.br/

[6]   Delucchi, M.A. and Lipman, T.E. (2010) Lifetime Cost of Battery, Fuel-Cell, and Plug-In Hybrid Electric Vehicles, Chapter 2. In: Pistoia, G. and Elsevier, B.V., Eds., *Electric and Hybrid Vehicles*: *Power Sources, Models, Sustainability, Infrastructure and the Market*, Elsevier, Amsterdam, 19-60.

[7]   Bottura, C.P. and Barreto, G. (1989) Electric Vehicles. UNICAMP, Campinas.

[8]   Larminie, J. and Lowry, J. (2003) Electric Vehicle Technology Explained. John Wiley and Sons Ltd., Chichester.

[9]   Baran, R. and Legey, L.F.L. (2011) Electric Vehicles: History and Prospects in Brazil. *BNDES*, **33**, 207-224.

[10]   Hoyer, K.G. (2008) The History of Alternative Fuels in Transportation: The Case of Electric and Hybrid Cars. *Utilities Policy*, **16**, 63-71. http://dx.doi.org/10.1016/j.jup.2007.11.001

[11]   Peças Lopes, J.A., Soares, F.J., Almeida, P.M. and Moreira da Silva, M. (2009) Smart Charging Strategies for Electric Vehicles: Enhancing Grid Performance and Maximizing the Use of Variable Renewable Energy Resources. EVS24, Stavanger.

[12]   Schiffer, M.B. (2010) Taking Charge—The Electric Automobile in America. Smithsonian Institution Press, Washington DC.

[13]   Struben, J.R. and Sterman, J. (2006) Transition Challenges for Alternative Fuel Vehicle and Transportation Systems. MIT Sloan Research Paper. http://ssrn.com/abstract=881800

[14]   DOE (2012) Department of Energy, Energy Efficiency and Renewable Energy. History of Electric Vehicles. http://www1.eere.energy.gov/vehiclesandfuels/avta/light_duty/fsev/fsev_history.html

[15]   Anderson, C.D. and Anderson, J. (2010) Electric and Hybrid Cars: A History. 2nd Edition, McFarland & Company, Inc., Jefferson.

[16]   Yergin, D. (1991) The Prize: The Epic Quest for Oil, Money, and Power. Free Press, Nova Iorque.

[17]   Valle Real, M. and Balassiano, R. (2002) Identify Priorities for Aadoção Mobility Management Strategies: The Case of

Dobrio City January. *Proceedings of the* 10*th Congress of Research and Training in Transportation*, ANPET, Natal. http://www.ivig.coppe.ufrj.br/ivig/Paginas/teses-dissertacoes-artigos.aspx

[18] Faia, S.M.R. (2006) Optimization of Vehicle Propulsion Systems for Fleet. Thesis of M.Sc., Instituto Superior Técnico, Lisboa.

[19] ANFAVEA (2011) National Association of Automobile Manufacturers, Statistical Yearbook of the Brazilian Automotive Industry.

[20] Dijk, M. and Yarime, M. (2010) The Emergence of Hybrid-Electric Cars: Innovation Path Creation through Co-Evolution of Supply and Demand. *Technological Forecasting and Social Change*, **77**, 1371-1390. http://dx.doi.org/10.1016/j.techfore.2010.05.001

[21] IEA (2011) International Energy Agency, Technology Roadmap: Electric and Plug-In Hybrid Electric Vehicles.

[22] Dettmer, R. (2001) Hybrid Vigour. *IEE Review*, **47**, 25-28. http://dx.doi.org/10.1049/ir:20010109

[23] Tomic, J. and Kempton, W. (2007) Using Fleets of Electric-Drive Vehicles for Grid Support. *Journal of Power Sources*, **168**, 459-468. http://dx.doi.org/10.1016/j.jpowsour.2007.03.010

[24] Kramer, B., Chakraborty, S. and Kroposki, B. (2008) A Review of Plug-In Vehicles and Vehicle-to-Grid Capability. National Renewable Energy Laboratory, 1617 Cole Blvd., Golden, CO 80401, USA.BEV-HEV-PHEV-FCEV.

[25] Broussely, M. (2010) Chapter 13—Battery Requirements for HEVs, PHEVs, and EVs: An Overview. In: Pistoia, G. and Elsevier B.V., Eds., *Electric and Hybrid Vehicles*: *Power Sources, Models, Sustainability, Infrastructure and the Market*, Elsevier, Amsterdam, 305-345. http://dx.doi.org/10.1016/B978-0-444-53565-8.00013-0

[26] Andersen, P.H., Mathews, J.A. and Rask, M. (2009) Integrating Private Transport into Renewable Energy Policy: The Strategy of Creating Intelligent Recharging Grids for Electric Vehicles. *Energy Policy*, **37**, 2481-2486. http://dx.doi.org/10.1016/j.enpol.2009.03.032

[27] Zeraoulia, M., Benbouzid, M.E.H. and Diallo, D. (2010) Electric Motor Drive Selection Issues for HEV Propulsion Systems: A Comparative Study. http://hal.inria.fr/docs/00/53/33/62/PDF/IEEE_VPPC_2005_ZERAOULIA.pdf

[28] Hermance, D. and Sasaki, S. (1998) Hybrid Electric Vehicles Take to the Streets. *IEEE Spectrum*, **35**, 48-52. http://dx.doi.org/10.1109/6.730520

[29] Hall, P.J. (2008) Energy Storage: The Route to Liberation from the Fossil Fuel Economy? *Energy Policy*, **36**, 4363-4367. http://dx.doi.org/10.1016/j.enpol.2008.09.041

[30] Reis, N.A.O. (2008) The Hybrid Car as a Supplier-Consumer Element of Electricity-Battery Modeling. M.Sc. Thesis, IST University, Lisbon.

[31] Junior, A.R.P. (2002) Regulation of Energy Demand in a Propulsion System for a Hybrid Electric Vehicle Series. M.Sc. Thesis, UFRGN, Natal.

[32] Bradley, T.H. and Frank, A.A. (2009) Design, Demonstrations and Sustainability Impact Assessments for Plug-In Hybrid Electric Vehicles. *Renewable and Sustainable Energy Reviews*, **13**, 115-128. http://dx.doi.org/10.1016/j.rser.2007.05.003

[33] Kiviluoma, J. and Meibom, P. (2011) Methodology for Modelling Plug-In Electric Vehicles in the Power System and Cost Estimates for a System with Either Smart or Dumb Electric Vehicles. *Energy*, **36**, 1758-1767. http://dx.doi.org/10.1016/j.energy.2010.12.053

[34] Hadley, W.S. and Tsvetkova, A. (2008) Potential Impacts of Plug-In Hybrid Electric Vehicles on Regional Power Generation. UT-Battelle, Oak Ridge National Laboratory, Oak Ridge.

[35] Green II, R.C., Wang, L. and Alam, M. (2011) The Impact of Plug-In Hybrid Electric Vehicles on Distribution Networks: A Review and Outlook. *Renewable and Sustainable Energy Reviews*, **15**, 544-553. http://dx.doi.org/10.1016/j.rser.2010.08.015

[36] Lin, S., He, Z., Zang, T. and Qian, Q. (2010) Impact of Plug-In Hybrid Electric Vehicles on Distribution Systems. *International Conference on Power System Technology* (POWERCON), Hangzhou, 24-28 October 2010, 1-5. http://dx.doi.org/10.1109/POWERCON.2010.5666508

14

Simplified Criterion of Steady-State Stability of Electric Power Systems

K. R. Allaev[1], A. M. Mirzabaev[2], T. F. Makhmudov[3], T. A. Makhkamov[4]

[1]Power Engineering Department, Tashkent State Technical University, Tashkent, Republic of Uzbekistan
[2]Mir Solar LLC, Tashkent, Republic of Uzbekistan
[3]Power Engineering Department, Tashkent State Technical University, Tashkent, Republic of Uzbekistan
[4]Research and Development Department, Tecon Groups, Moscow, Russian Federation
Email: solarmir@mail.ru, temur.ma@gmail.com

Abstract

In the paper the simplified criterion of a steady-state stability of electric power systems (EPS) is justified on the basis of Lyapunov functions in a quadratic form ensuring necessary and sufficient conditions of its performance. Upon that, the use of the node-voltage equations allows reducing study of a steady-state stability of complex EPS to study of the generator-bus system. The obtained results facilitate studies of a steady-state stability of the complex systems and have practical importance.

Keywords

Steady-State Stability, Matrix Method, Lyapunov Function, Node-Voltage Equations

1. Introduction

Study of EPS stability at small disturbances is based on known classical concepts of the General Theory of Stability of Motion [1]-[8].

As is known [1]-[4] [7], features of EPS are their continuous flow process, complexity, multiple connection of system of facilities and their control devices. With increasing the capacity and complexity of EPS the problem of automation of system state control both in normal operating conditions and during disturbances becomes more acute. Development of the automated dispatch control systems (ADCS) for EPS and automatic control systems require the new approach to development of the simplified mathematical models and the algorithms meeting the practical requirements that form the basis of complex EPS control [1] [4] [7] [8].

There is a real-time calculation problem when the issue to ensure maximum operativeness of a current operating conditions stability calculation and timely definition of EPS approaching to stability limits is critical. The most acceptable for their solution are the matrix computational methods grounded on use of the newest methods to discover structural properties of dynamic systems which include modern electric systems [7]-[11].

Steady-state stability study of electric power systems consists in definition of a possibility of steady-state operation under given values of electric power system's parameters, an electricity generating sources operation state, loads in nodal points and settings of automatic control equipment [1] [2]. The problem is usually solved for the determined conditions.

In most cases the complex power system steady-state stability analysis is carried out under the supposition of the lack of a self-oscillation in the electric power system considering that this requirement is ensured by appropriate setting of automatic regulators [3]. In this case the problem becomes simpler and is reduced to study of an aperiodic steady-state stability of the system, *i.e.* to definition of dependence and a sign for the constant term of the characteristic equation of the system upon the continuous variation of any parameter of the operation condition.

Now, except for classical, direct solution methods for calculation of a matrix spectrum of system are also used for complete steady-state stability analysis, including an estimate of lack of eigenvalues in the right half-plane by indirect criteria, dynamic simulation methods, etc. [11].

In the practice of calculations there is the popular method of so-called critical sections when the critical line (section) is determined by high probability of loss of EPS stability because of overloading. Upon that, finally it is not specified what generator or station are referred to the critical section, therefore definition of critical section is ambiguous [12].

The devices of the synchronized phasor measurement units (SPMU) being developed create premises for development of new analysis methods and algorithms of an electric power system state at small disturbances including methods to detect weak tie-lines and critical sections that are based exclusively on operatively obtained values of operating condition parameters [7].

2. Lyapunov Function in a Quadratic Form as the Tool for the Full Study of Steady-State Stability in EPS

One of the most fruitful methods for study of EPS stability is application of a direct (second) Lyapunov's method which require selection of special Lyapunov functions and obtaining of their derivatives taking into account the perturbation equations [5] [6].

According to Lyapunov's direct method which is applied to study of a dynamic systems stability including electrical power systems it is generally supposed definition of special sign-definite function of state variables $V(x_1, x_2, x_3 \cdots x_n)$ which derivative dV/dt taken on account of system of differential equations describing dynamics of the system, should be sign-definite with an opposite sign to V or be identically zero or strictly sign-definite with an opposite sign to V. Under these requirements the system is, accordingly, stable or asymptotically stable [5] [6]. Construction of V for nonlinear systems is generally performed by a trial method and obtained results ensure only sufficient conditions of stability for the explored system.

At the same time, in the case of linear autonomous systems there is a Lyapunov function in a quadratic form ensuring both necessary and sufficient conditions of its stability [5]. Modern counting machines offer ample opportunities for a solution of the higher order equations [13] and, accordingly, successful application of Lyapunov functions in a quadratic form for study of a steady-state stability for complex EPS [14].

The essence of the method consists in the following. Let's consider the linear time-invariant system described in a state space by the system of differential equations:

$$dx/dt = AX \tag{1}$$

where A is a square matrix with constant elements; X is a n—dimensional column vector with coordinates $x_1 - x_n$.

Following Lyapunov, we will define for this system function V in a quadratic form [5] [6]:

$$V = \sum_{i,j=1}^{n} q_{ij} x_i x_j = X^{\mathrm{T}} QX, \quad i, j = 1, 2, \cdots, n \tag{2}$$

where Q is yet unknown square matrix of coefficients of a quadratic form; X^{T} is transposed X (row-vector).

Owing to (1) the total derivative of V with time looks like:

$$dV/dt = X^{\mathrm{T}} \left(A^{\mathrm{T}} Q + QA \right) X \tag{3}$$

Let's require that V should satisfy the condition:

$$dV/dt = -W \tag{4}$$

where W is arbitrarily prescribed quadratic form of state variables.

Let's denote:

$$A^{T}Q + QA = -C \tag{5}$$

The main result consists in that the system (1) is asymptotically stable in only case when the (2) has positive definite solutions Q at any positive definite matrix C [5].

The Equation (5) puts in correspondence to any symmetric matrix Q a matrix C and *vise versa*, and this correspondence is linear [5] [17]. Elements of matrix Q are determined from (5) by a solution of $n(n+1)/2$ equations where n is a number of initial differential equations. If to set a positive definite symmetric matrix C (where the determined from (5) matrix Q will be also positive definite) then due to linearity and stationarity of system (1), according to the Lyapunov's theorem, we will obtain an asymptotical stability of its equilibrium state. Upon that, stability conditions should be strictly equivalent to the obtained on the basis of Routh-Hurwitz criterion [2].

There is a close connection between the Lyapunov's theorem and other algebraic stability criteria: the Routh-Hurwitz criterion [15], the Hermite stability criterion [16], the Shur-Kon criterion [17], and the constituent matrix method [18] [19]. The main advantage of the Lyapunov's second method for stability when studying stability conditions is related to a possibility to operate in calculations with elements of a matrix A omitting calculations of coefficients of a characteristic polynomial for this matrix.

On the basis of Lyapunov's functions in a quadratic form, we will carry out computational-experimental research of a steady-state stability of both simplex and complex EPSs and compare results for them with the results obtained conventionally on the basis of the Routh-Hurwitz criterion.

2.1. Simplex EPS

The characteristic equation at small fluctuations of operating condition parameters at study of a steady-state stability of the simplex uncontrolled EPS (**Figure 1(a)**) taking into account transients in a field winding looks like [2] [14]:

$$a_0 p^3 + a_1 p^2 + a_2 p + a_3 = 0 \tag{6}$$

where a_0, a_1, a_2, a_3 are the coefficients of a characteristic equation which are functions of operating condition parameters and EPS.

In our case Lyapunov function in a quadratic form at $n=3$ according to (2) looks like:

$$V = X^{T}QX = q_{11}x_1^2 + q_{22}x_2^2 + q_{33}x_3^2 + 2q_{12}x_1x_2 + 2q_{13}x_1x_3 + 2q_{23}x_2x_3 \tag{7}$$

Let set C in the form of an identity matrix. Then

$$W = x_1^2 + x_2^2 + x_3^2 \tag{8}$$

Upon that, the matrix of coefficients of a quadratic form (7) according to (2) looks like:

$$Q = \begin{bmatrix} q_{11} & q_{12} & q_{13} \\ q_{21} & q_{22} & q_{23} \\ q_{31} & q_{31} & q_{33} \end{bmatrix} \tag{9}$$

We have set negatively definite derivative W (positively definite symmetric C). If upon that the positive definiteness requirements to the matrix Q of a quadratic form (7) are satisfied then initial system equilibrium state asymptotic stability conditions (1) will be obviously provided.

For the positive definiteness of a quadratic form (7) according to Sylvester's criterion [5] [6] it is necessary and sufficient that principal diagonal minors of matrix Q were positive:

$$\Delta_{n1} = q_{11} > 0; \quad \Delta_{n2} = \begin{bmatrix} q_{11} & q_{12} \\ q_{21} & q_{22} \end{bmatrix} > 0; \quad \Delta_{n3} = \begin{bmatrix} q_{11} & q_{12} & q_{13} \\ q_{21} & q_{22} & q_{23} \\ q_{31} & q_{31} & q_{33} \end{bmatrix} > 0 \tag{10}$$

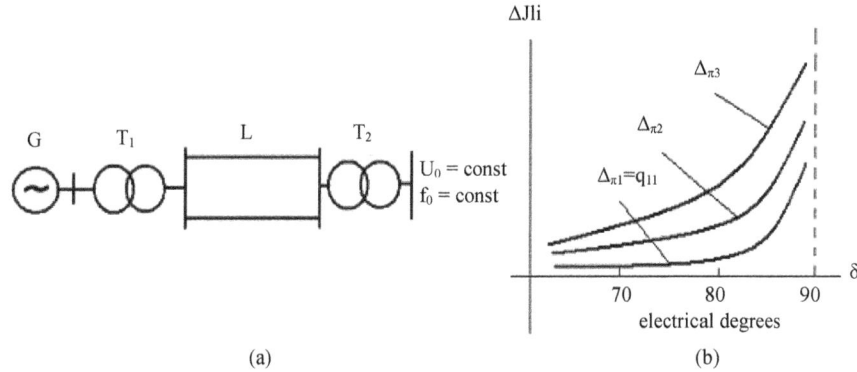

Figure 1. Behavior of minors of a quadratic form for the matrix Lyapunov equation (b) for the EPS circuit design (a).

Let's start checking violation of positivity of these minors with the first $\Delta_{\pi 1} = q_{11}$. As disclosing $\Delta_{\pi 1}$ shows, the minor can become negative if following requirements [14] are broken:

$$C_1 = \frac{\partial P_{E_q}}{\partial \delta} > 0 \tag{11}$$

$$X_d > X_c > X_d' \tag{12}$$

$$\delta > a_{ii} - \arcsin \frac{2 E_q \sin a_{ii}}{U} \tag{13}$$

where P_{E_q} is the real output power of the generator at $E_q = $ const ., U is the terminal voltage of the generator, E_q is EMF of the generator, X_d, X_d', X_c are synchronous and transient reactance of the synchronous generator and reactance of the transmission line, accordingly, δ is the torque angle of the generator that determines stability of the generator and therefore, EPS, α_{ii} is a complementary angle [2] [3] [14].

The requirement (11) can be broken only in the case of overload which may cause an aperiodic instability of EPS.

Violation of the inequality (12) is possible in the case of overcompensation of an reactance of the power line by the direct compensation plants (series capacitances) that leads to electromagnetic instability in a power system (self-excitation of generator).

The requirement (13) can be broken in operating conditions close to light load conditions and synchronous generator operation with a transmitting line with appreciable resistances and thereby a stability violation process has oscillatory behavior and is observed in the form of electromechanical oscillation of the generator rotor (self-oscillation).

In other words, requirements to violation of positivity of the first minor in a square matrix (10) imply all possible conditions that may lead to violation of an electrical power system steady-state stability, *i.e.*, the complete problem to define conditions of EPS instability is solved "in the small".

The analysis has shown [14] [20] that positivity of $\Delta_{\pi 2}$, $\Delta_{\pi 3}$ is reduced to satisfaction of the same requirements (11)-(13).

So, conditions that may lead to violation of an electrical power system steady-state stability obtained by Lyapunov's second method coincide with earlier discovered on the basis of the generalized Routh-Hurwitz conditions.

It is necessary to note that for the first time requirements of adequacy of the results obtained on the basis of Lyapunov functions in a quadratic form and a Routh-Hurwitz criterion for an electrical power system have been obtained in the work [20].

With a view of checking theoretical rules there were carried out computational-experimental researches of violation of principal minors positivity in a quadratic matrix (10). Calculations were carried out for the simplex and complex EPS [14]. Calculations were also carried out on the basis of the Routh-Hurwitz criterion aor the purpose of comparison.

Figure 1(b) shows variations of minors (10) at gradual increasing the load in EPS (increasing the transmitted real power). The analysis shows that upon increasing the operation condition loads variations for all minors $\Delta_{\pi i}$ from the matrix of a quadratic form of Lyapunov function (10) have the equal character, while variations of characteristic equation coefficients (6) and Hurwitz determinants are absolutely different [14].

The result allows using positivity of the first minor, *i.e.* $q_{11} > 0$, for complete analysis of a steady-state stability of the electrical power system because q_{11} contains all information on possible kinds of EPS instability "in the small". Upon that, positivity of the higher minors, *i.e.* $\Delta_{\pi i} > 0$ ($i = 2, \cdots, n$; where n is the order of a differential equation of initial EPS) may not be considered in the first approximation. Hence, it is possible to state simplified (practical) criterion of a steady-state stability $q_{11} > 0$ which gives both necessary and sufficient conditions of its performance. Traditionally [2], these requirements are obtained on the basis of positivity of characteristic equation coefficients for the system (the necessary condition) and positivenesses of Routh-Hurwitz matrix determinants (the sufficient condition).

2.2. Complex EPS

The calculation analysis of a steady-state stability for EPSs of various complexity shows [4] that the most strict in theoretical aspect, convenient in computing aspect and effective aspect by the obtained results is use of two fundamental methods: the method of Lyapunov functions in a quadratic form and the method of the nodal equations [14] [21]. The proposed stability research technique "in the small" has been suggested for the first time in the work [22], and its essence consists in the following.

When studying a steady-state stability of complex EPS, calculation of the steady-state condition on the basis of the node-voltage equations is carried out at first, voltages U_k for each node k and their arguments δk, and further for each j-th generator are determined; then positivity of the first minor q_{11j} (10) of the matrix of quadratic form Q is checked using these data. Thereby the stability of the generator which is the fastest to come closer to a limit at the given load is determined. In essence, stability study "in the small" of the complex EPS using the proposed method turns to study of the generator-bus system.

Let's consider steady-state stability conditions of complex EPS by the example of three-generator system (**Figure 2**) since such model of the electrical power system as a whole adequately reflects properties and performances of the complex electrical system [4] [23] [24].

Figure 3 represents characters of variations q_{11j} (where $j = 1-3$) of the first elements of minors for matrixes of quadratic forms Q_j for each generator of the system in question.

Changes (increasing) of minors of a quadratic form ξ_I for different generators in the case if loading of operating conditions of the electrical power system increases are different, that is clear from **Figure 3**.

Upon condition if the small deviations of EPS operating condition parameters are constant ($\Delta\Pi$ = constant) it is possible to write:

$$\xi_3 \succ \xi_2 \succ \xi_1 \text{ and } \Delta h_3 \succ \Delta h_1 \succ \Delta h_1 \tag{14}$$

where Δh_j are the incremental rates of first minors of quadratic form matrixes for each generator showing how quickly a generator comes closer to the stability limit. The analytical expression Δh_j, for example, for the second generator at i-th step looks like:

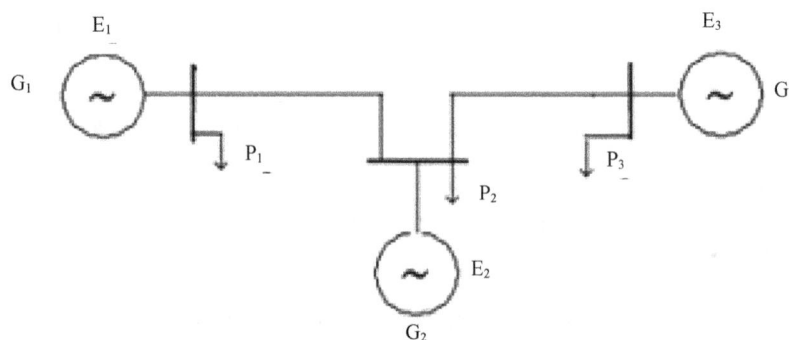

Figure 2. The circuit design of three-generator electrical power system.

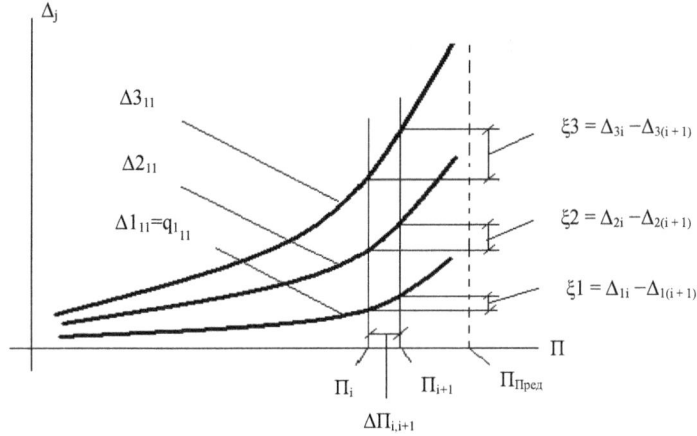

Figure 3. Definition of the steady-state stability simplified criterion for the complex EPS (Π is the operating condition parameter of EPS).

$$\Delta h_2 = \frac{\xi_2}{\Delta \Pi_{2(i,i+1)}} = \Delta \left(\frac{\Delta_{2,i} - \Delta_{2(i+1)}}{\Pi_{2,i} - \Pi_{2(i+1)}} \right) \tag{15}$$

Generally (15) looks like:

$$\Delta h_j = \frac{\xi_j}{\Delta \Pi_{ji,j(j+1)}} = \Delta \left(\frac{\Delta_{ji} - \Delta_{j(i+1)}}{\Pi_{ji} - \Pi_{j(i+1)}} \right) \tag{16}$$

where $\Pi \to U, f, I, \delta$, etc. are the operating condition parameters of EPS by which variations of minors for generators of the explored system as a result of increasing the loads of operating conditions can be determined, j is the generator for which (16) is calculated, i is a current step of increasing the loads for the given EPS operating condition parameter.

Let's transfer to differentials under condition of small increments:

$$\Delta h_j = d\left(\Delta h_j \right) \tag{17}$$

The condition $\xi_l > 0$ for a quadratic form is always satisfied, therefore it is possible to be restricted to study of strict performance of the inequality:

$$\Delta h_j = \Delta \left(\frac{\Delta_{ji} - \Delta_{j(i+1)}}{\Pi_{ji} - \Pi_{j(i+1)}} \right) = \frac{dq_{11,j}}{d\Pi_i} \succ 0 \tag{18}$$

The condition (18) means that in order to provide a steady-state stability of j-th generator and, hence, EPS, the fulfillment of the following condition is required:

$$\frac{dq_{11,j}}{d\Pi_i} \succ 0 \tag{19}$$

On the basis of the obtained results it is possible to propose the following algorithm for studies of a steady-state stability of the complex electric power systems.

With increasing the load by the given parameter of operating condition Π at each step i and for each generator or the selected groups of generators the condition (19) is checked and compared with other similar conditions, i.e., fulfillment of conditions is checked:

$$\frac{dq_{11,1}}{d\Pi} \succ \cdots \succ \frac{dq_{11,1}}{d\Pi} \succ \cdots \succ \frac{dq_{11,n}}{d\Pi} \tag{20}$$

where n is a number of generators or stations checked on a steady-state stability. The generator which first minor variation in a matrix of coefficients of a quadratic form is maximum will be the most critical from the point of view of steady-state stability violation:

$$\frac{dq_{11,j}}{d\Pi} \to \max \qquad (21)$$

for the considered series of generators.

Hence, j-th generator for which $dq_{11,j}/d\Pi \to \max$ will be the first which comes to a steady-state stability limit of EPS.

Thus, that generator which tends to steady-state stability violation, and also possible sections (lines) which represent the greatest danger from this point of view should be determined at first. The given factor is valuable also in that it allows to define the corresponding parameter of operating conditions which is most preferable to control the transient behavior of EPS in the case of regulation. This makes it possible to organize control of transient process of the generator using automatic excitation control, automatic speed control and other control systems and proactively ensure its steady-state stability. The significance of this result is obvious to practice of maintenance of electric power systems.

This data imply efficiency of common application of the nodal equations and Lyapunov functions in a quadratic form and prospectivity of the proposed method for study of a steady-state stability of complex EPS.

3. Conclusions

The obtained theoretical and computational results confirmed for systems of various complexity allow checking stability of EPS "in the small" by study of positivity condition for the first minor of a Lyapunov function matrix in a quadratic form $q_{11j} > 0$ and to consider it as the practical (simplified) criterion of EPS steady-state stability providing its both necessary and sufficient conditions. Definition of the requirement $dq_{11,j}/d\Pi \to \max$ allows the generator to reveal that represents the greatest danger from the point of view of stability violation. Upon that, the study of a steady-state stability of the complex EPS turns to study of the "generator-bus" circuit design that makes it possible to determine the particular generator or station which leads to violation of a system stability and an asynchronous condition in system.

Thus, joint use of Lyapunov functions in a quadratic form and the node-voltage equations allows us to the fullest extent to explore a steady-state stability of the complex electrical power system including both its electromechanical and electromagnetic violations.

References

[1]   Rudenko, Y.N. and Semyonova, V.A. (2000) Automation of Dispatch Control System in Electrical Power System. MPEI, Moscow.

[2]   Venikov, V.A. (1964) Transient Phenomena in Electrical Power System. Pergamon Press, Macmillan, New York.

[3]   Barinov, V.A. and Sovalov, S.A. (1990) Electrical Power Systems Operating Conditions: Analysis and Control Method. Energoatomizdat, Moscow.

[4]   Anderson, P.M. and Fouad, A.A. (2003) Power System Control and Stability. 2nd Edition, IEEE, Wiley-Interscience, USA.

[5]   Barabashin, E.A. (1970) Lyapunov Function. Science, Moscow.

[6]   Andreev, Y.N. (1976) Finite-Dimensional Linear Object Control. Science, Moscow.

[7]   Misrikhanov, M.Sh. and Ryabchenko, V.N. (2006) Solution of Algebraic Systems with Local Variable Matrix for Dispatching Control in EPS. *A&T*, **5**, 131-141.

[8]   Vasin, V.P., Starshinov, V.A. and Domnina, V.V. (2006) Application of New Processing Technology in Electrical Power System Control. MPEI.

[9]   Bukov, V.N. (2006) System's Embedding. Analytical Approach to Analysis and Design of Matrix Systems. Kaluga.

[10] Asanov, A.Z. (2007) System's Embedding Method and Its Application. USATU, Ufa.

[11] Misrikhanov, M.Sh. and Ryabchenko, V.N. (2006) Quadratic Problems of Eigenvalue in EPS. *A&T*, **5**, 24-47.

[12] Narovlyanskiy, V.G. and Kurmak, V.V. (2012) The Critical Section Determination Algorithm in EPS in the Design of Emergency Automatics. *Electric Power Plant*, **3**, 48-51.

[13] Alalishev, O.S. (2004) Super Computers: Field of Application and Performance Requirements. *Electronics*, **1**.

[14] Allaev, K.R. and Mirzabaev, A.M. (2011) Small Fluctuations of Electrical Power System. Matrix Approach. Science and Technologies, Academy of Sciences Republic of Uzbekistan, Tashkent.

[15] Kalman, R.E. (1695) On the Hermite-Fujiwara Theorems in Stability Theory. *Quarterly of Applied Mathematics*, **23**, 279-285.

[16] Kalman, R.E. and Bertram, I.E. (1960) Control System Analysis and Design via the "Second Method" of Lyapunov. 1. Continuous-Time Systems. *Transactions of the ASME. Series D, Journal of Basic Engineering*, 371-393. http://dx.doi.org/10.1115/1.3662604

[17] Parks, P.C. (1962) A New Proof of the Routh-Hurwitz Stability Criterion Using the Second Method of Lyapunov. *Proc. Camb. Phil. Soc. Math. Phys. Sci.*, **58**, 694-702. http://dx.doi.org/10.1017/S030500410004072X

[18] Andreev, Y.N. (1977) Algebraic State-Space Methods in Linear Objects Control Theory. *A&T*, **3**, 5-50.

[19] Power, H.M. (1967) The Companion Matrix and Lyapunov Functions for Linear Multivariable Time-Invariant System. *Journal of the Franklin Institute*, **283**, 214-234. http://dx.doi.org/10.1016/0016-0032(67)90025-7

[20] Allaev, K.R. (1973) Lyapunov Function in a Quadratic Form Applied to Study Steady-State Stability in EPS. *Science and Technologies*, **5**, 13-17.

[21] Fazilov, Kh.F. (1964) Network Calculations Methods. Science, Tashkent.

[22] Allaev, K.R. (1984) Lyapunov Function in a Quadratic Form as the Tool for Study of Steady-State Stability in EPS. *Science and Technologies*, **5**, 13-17.

[23] Sovalov, S.A. (1983) Operating Conditions of Unified Energy System. Energoatomizdat, Moscow.

[24] Bukov, V.N. and Misrikhanov, M.Sh. (2003) About Matrices and Modeling Transformation in EPS. Energoatomizdat, Moscow.

15

Generation of Electric Power from Domestic Cooking System

Syed Ali Raza Shah[1], Zahoor Ahmed[2], Bashir Ahmed Leghari[1], Wazir Muhammad Laghari[2], Attaullah Khidrani[2]

[1]Department of Mechanical Engineering, Balochistan University of Engineering and Technology, Khuzdar, Pakistan
[2]Department of Electrical Engineering, Balochistan University of Engineering and Technology, Khuzdar, Pakistan
Email: razadopasi@gmail.com, zahoor@buetk.edu.pk, baleghari@gmail.com, niaz1111@yahoo.com, atta_khidrani@yahoo.com

Abstract

This study work related with floating of an idea about conversion of reclaimed thermal energy from domestic cooking system into the electrical power. There were different techniques in use worldwide for harnessing the energy into the appropriate useful work and also to create efficient system for the energy conversion process. The ignorance in this regard might be due to the reason that this wastage did not cost too much for a single home on per day or per month basis, but it could be an ample amount of cost if integrated this loss for a whole city or on yearly bases for a single home. The idea in this work depended upon the recovery of waste heat from Pressure Cookers used in the houses for the domestic cooking purposes, and optimized the reclaimed thermal energy for the conversion into electric power. This research work related with losses of energy discussed and analyzed on the basis of thermo dynamically regarding (a) the wastage of thermal energy escaped through the system due to the spreading of exhaust vapors and taking away significant amount of thermal energy; (b) losses of enthalpy through the dissipate steam; (c) heat losses in the tubing from the Pressure Cooker to the turbine; (d) electric power produces from the system. In this work, new methods were advised, in order to reduce the losses of thermal energy from the system. It would open the venue for researchers to promote this new idea in near future.

Keywords

Waste Heat, Pressure Cooker, Thermal Energy, Electric Power, Enthalpy Losses, Steam Turbine

1. Introduction

The every nation desires for self-sufficient in the sector of energy, as it plays a vital role in the economic development either in industrial area or in the domestic use. More over the rising in grassroots population throughout the world, it will adversely affect the resources and develops pressure on the basic amenities. They also rely on the energy sector for their likely-hood in future. Hence, demand of energy is increasing day by day. In order to overcome this issue and fulfill the requirement of energy demand in future, proper planning and development activities can be started in every sector immediately. The idea in this work depends upon the recovery of waste heat from the Pressure Cooker which uses in the houses for daily domestic cooking purpose, and optimizes the reclaimed thermal energy for the electric power [1]. Pressure Cooker dissipates the high temperature steam continuously during the entire process. The exhausted steam is more appropriately used to convert the enthalpy of the steam into electrical power energy through De-Laval's turbine (a single stage, single nozzle steam turbine), and it is used for convenience in manufacturing. This set-up can be applied to the systems where cooking done on larger scales continuously like hotels and food companies, but it is not feasible to be adopted at domestic level because turbine alternator set-up makes the system complex and bulky [2].

2. Design Procedure

The waste exhausted heat recovered from the Pressure Cooker was used for optimizing the efficiency of system and utilized the reclaimed thermal energy for conversion in to the electric power for specified given data. In this work study was focused upon the Pressure Cooker used for cooking food stuff or heating the potable water in the house daily. The calculation related with its design procedure is discussed as under.

2.1. Energy Available in the Fuel Gaseous

The fuel gas liberated an amount of energy in the form of fuel consumption rate by unit time from natural gas to the burner of Pressure Cooker. The measuring unit of energy is Joule or KJ and in case of gas consumed per unit time then it will be considered as KW [2]:

$$Q^{\cdot} = m^{\cdot} \times V_c \qquad (1)$$

In the above equation $Q^{\cdot}$ is equivalent to the rate of fuel consumption per time and measured in KW. While: $m^{\cdot}$ = Mass flow rate in Kg/sec:

$$V_c = \text{Calorific value in KJ / Kg}$$

The calorific value of any substance is to be considered to the heating value of that material. The quantity of heat liberated during the process of combustion for any fuel; and measuring unit for energy/unit of time for the particular substance such as MJ/Kg. The various fuels have different calorific values as indicated in **Table 1** [3].

The Natural gas has methane (CH4) 85% to 90% as an ingredient, while the remaining other particles depends upon ethane, butane, propane, nitrogen, and CO_2 etc. Hence it is clear from **Table 1** that the Natural gas has Calorific value near about (55.5 MJ/Kg) as per consideration on the basis of methane. Since natural gas is not available in pure form and having sufficient amount of impurities and also keeping the other losses under consideration. Then it will be considered that the natural gas about 70% of its original mass will be available for the design calculation, hence the calorific value of natural gas is as under:

Table 1. Calorific values of common fuels.

Type of Fuel	MJ/Kg
Hydrogen	141.8
Methane	55.5
Ethane	51.9
Propane	50.35
Butane	49.5
Gasoline	47.3
Diesel	44.8
Coal	15 - 27

$$V_c = \left[(55.5 \times 1000) \times \left(\frac{70}{100} \right) \right] = 38850 \, \text{KJ/Kg}$$

The study was carried out through practical conduction on the gas cylinder. The gas cylinder was placed over the digital weight meter and fuel consumptions were observed after expiry of ten 10 seconds. The data was analyzed as per consumed mass of fuel per unit time. Hence the rate of fuel consumption was the actual power produced by the gas fuel for the burner. The $m = 0.00022 \text{Kg/sec}$ was recorded during practical. Putting the calculated values in Equation (1), then the energy in the form of power calculated as:

$$Q = 8.547 \frac{\text{KJ}}{\text{Second}} = \text{KW}$$

2.2. Losses of Energy from the System

It was apparently revealed that without the economizing arrangement; the small amount of thermal energy was lost due to the absorption by the system and operating pressure vary from 3 - 9 KPa during the process. It is clear from **Table 2** that the power released by the fuel gas supplied to the furnace amounting to (8.574 KW). Similarly on other hand the power losses by evaporation was (2.136 - 2.406 KW) as shown in **Table 2**.

The efficiency in percentage wise for the system without doing any economic planning can be estimated as mentioned below [3].

$$\% = \frac{\text{output}}{\text{input}} \times 100 \tag{2}$$

$$\% = \left[\left(\frac{2.406}{8.574} \right) \times 100 \right] = 28.15\%$$

Hence, losses of thermal energy/second will be:

$$Q = 8.574 - 2.406 = 6.168 \text{KJ}$$

It is clear from the above discussion that 6.168 KJ of energy lost by each second from the system; therefore efficiency reduced to 28.15%, as shown in **Figure 1**, but that is very low, so it can be enhanced.

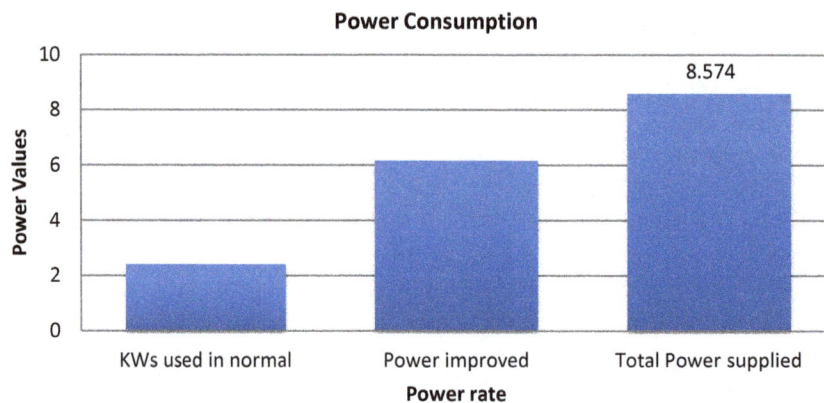

Figure 1. Power consumption rate.

Table 2. Thermal properties of steam.

Input Q_{gas} KW	Pressure P_{gauge} KP$_a$	Pressure P_{abs} KP$_a$	Saturation Température T_{sat} °C	Enthalpy Vapor H_g KJ/K$_g$	Mass Flow m K / s	Power of Steam Q_{steam} KW
8.574	9	96.469	98.63	2673.59	0.0009	2.406
	3	90.46	96.853	2670.75	0.0008	2.136

3. Steps Suggested for Improvement of Efficiency

In order to improve the efficiency of the system; the under mentioned steps are suggested keeping some economical arrangements in the system.

3.1. Insulation of the System

It is proposed that to collect the energy of flue gases from the domestic Pressure Cooker at home then insulation may be provided in the system. It was suggested that a metallic shell with area considerably greater than the covered area of the burner including Pressure Cooker be fabricated around it. The maximum temperature will be in the space within the shell and between the walls of the Pressure Cooker, that will helps to raise the temperature up to the maximum level. The energy of flue gases can be trapped and utilized for heating the cooker. The external insulation of burner including pan with shell will prevent the loss of heat transfer through mode of conduction and convection to atmosphere as seen in **Figure 2** [4].

The influence of temperature and color of flame vary with type of fuel involved for combustion process. The variable absolute temperature values for different fuels react with air or O_2 during the combustion process are mentioned below in **Table 3**.

The color of flames are depends upon the molecular constituents of the fuels. It is not necessary that the blue colored flames are every time considered to be scorching than yellow colored flames [5]. A light yellow flame causes the losses of heat to some level instead of blue flame, in order to overcome the losses and try to make the system efficient; then make arrangement for a decent sharp blue colored flame for the system. The orifice of nozzles for the burner must be fabricated considering the back pressure including the feeding rate of the fuels. In order to accomplish the good hot flame for the system; then make arrangement for efficient supply of fuel. Then domestically system became efficient by adopting these all above mentioned arrangement collectively (**Figure 3** and **Figure 4**).

Figure 2. View of metallic shell for burner.

Table 3. Common temperature (K) values for different fuels.

Types of Fuels	Temperature in Kelvin
Acetylene in oxygen	3410
Heptane in oxygen	3100
Hydrogen in oxygen	3080
Methane in oxygen	3053
Acetylene in air	2600
Hydrogen in air	2400
Heptane in air	2290
Methane in air	2232

Figure 3. Light yellow flame.

Figure 4. Blue flames.

3.2. Measuring Vapor Pressure

Here in this case the fuel gas supplied to the burner for providing the heat energy to the Pressure Cooker for evaporation process. A vapor pressure was measured through pressure gauge installed over the Pressure Cooker for getting the gauge pressure instantly. The Pressure Cooker shut down during the process of recording the value of gauge pressure. These recorded readings of pressure gauge were used for analysis the efficiency of the system. The online SPIRAX SPARCO calculator was used to find out other related properties such as flow rate of mass and power output etc. for steam tables as given in **Table 4** [6].

4. Improvement in Efficiency of the System

The Pressure Cooker is used for cooking the food stuff everywhere including towns as well as in big cities. In these systems there are many chances for losses of thermal energy from the surrounding and system became inefficient and uneconomical for the use of domestic purposes. The system has efficiency within the range of 20% - 30%. The economizing arrangements were made and calculated as shown in **Table 4**. The output power without the economizing arrangement was ranging from 2.136 to 2.406 KW as calculated power supplied was 8.547 under operating pressures between 3 - 9 KPa [7]. The exact calculation of reclaimed thermal energy is almost impossible due to the involvement of analysis of flue gases with variation in flame shapes. Therefore readings were directly obtained through software and observed with economizing arrangement under operating pressure that was 50.5 to 60.5 KPa having corresponding output powers between 4.303 - 4.847 KW. The new efficiency of the system can be improved up to:

$$\eta = \frac{4.847}{8.574} \times 100 = 56.53\%$$

5. Energy Analyses

The energy produced was not much enough in amounts. However, it was a useful for analyzing steam conditions

Table 4. Net effect after making the economizing arrangement.

S. No.	A	B	C
	Gauge Pressure P_{gauge}	Steam Mass Flow m·Kg/Sec	Steam Power Q_{steam} KW
1	275.00	0.0052	14.22
2	220.00	0.0042	11.45
3	170.00	0.0035	9.514
4	120.50	0.0028	7.583
5	100.50	0.0024	6.488
6	60.50	0.0018	4.847
7	50.50	0.0016	4.303
8	25.00	0.0012	3.216
9	20.00	0.0011	2.946
10	10.50	0.0010	2.674
11	9.00	0.0009	2.406
12	3.00	0.0008	2.136

for requirement of turbine design [8]. Moreover, the Wattmeter was attached with the alternator that provided the output parameters, which shows the effectiveness of the arrangements. The initial reading on the pressure gauge with the final reading on the wattmeter was used for calculations and it was included in **Table 3**. Since frictional and conductional losses from the exit point of steam from the Pressure Cooker to the turbine and alternator were neglected. Energy consumed to form steam by evaporating water was treated as the useful work of the system. Steam exhausted from the Pressure Cooker was the only output parameter [9]. The amount of the heat absorbed by the system and the efficiency of the system was improved by adopting the economizing arrangement on it. The study required to check the possibility of converting the steam enthalpy into electrical power by using a single nozzle De-Laval's Turbine as shown in **Figure 5**. Laval developed a nozzle in 1890 in order to increase the working of the kinetic energy of the steam, rather than its pressure for steam jet to the supersonic speed.

The wattmeter reading practically finds out the efficiency of the whole mechanism. At the peak pressure conditions, the theoretical power output of the turbo-generated was estimated 347.10 W; but practically it came out 131.898W on the wattmeter as shown in **Table 5**.

The overall efficiency of tubing, nozzle, turbine and alternator calculated as under

$$\eta = \frac{131.898}{347.10} \times 100 = 38\%$$

This percentage is applied to all the values of the shaft power to calculate the actual output. The result shows that loss of 62% of available energy across the power generation setup that was obviously a result of non-professionalism and manufacturing limitations within the remote area.

6. Design Configuration of Steam Turbine

1. Power supplied by the Gaseous Fuel $Q = 8.547$Kw
2. Absolute pressure $P_{abs} = P_{gauge} + P_{atm} = 275 + 87.4699 = 362.469KP_a$
3. Saturation temperature $T_{sat} = 140.10°C$
4. Mass flow rate of steam $(m\dot{\ }) = 0.0052$ Kg/Sec
5. Enthalpy of saturated vapor $H_g = 2734.1$KJ/Kg
6. Specific volume of saturated vapor $V_p = 0.50736$m^3 / Kg
7. Velocity of the steam outlet $V_{steam} = 190.530$ m/sec

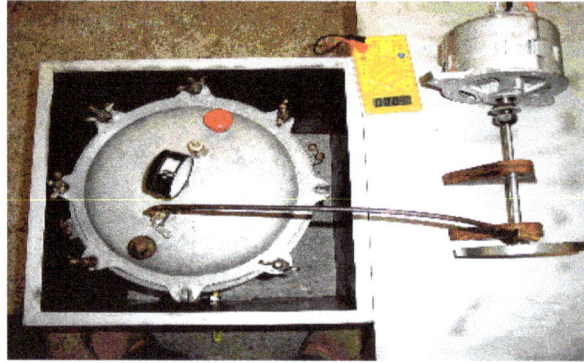

Figure 5. Energy conversion arrangements.

Table 5. Power output over shaft and in watt-meter.

Final Pressure P_2 KP_a	Final Température T_2 °C	Final Enthalpy H_{g2} KJ/Kg	Power at Shaft KW	Power on Watt-meter KW
182.092	117.30	2682.52	347.10	131.898
171.738	115.49	2681.70	271.75	103.265

8. Diameter of Jet $d = 4.2\text{mm} = 0.0042\text{m}$

9. Area of Jet $A = \dfrac{\pi}{4} d^2 = 13.847 \times 10^{-6} \text{m}^2$

10. Thermal power at steam outlet $Q = 14.21732\text{KW}$

11. Specific heat at constant pressure of steam $C_p = 2.23375 \dfrac{\text{KJ}}{\text{Kg}} K \left(28.57\% \, 4.2 \text{ mm to 3 mm}\right)$ convergent nozzle used

12. Kinetic energy of the steam before the nozzle: $K \cdot E_1 = 8150.841 \dfrac{\text{m}^2}{\text{sec}^2} = 18.150\text{KJ/Kg}$

13. Initial Mach number: $\left(M_1\right) = 0.338$

14. Velocity of steam beyond nozzle (28.57% converged): $V_2 = 373.42\text{m/s}$

15. Kinetic energy of the steam beyond the nozzle $K \cdot E_2 = 69.721\text{KJ / Kg}$

16. Enthalpy of the steam beyond the nozzle $H_2 = 2682.527\text{KJ / Kg}$

17. Temperature of the steam beyond the nozzle $T_2 = 390.3K = 117.3°C$

18. Pressure drop through the nozzle $\left(P_2\right) = 182.092\text{KP}_a$

19. Mach number of the steam beyond the nozzle $M_2 = 0.775$

20. Velocity of blade to the velocity of steam jet $V_b = 182.60\text{m/sec}$

21. Angle of blade $\theta = 12°$

22. Specific heat ratio $\gamma = 1.237$

23. Volume of the rim $V = 0.000113 \text{ m}^3$

24. Weight of the rim: $W = 564.6723 \text{ gm}$

25. Blade height: h = 7 mm

26. Inlet angle $\theta_1 = 23°$

27. Outlet angle $\theta_1 = 23°$

28. Blade width: = 10 mm

29. Radius of curvature of the blade $R = 5.4318\text{mm}$

30. Third angle of blade $\theta_3 = 134° OR \, 2.33 \text{ radians}$

31. Area of blade $\Rightarrow$ A = 12.703 mm

32. Circumference of the Rim: $C = \pi \times 120 = 376.8$mm
33. Number of blades = 45
34. Total mass of blades = 59.861 gm
35. Total mass of the turbine = 624.533 gm
36. Corresponding power at the shaft = 347.10 W
37. Diameter of shaft = 2.269 mm
38. Permissible Shear stress $\sigma_{ms} = 42$MP$_a$
39. Bending stress (τ_{ms}) for the Shaft = 80 MP$_a$
40. Length of shaft = 0.015 mm

7. Conclusion

This study work will address the problem for reclaiming of waste heat energy from the domestic Pressure Cooker used for cooking system and converts the same into the electric power. The efficiency of system is improved through proposed process and suggests to optimize the domestic cooking system for achieving the better efficiency and to save the thermal energy for future requirement.

Acknowledgements

The authors acknowledge the chairman and Incharge of Laboratories section, Mechanical Engineering Department, Balochistan University of Engineering & technology, Khuzdar, for providing the laboratory facilities.

References

[1]   Domestic Waste Water Heat Recovery Device.
       http://www.pera.com/website/clientsandcasestudies/creatingnewproductideas/

[2]   Liu, L.B., Fu, L. and Jiang, Y. (2010) Application of an Exhaust Heat Recovery System for Domestic Hot Water. *Journal of Energy*, **35**, 1476-1413. http://dx.doi.org/10.1016/j.energy.2009.12.004

[3]   De Paepe, M. and Theun, E. (2003) Heat Recovery System for Dishwashers. *Journal of Applied Thermal Engineering*, **23**, 743-756. http://dx.doi.org/10.1016/S1359-4311(03)00016-4

[4]   Lukitobudi, R., Akbarzadeh, A., Johnson, P.W. and Hendy, P. (1995) Design, Construction and Testing of a Thermo syphon Heat Exchanger for Medium Temperature Heat Recovery in Bakeries. *Journal of Heat Recovery Systems and CHP*, **15**, 481-491. http://dx.doi.org/10.1016/0890-4332(95)90057-8

[5]   Colangelo, G., de Risi, A. and Laforgia, D. (2006) Experimental Study of a Burner with High Temperature Heat Recovery System for TPV Applications. *Journal of Energy Conversion and Management*, **47**, 1192-1206. http://dx.doi.org/10.1016/j.enconman.2005.07.001

[6]   www.SPIRAXSPARCO.com

[7]   Noie-Baghban, S.H. and Majideian, G.R. (2000) Waste Heat Recovery Using Heat Pipe Heat Exchanger (HPHE) for Surgery Rooms in Hospitals. *Journal of Thermal Engineering*, **20**, 1271-1282. http://dx.doi.org/10.1016/S1359-4311(99)00092-7

[8]   El-Baky, M.A.A. and Mohamed, M.M. (2007) Heat Pipe Heat Exchanger for Heat Recovery in Air Conditioning. *Journal of Applied Thermal Engineering*, **27**, 795-801. http://dx.doi.org/10.1016/j.applthermaleng.2006.10.020

[9]   W.J. Kearton in His Book "Steam Turbine Theory and Practice".

16

Energetic Macroscopic Representation of an Electrically Heated Building with Electric Thermal Storage and Heating Control for Peak Shaving

Cristina Guzman, Kodjo Agbossou, Alben Cardenas

Hydrogen Research Institute and The "Département de Génie Électrique et Génie Informatique", Université du Québec à Trois-Rivières, Trois-Rivières, Canada
Email: guzman@uqtr.ca, kodjo.agbossou@uqtr.ca, cardenas@uqtr.ca

Abstract

As a part of the Smart Grid concept, an efficient energy management at the residential level has received increasing attention in lately research. Its main focus is to balance the energy consumption in the residential environment in order to avoid the undesirable peaks faced by the electricity supplier. This challenge can be achieved by means of a home energy management system (HEMS). The HEMS may consider local renewable energy production and energy storage, as well as local control of some particular loads when peaks mitigation is necessary. This paper presents the modeling and comparison of two residential systems; one using conventional electric baseboard heating and the other one supported by Electric Thermal Storage (ETS); the ETS is employed to optimize the local energy utilization pursuing the peak shaving of residential consumption profile. Simulations of the proposed architecture using the Energetic Macroscopic Representation (EMR) demonstrate the potential of ETS technologies in future HEMS.

Keywords

Home Energy Management, Energetic Macroscopic Representation, Building Modeling, Electric Thermal Storage

1. Introduction

Smart grid concept brings within the scope of implementing more efficient and performing technologies to improve the electricity system. Modern and attractive technologies would also motivate residential clients to become managers of their own consumption as a better way for the energy use [1].

Local energy production systems, local energy storage and residential load management are new challenges pursuing the efficiency improvement of the grid without the stress engendered by the unbalance between energy production and consumption. Among the many initiatives to improve the residential energy efficiency and commercial buildings, some of them propose local renewable electricity generation as photovoltaic, and electrical or thermal local storage systems [2]-[4].

Electric Thermal Storage (ETS) systems are inspired from an ancient technology used by aboriginal arctic populations, which employed heated stones as storage medium. This technology has lately emerged as a good option to improve the energy utilization, and can be employed for water-heating systems, space heating/cooling systems of buildings and off-peak electricity storage systems [5]-[8]. In the residential environment this option is quite easy to carry out, however some limitations and rejection arise taking into account, the space and the quite high weight of such storage systems. In sensible heat technologies [9] [10], when water heater is employed as storage medium, the temperature must remain between 60°C and 82°C (140°F - 180°F). Otherwise when solid materials are employed, as ceramic and brick, temperatures can go up to 700°C (1300°F). In latent heat technologies phase change materials (PCM) are employed which permit to store heat using a narrow temperature range without employing high temperatures [11]-[13].

Electric Thermal Storage (ETS) typically use high density ceramic bricks as storage element. The bricks are placed inside a cabinet which can weight hundreds of pounds [14]. For residential use, ETS is well adapted to be supplied by renewable energy local generation (as photovoltaic or wind power) or taking advantage of cheap off-peak power [15]. The use of ETS might be a good alternative to shave the unwanted peak of energy consumption, when this peak is mostly caused by the heating utilization and, can be appropriate for cold climate countries like Canada. As matter of facts, it could be a practical choice taking into account that heating power represents more than 60% of energy consumption in Canadian residential buildings.

The Energetic Macroscopic Representation (EMR) has been introduced as a graphical descriptor approach offering a macroscopic energetic view of multiphysics system [16]-[18]. EMR offers an inversion-based control contributing to the controller design and the implementation of energy management strategies.

This paper presents the development of the EMR of a residential system with electric thermal storage. The main parts of the system have been represented and a simulation model has been implemented by using MATLAB/Simulink. Simulation results using a classic management strategy show the potential of the electric thermal storage for peak shaving during winter grid peak periods. The remainder of this paper is organized as follows, Section 2 presents the EMR of the studied system; Section 3 presents the simulation results for a system with and without electric thermal storage; and the conclusion of the paper will be presented in Section 4.

2. EMR of Residential System with Electric Thermal Storage

2.1. General Description of Studied System

The studied system consists of an electrically heated detached house. In the base case, the system comprises the Electric Heating System (EHS) composed of resistive baseboards, represented as a single aggregated unit and the critical loads. A second case considers additionally the use of Electric Thermal Storage (ETS) as illustrated in **Figure 1**. The modeling of the thermal envelope of the building and the behavior of the external and the internal temperatures are important features in the management of this kind of systems which become essentials for the study.

Figure 2 represents a simplified view of the energetic macroscopic representation of the residential system, which is coupled to the utility grid as the main Energy Source (ES). Through the distribution panel three subsystems are connected: the Space Heating System (SHS), the Electric Thermal Storage (ETS) and the Critical Loads (water heater, lighting and home appliances). The subsystems are denoted DS-1, DS-2 and DS-3 respectively.

The utility grid is represented as the electric source ES, the utility voltage and main current are denoted as U_G and I_G respectively. The main current of the system is represented by:

$$I_G = I_L + I_H + I_S \tag{1}$$

where, I_L is the current of critical loads, I_H is the current of the SHS and I_S is the current of the ETS. As il-

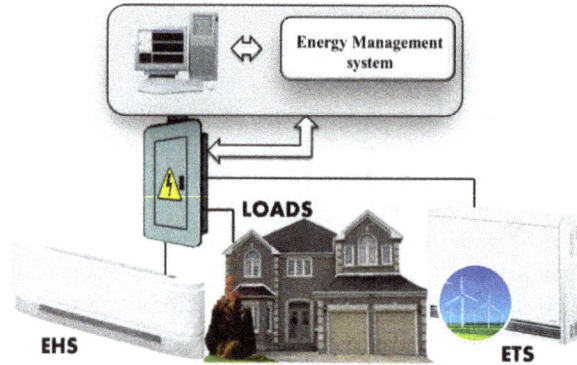

Figure 1. Simplified EMR representation of residential system with electric thermal storage.

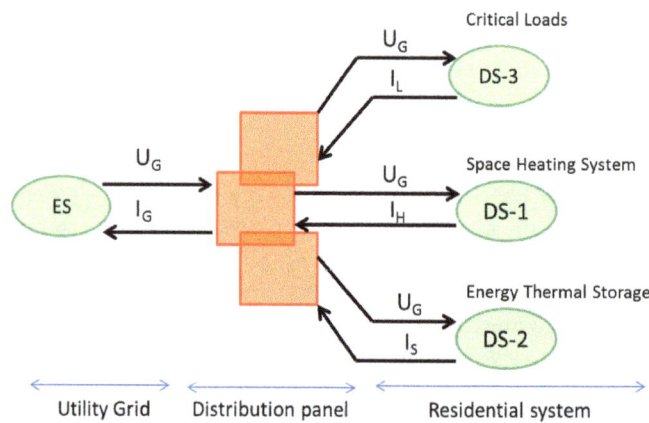

Figure 2. Simplified EMR representation of residential system with electric thermal storage.

lustrated in **Figure 2** the distribution panel is modeled as a coupling block, and the voltages applied to each subsystem are the action variables, in counterpart the currents are the reaction variables.

2.2. Space Heating System

Residential heating systems in Quebec are essentially electric systems, and in this study it is supposed that all heating needs are supplied from the electric source. The EMR representation of the SHS is depicted by **Figure 3**, there, the Electric Heater, normally an assembly of baseboard systems, is controlled by means of a thermostatic control which is connected to the voltage source. The control signal of the thermostatic control is called m_1 and permits to modulate the power sent to the baseboard assembly. The mean voltage applied to the electric heater (U_{SH}) is defined by

$$U_{SH} = m_1 \cdot U_G \tag{2}$$

The reaction to the applied voltage is the current I_{SH} which is defined by

$$I_{SH} = \frac{U_{SH}}{R_{SH}} \tag{3}$$

where, R_{SH} is the equivalent electric resistance of the electric heater assembly. Then, the current from the main source (I_H) can be defined by

$$I_H = I_{SH} \cdot m_1 \tag{4}$$

Notice that the electric heater is represented by a multi-physical conversion element. This representation because, the left side there is an electric system and at the right side there is a thermal system. According to the

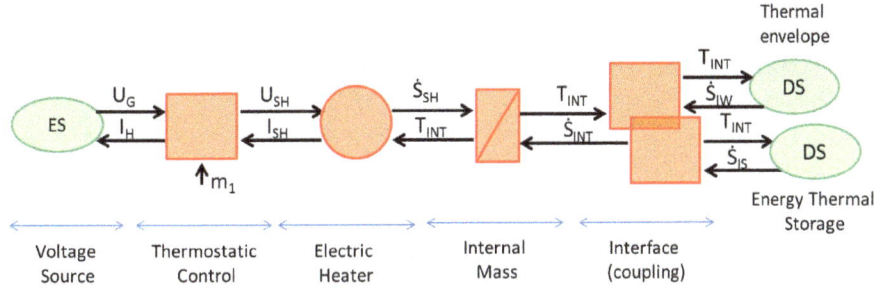

Figure 3. EMR representation of residential space heating system (SHS).

second law in thermodynamics conservation the entropy flow rate $\dot{S}$ is found [19]. Then, the power at each side of this block can be computed as

$$I_{\text{SH}} \cdot U_{\text{SH}} = \dot{S}_{\text{SH}} \cdot T_{\text{INT}} \tag{5}$$

where, $\dot{S}_{\text{SH}}$ is the entropy flow rate of heating in [W/°K] and T_{INT} is the internal temperature of the residential building in [°K].

An energy storage element is employed to represent the internal thermal mass of the building (internal walls, air mass and home furniture). The differential equation of the internal thermal mass and the internal temperature $\left(T_{\text{INT}} \right)$ can be written as

$$T_{\text{INT}} \left(\dot{S}_{\text{SH}} + \dot{S}_{\text{INT}} \right) = mCp_{\text{IM}} \cdot \dot{T}_{\text{INT}} \tag{6}$$

where

$\dot{T}_{\text{INT}} = \dfrac{\text{d}}{\text{d}t} T_{\text{INT}}$, and mCp_{IM} is the heat capacity of the internal thermal mass in [W·s/°K]. The output entropy flow rate is defined by

$$\dot{S}_{\text{INT}} = \dot{S}_{\text{IW}} + \dot{S}_{\text{IS}} \tag{7}$$

where $\dot{S}_{\text{IW}}$ is the entropy flow rate from the internal side of the walls (coming from the thermal envelope), and $\dot{S}_{\text{IS}}$ is the entropy flow rate from the energy thermal storage system.

It is to remark that in the coupling system three entropic flow rates are defined, the first for the internal thermal mass, the second for the thermal envelope and the third one for the ETS. A common temperature is at the three sides of coupling block corresponding to the internal temperature of the residential building.

2.3. Building Thermal Envelope

Canadian homes are built using materials with high thermal resistance looking to reduce the heat losses during winter time. These construction characteristics impose a modeling of the walls as a resistive part (thermal insulation) and a thermal capacity part. The EMR of the building envelope is depicted in **Figure 4**. There, the internal layer of walls is represented by a mono-physical conversion element connecting the interior of the building (internal thermal mass) and the thermal mass of the walls. The power at each side of the conversion block are defined by

$$\dot{S}_{\text{IW}} \cdot T_{\text{INT}} = \dot{S}_{W} \cdot T_{\text{WM}} \tag{8}$$

$$\dot{S}_{\text{IW}} \cdot T_{\text{INT}} = \frac{\left(T_{\text{WM}} - T_{\text{INT}} \right)}{R_{\text{IW}}} \tag{9}$$

$$\dot{S}_{W} \cdot T_{\text{WM}} = \frac{\left(T_{\text{INT}} - T_{\text{WM}} \right)}{R_{\text{IW}}} \tag{10}$$

where, $\dot{S}_{\text{IW}}$ is the entropy flow rate from internal layer of walls to the internal thermal mass; $\dot{S}_{W}$ is the entropy flow rate to the thermal mass of the walls from the internal layer of the walls. R_{IW} is the equivalent thermal resistance of the internal side layer of walls.

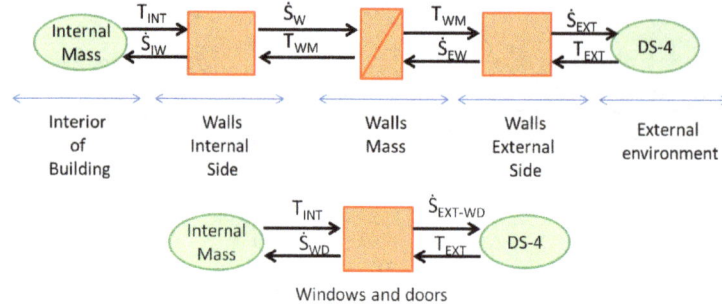

Figure 4. EMR representation of building thermal enveloppe.

The thermal mass of the walls is represented by an energy storage element connected to the internal and external layers of the walls; the differential equation of this storage element can be written as

$$T_{WM}\left(\dot{S}_W + \dot{S}_{EW}\right) = mCp_{WM} \cdot \dot{T}_{WM} \tag{11}$$

where, $\dot{S}_{EW}$ is the entropy flow rate from external layer of the walls, T_{WM} is the temperature inside the walls; mCp_{WALL} is the heat capacity of the walls mass.

The external layer of the walls, like the internal one, is also represented by a mono-physical conversion element with an equivalent thermal resistance denoted R_{EW}. This element connects the thermal mass of the walls with the external environment. The external environment which imposes a variable temperature $\left(T_{EXT}\right)$ is denoted as DS-4 in **Figure 3**. Then, the power at each side of the conversion element can be written as

$$\dot{S}_{EW} \cdot T_{WM} = \dot{S}_{EXT} \cdot T_{EXT} \tag{12}$$

The entropy flow rate to the environment $\left(\dot{S}_{EXT}\right)$ is defined by the difference of temperature between the walls and the environment, and can be written as

$$\dot{S}_{EXT} = \frac{\left(T_{WM} - T_{EXT}\right)}{T_{EXT} \cdot R_{EW}} \tag{13}$$

The windows and doors are represented as a mono-physical conversion element, as the resistive layers of the walls. The power at each side of the conversion element can be written as

$$\dot{S}_{WD} \cdot T_{INT} = \dot{S}_{EXT\text{-}WD} \cdot T_{EXT} \tag{14}$$

The entropy flow rate to the environment $\left(\dot{S}_{EXT\text{-}WD}\right)$ is defined by the difference of internal mass temperature and the environment, and can be written as

$$\dot{S}_{EXT\text{-}WD} = \frac{\left(T_{INT} - T_{EXT}\right)}{T_{EXT} \cdot R_{WD}} \tag{15}$$

where R_{WD} corresponds to the equivalent thermal resistance of the doors and windows.

2.4. Electric Thermal Storage System

The electric thermal storage system normally consists of a heat storage control, an electric heater (heating element), a thermal mass, and a heating blower (forced air heating). The EMR of this subsystem is depicted in **Figure 5**. The heat storage control permits to modulate the voltage applied to the electric heater according to a modulation index m_2. In most of cases, m_2 is considered a binary input which imposes an ON/OFF control of the electric heater. The voltage and currents of the control block can be defined by

$$U_{ETS} = m_2 \cdot U_G \tag{16}$$

$$I_{ETS} = \frac{U_{ETS}}{R_{ETS}} \tag{17}$$

$$I_S = I_{ETS} \cdot m_2 \tag{18}$$

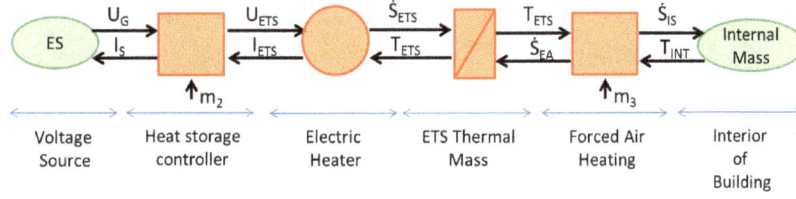

Figure 5. EMR representation of the electric thermal storage system (ETS).

As in the space heating system case, the electric heater is modeled as a multi-physical conversion element. The differential equation of this element can be written as

$$T_{\text{ETS}}\left(\dot{S}_{\text{ETS}} + \dot{S}_{\text{EA}}\right) = mCp_{\text{ETS}} \cdot \dot{T}_{\text{ETS}} \tag{19}$$

$$\dot{T}_{\text{ETS}} = \frac{\text{d}}{\text{d}t} T_{\text{ETS}} \tag{20}$$

where mCp_{ETS} is the heat capacity of the thermal mass of the storage system. $\dot{S}_{\text{ETS}}$ is the entropy flow rate from the electric heater; $-\dot{S}_{\text{EA}}$ is the entropy flow rate from the storage system (transferred to the internal mass of the building).

The forced air heating system is modeled similarly to the heat storage control, that's as a mono-physical conversion element. The input and output powers can be written as

$$\dot{S}_{\text{EA}} \cdot T_{\text{ETS}} = \dot{S}_{\text{IS}} \cdot T_{\text{INT}} \tag{21}$$

The heating power transferred from the ETS to the internal mass of the building is controlled by means of the input m_3 and the nominal heating power $\left(P_{\text{BH}}\right)$; a constant power representing the heat leaks of the storage system $\left(P_{\text{HL}}\right)$ is also included using

$$\dot{S}_{\text{IS}} \cdot T_{\text{INT}} = P_{\text{HL}} + m_3 P_{\text{BH}} \tag{22}$$

2.5. Critical Loads

As illustrated in **Figure 2** critical loads are considered in the system. They are simulated using hourly typical residential profiles of energy consumption, measured from a Quebec residence by means of a commercial energy monitoring system [20]. The total current of critical loads I_L is computed using the load profile and the input voltage. The complete EMR of the studied residential system with ETS is presented in **Figure 6**, where all the explained modeled subsystems are coupled.

3. Simulation Results

The system described in previous section has been implemented in MATLAB/Simulink. Two systems have been implemented, one including the ETS and the other without it, as illustrated in **Figure 7**, the two systems have similar characteristics and their parameters are presented in **Table 1**. These parameters correspond to a single family house (detached house) of Quebec as explained before. It is to remark that only the temperature effect has been studied being the most important parameter for this Nordic locations; e.g. solar radiation gains and internal gains have been neglected.

For simulation purposes a classic proportional and integral (PI) controller is employed to control the internal temperature which is set to 21°C during the day period (from 6H to 22H) and at 19.5°C at night. The proportional and integral gains of controllers are also presented. The external temperature used in this study has been obtained from the SIMEB Hydro-Quebec website [21] for the winter period of 2013-2014 (November 2013 to April 2014).

The ETS is controlled considering the "heat storage" at "off-peak periods" and the "heat discharge" at peak period; in the "off-peak period" the load current at the distribution panel is measured and used to obtain the total power of the residential building, then the storage power is computed according to the actual power not exceeding 6 kW. The storage is enabled from November to March, months during which larger amount of heating is usually required. The "peak period" is considered fixed for weekdays between 6H and 10H and between 16H

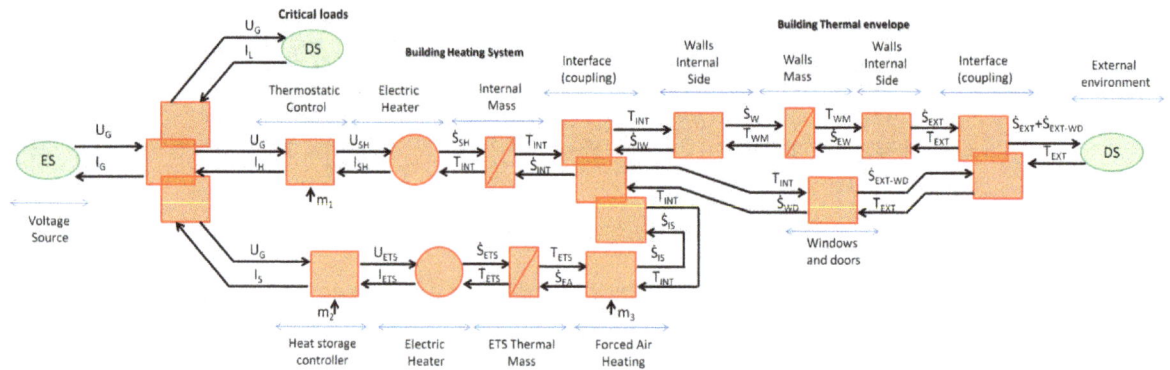

Figure 6. EMR representation of residential system with electric thermal storage.

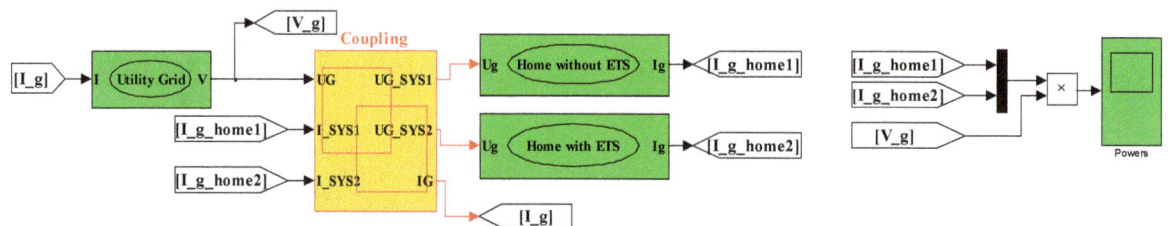

Figure 7. Model implemented in Simulink/MATLAB.

Table 1. System parameters.

Description	Value	Units
Voltage source	240	V
Building envelope		
mCp_{WALL}	107.53	MW·s/°K
R_{IW}	0.0127	°K/W
R_{EW}	3.719e5	°K/W
R_{WD}	0.0301	°K/W
Electric thermal storage		
mCp_{ETS}	12,240	kW·s/°K
P_{HL}	$2 \times 3400 \times 0.1$	W
P_{BH}	$2 \times 3400 \times 0.9$	W
R_{ETS}	8.47	Ω
Space heating system		
mCp_{IM}	6904.7	kW·s/°K
P_{SH}	10	kW
R_{SH}	5.76	Ω
Temperature controller		
Kp	852	W/°K
Ki	8.52	W/°K rad/s

and 20H, weekend period is considered off-peak. During the "peak period" the internal temperature of the building is controlled by means of two PI controllers, one driving the ETS and the other one driving the classic space heating system.

Simulation results are plotted in **Figures 8-10**, where **Figure 8** presents the internal and external temperature during the simulation period (November 2013 to April 2014) and the total power of the two systems; it is to re-

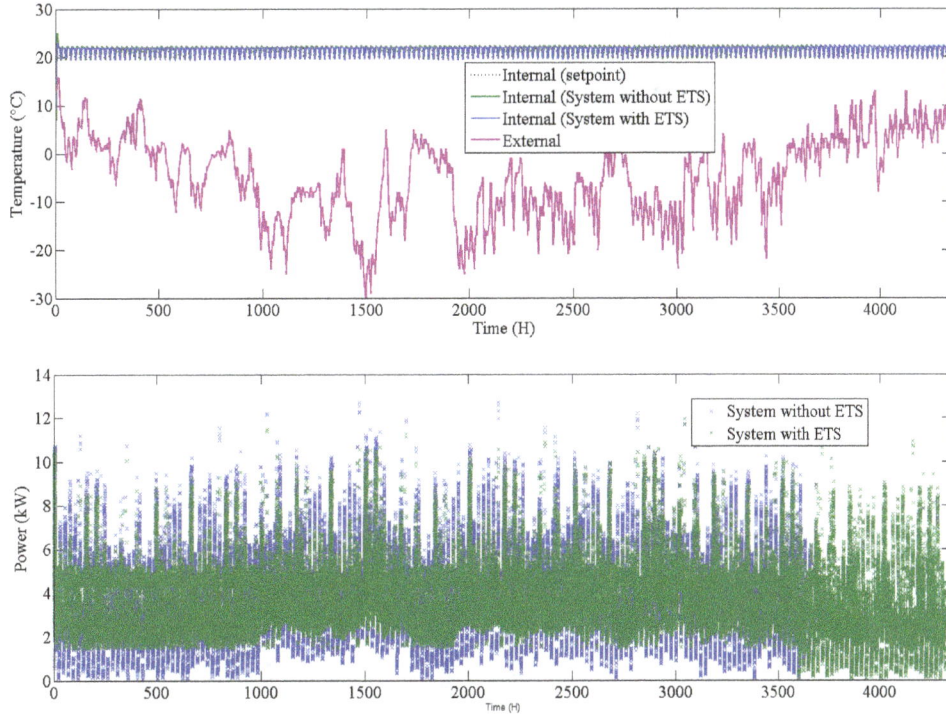

Figure 8. External temperature at Trois-Rivieres Canada and instantaneus power of each simulated residential building (from November 2013 to April 2014).

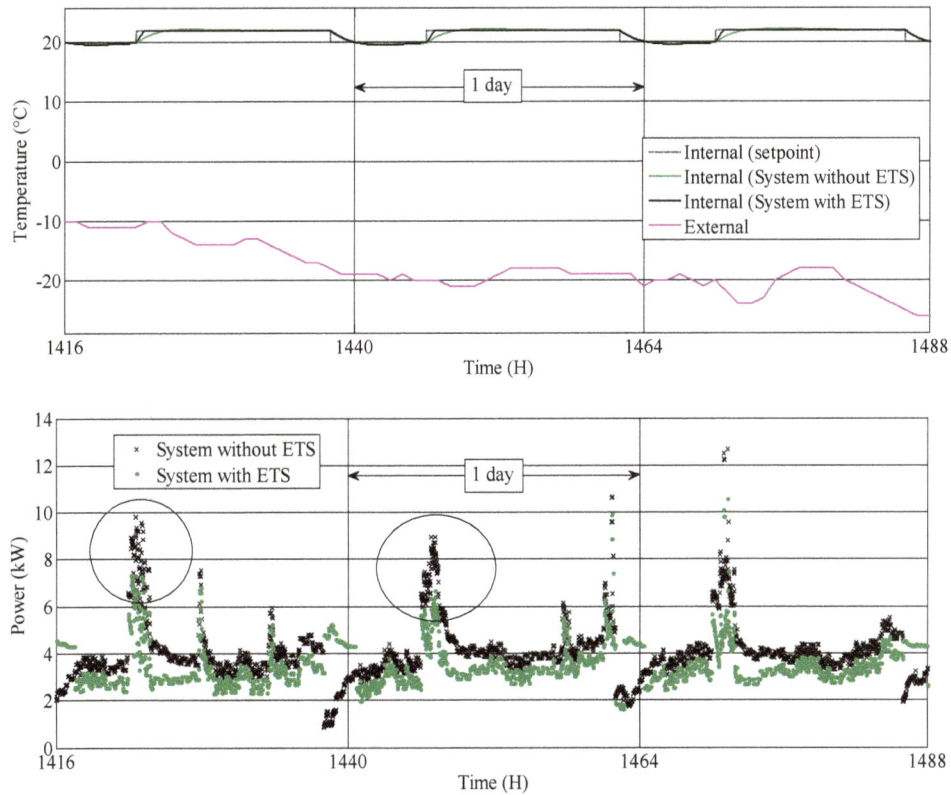

Figure 9. External temperature at Trois-Rivieres Canada and instantaneus power of each simulated residential building (3 days on January 2014).

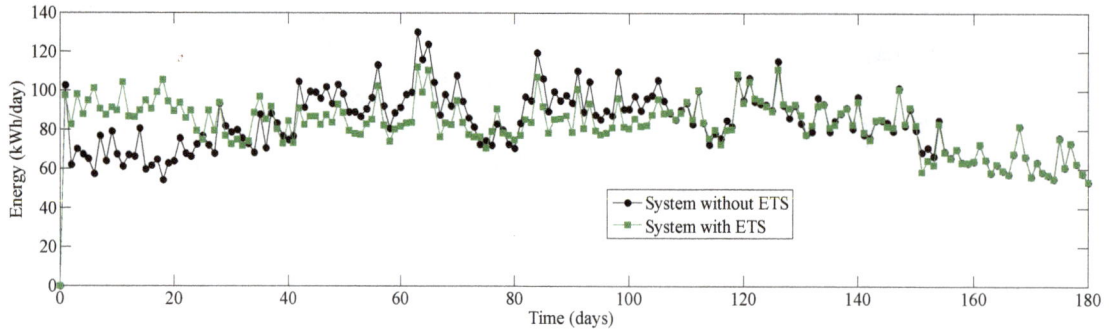

Figure 10. Energy consumption of each simulated residential building (from November 2013 to April 2014).

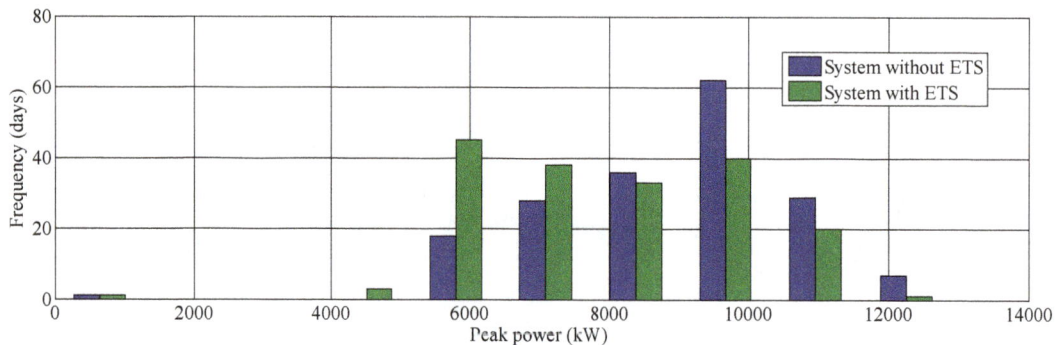

Figure 11. Distribution of peak power for each simulated residential building (from November 2013 to April 2014).

mark that the coldest period corresponds to the January and February months. Results corresponding to 3 days of cold period (January 2014) are plotted in **Figure 9**. It is observed that the peak of consumption is shaved using the electric thermal storage while the control of temperature is maintained. Morning peaks are encircled in **Figure 9** in order to better illustrate the peak shaving. In **Figure 10** the total energy consumption in kWh/day is plotted, showing that the peak of daily energy in the coldest period is also smoothed. Notice that the two systems use the same energy during the simulation period. The distribution of peak power per day for each system is presented in **Figure 11**, where the maximal frequency appears at 9.67 kW (62 days) and at 5.8 kW (45 days) for the system using classic heating and for the one with ETS respectively.

4. Conclusion

This paper has presented the modeling of two residential systems: one using conventional electric baseboard heating and the other one supported by Electric Thermal Storage. The models have been built using the Energetic Macroscopic Representation (EMR) which permits the description of multiphysic systems using simple blocks interconnected with power exchanges links. The implemented models, using MATLAB, have permitted to validate the concept of Electric Thermal Storage and its potential for peak shaving; in fact peak power reduction without negative effects on the temperature regulation has been obtained by simulations. This model could be employed to study control strategies for Demand Side Management (DSM).

Acknowledgements

This work was supported in part by the LTE Hydro-Québec, the "Bureau de l'Efficacité et de l'Innovation Énergétiques du Québec" and Natural Science and Engineering Research Council of Canada.

References

[1]   Kok, K., Karnouskos, S., Ringelstein, J., Dimeas, A., Weidlich, A., Warmer, C., Drenkard, S., Hatziargyriou, N. and Lioliou, V. (2011) Field-Testing Smart Houses for a Smart Grid. 21*st International Conference on Electricity Distribution (CIRED)*, Frankfurt, 6-9 June 2011, 6-9.

[2]   Van Roy, B.J., Verbruggen, B. and Driesen, J. (2013) Ideas for Tomorrow. *IEEE Power and Energy Magazine*, **11**, 75-81. http://dx.doi.org/10.1109/MPE.2013.2268815

[3]   Molderink, A., Bakker, V., Bosman, M.G.C., Hurink, J.L. and Smit, G.J.M. (2010) Management and Control of Domestic Smart Grid Technology. *IEEE Transactions on Smart Grid*, **1**, 109-119. http://dx.doi.org/10.1109/TSG.2010.2055904

[4]   Davito, B., Tai, H. and Uhlaner, R. (2010) The Smart Grid and the Promise of Demand-Side Management. McKinsey on Smart Grid. McKinsey & Company, Atlanta, 38-44.

[5]   Craven, C. and Grunau, B. (2013) Thermal Storage Technology Assessment. Cold Climate Housing Research Center (CCHRC), Alaska Housing Finance Corporation & the Alaska Department of Commerce, Community, and Economic Development, Fairbanks.

[6]   Sharma, A., Tyagi, V.V., Chen, C.R. and Buddhi, D. (2009) Review on Thermal Energy Storage with Phase Change Materials and Applications. *Renewable and Sustainable Energy Systems*, **13**, 318-345.

[7]   Science Applications International Corporation (SAIC Canada) (2013) Compact Thermal Energy Storage Technology Assessment Report. Presented to City of Pickering and Natural Resources Canada.

[8]   Dincer, I. (2002) On Thermal Energy Storage Systems and Applications in Buildings. *Energy and Buildings*, **34**, 377-388. http://dx.doi.org/10.1016/S0378-7788(01)00126-8

[9]   Pérez-Lombard, L., Ortiz, J., Coronel, J.F. and Maestre, I.R. (2011) A Review of HVAC Systems Requirements in Building Energy Regulations. *Energy and Buildings*, **43**, 255-268. http://dx.doi.org/10.1016/j.enbuild.2010.10.025

[10]   Kulkarni, M.R. and Hong, F. (2004) Energy Optimal Control of a Residential Space-Conditioning System Based on Sensible Heat Transfer Modeling. *Building and Environment*, **39**, 31-38. http://dx.doi.org/10.1016/j.buildenv.2003.07.003

[11]   Kiziroglou, M.E., Wright, S.W., Toh, T.T., Mitcheson, P.D., Becker, T. and Yeatman, E.M. (2014) Design and Fabrication of Heat Storage Thermoelectric Harvesting Devices. *IEEE Transactions on Industrial Electronics*, **61**, 302-309. http://dx.doi.org/10.1109/TIE.2013.2257140

[12]   Rousse, D.R., Ben Salah, N. and Lassue, S. (2009) An Overview of Phase Change Materials and Their Implication on Power Demand. 2009 *IEEE Electrical Power & Energy Conference* (*EPEC*), Montreal, 22-23 October 2009, 1-6. http://dx.doi.org/10.1109/EPEC.2009.5420979

[13]   Kaplan, F., De Vivero, C., Howes, S., Arora, M., Homayoun, H., Burleson, W., Tullsen, D. and Coskun, A.K. (2014) Modeling and Analysis of Phase Change Materials for Efficient Thermal Management. 32*nd IEEE International Conference on Computer Design* (*ICCD*), Seoul, 19-22 October 2014, 256-263.

[14]   Steffes Corporation, Owner's and Installer's Manual for Room Heating Units-2100 Series. http://www.steffes.com

[15]   Armaroli, N. and Balzani, V. (2006) The Future of Energy Supply: Challenges and Opportunities. *Angewandte Chemie International Edition*, **46**, 52-66. http://dx.doi.org/10.1002/anie.200602373

[16]   Chen, K. (2010) Common Energetic Macroscopic Representation and Unified Control Structure for Different Hybrid Electric Vehicles. PhD Dissertation, École Doctorale des Sciences pour l'Ingénieur, Université Lille 1.

[17]   University of Lille 1, Energetic Macroscopic Representation Web Site. http://www.emrwebsite.org/energetic-macroscopic-representation.html

[18]   Horrein, L., Bouscayrol, A. and El-Fassi, M. (2012) Thermal Energetic Model of an Internal Combustion Engine for Simulation of a Thermal Vehicle. 2012 *IEEE Vehicle Power and Propulsion Conference* (*VPPC*), Seoul, 9-12 October 2012, 978-983. http://dx.doi.org/10.1109/VPPC.2012.6422768

[19]   Dong, Y., El-Bakkali, A., Descombes, G., Feidt, M. and Périlhon, C. (2012) Association of Finite-Time Thermodynamics and a Bond-Graph Approach for Modeling an Endoreversible Heat Engine. *Entropy*, **14**, 642-653. http://dx.doi.org/10.3390/e14040642

[20]   Technical Specifications of TED Pro Energy Monitoring and Control System, Rev 7.1, TED the Energy Detective. http://www.theenergydetective.com/5000docs

[21]   SIMEB Web Site, Simulation énergétique des bâtiments. https://www.simeb.ca

17

Dynamic State Forecasting in Electric Power Networks

Sideig A. Dowi, Amar Ibrahim Hamza

School of Electrical and Electronics Engineering, North China Electric Power University, Beijing, China
Email: sideigdowi@yahoo.com, amarhamaza2010@hotmail.com

Abstract

The real time monitoring and control have become very important in electric power system in order to achieve a high reliability in the system. So, improvement in Energy Management System (EMS) leads to improvement in the monitoring and control functions in the control center. In this paper, DSE is proposed based on Weighted Least Squares (WLS) estimator and Holt's exponential smoothing to state predicting and Extended Kalman Filter to state filtering. The results viewing the dynamic state the estimator performance under normal and abnormal operating conditions.

Keywords

Dynamic State Estimation; Extended Kalman Filter; State Estimation; Electric Power Flow

1. Introduction

Recently, the power system has begun to grow very largely and more complex, so real time monitoring and control become very important in order to fulfill a reliable operation. Energy Management System (EMS) is responsible for this mission, and it forms the basis for efficient operating and control. State Estimation (SE) forms the spine of the EMS by providing the information of the real time state of the system which can be used in other EMS functions. Hence, an accurate and efficient state estimation is necessary for a reliable and efficient operation of the power system [1] [2].

The state estimator computes the voltage magnitudes and voltage angles at the buses of the power system. We know that, power system is not a static system, but it changes very slowly with time and continuously. That means, when the load on the buses changes, the generations also have to change to overcome these changes in load. This in turn causes the change in power flows and injections at the buses, also leads to change in voltage angular at the buses and perhaps change in voltage magnitude at some buses depending on the size of this change; therefore, change the nature of the power system from static state to dynamic state nature. These dynamic behaviors of the power system are difficult to overcome by the conventional Static State Estimation (SSE). This led to the development of a new algorithm called Dynamic State Estimation (DSE) [1]-[5]. The DSE technique possesses a mathematical model for the time variation of the power system. The DSE uses this mathematical model depending on the previous state of the system at time (*t*), to predict the state vectors at the next state of the system (one step ahead) at time (*t* + 1). This capability of predicting the state one step ahead is a very important advantage in the control center, because the state forecasting gives a longer decision time to the system operator, and because the security assessment, economic dispatching and the other functions can be performed in advance. So, Dynamic State Estimation has the ability to represent an important role in the modern-day control center [1].

In this paper, we will not only describe the dynamic model for the time behavior of the system state, but will show more details about the DSE, mainly the state predicting and state filtering. When the state variables are estimated at time k by state estimation technique, we will use these state variables to forecast the state vectors at time $k + 1$ using linear exponential smoothing. The state vectors are filtered based on Extended Kalman Filter and weighted least squares method. The proposal is tested using IEEE 14 bus test system. The test includes normal and abnormal operations.

2. Mathematical Models

The measurement vector consists of active power and reactive power flows and injection's power as well as some voltage magnitudes, is denoted by an m-dimensional vector *z*. The power equation s is expressed by [6] [7].

$$P_i = V_i \sum_{j=1}^{n} V_j \left(G_{ij} \cos \theta_{ij} + B_{ij} \sin \theta_{ij} \right) \tag{1}$$

$$Q_i = V_i \sum_{j=1}^{n} V_j \left(G_{ij} \sin \theta_{ij} - B_{ij} \cos \theta_{ij} \right) \tag{2}$$

$$P_{ij} = V_i^2 \left(g_{si} + g_{ij} \right) - V_i V_j \left(g_{ij} \cos \theta_i + b_{ij} \sin \theta_{ij} \right) \tag{3}$$

$$Q_{ij} = V_i^2 \left(b_{si} + b_{ij} \right) - V_i V_j \left(g_{ij} \sin \theta_i - b_{ij} \cos \theta_{ij} \right) \tag{4}$$

The measurement and state variable z_k and x_k at time instant *k* are given by equation

$$Z_k = h(x_k) + v_k \tag{5}$$

where z_k is measurement vector m-dimensional $(m \times 1)$, x_k is state variables dimensioned $(n \times 1)$, $h(x_k)$ is nonlinear function relating measurement to the state vector .dimensioned $(m \times 1)$ and v_k is the measurements error with zero mean and standard deviation of *R*.

where $R = \text{diag}\left[\sigma_i^2, \sigma_2^2, \cdots, \sigma_n^2 \right]$, σ is the standard deviation of error.

$$Z_k = H x_k + v_k \tag{6}$$

where $H = \dfrac{\partial h}{\partial x}\bigg|_{x=x^0}$ is Jacobian matrix dimensional $(m \times n)$, *m* is number of measurements and *n* is number of state vectors.

The general model for DSE is given by.

$$x_{k+1} = F_k x_k + G_k + w_k \tag{7}$$

where x_k and x_{k+1} are the state vector at instants *k* and *k*+1 respectively, F_k is nonzero diagonal matrix dimensioned $(n \times n)$, a function represent the state transition between two instant of time, G_k is nonzero vector associated with trend behavior of the state trajectory dimensional $(n \times 1)$ and w_k is white Gaussian noise with zero mean and covariance matrix *Q* [1]-[4].

2.1. Parameters Identification

The parameters F_k and G_k are identified using Holt's two-parameter linear exponential smoothing method [1], [8]-[10]. This method is very simple, used when the data shows a trend. In this method, two components must be updated each period of time, level and trend.

- The level is a smoothed estimate of the value of the data at the end of each period represented by a_k. as shown in Equation (8).
- The trend is a smoothed estimate of average growth at the end of each period. represented by b_k .as shown in Equation (8).

The specific model for simple exponential smoothing is written as:

$$\tilde{x}_{k+1} = a_k + b_k \tag{8}$$

where $a_k = \alpha \hat{x} + (1-\alpha)\tilde{x}_k$ & $b_k = \beta(a_k - a_{k-1}) + (1-\beta)b_{k-1}$ $0 < \beta < 1, 0 < \alpha < 1$

$$F_k = \alpha(1+\beta) \tag{9}$$

$$G_k = (1+\beta)(1-\alpha)\tilde{x}_k - \beta a_{k-1} + (1-\beta)b_{k-1} \tag{10}$$

α and β represent the smoothing parameters [2], [8]-[10].

2.2. State Predicting or State Forecasting

Let x_k and Σ_k be the estimated state and estimated state covariance respectively at time k the forecasted state vector and its covariance matrix is obtained by

$$\tilde{x}_{k+1} = F_k \hat{x}_k + G_k \tag{11}$$

$$M_{k+1} = F_k \Sigma_k F_k^T + Q_k \tag{12}$$

where M_{k+1} is covariance matrix of the forecasted state.

2.3. State Filtering

The forecasted state would use to forecast new measurements Z_{k+1} at the time $k + 1$ based on the data at instant k; the predicted state vector at $k + 1$ will be filtered to obtain new estimates (filtered states) x_{k+1} with its error's covariance matrix Σ_{k+1}. Then, the objective function for the filtering process at the instant of time $(k + 1)$ is

$$J(x) = \left[Z - h(\tilde{x})\right]^T R^{-1}\left[Z - h(x)\right] + \left[x - \tilde{x}\right]^T M^{-1}\left[x - \tilde{x}\right] \tag{13}$$

Note that, the time index $(k + 1)$ has been omitted.

Extended Kalman Filter (EKF) used for minimizing the objective function and getting the final filtering state.

$$\hat{x}_{k+1} = \tilde{x}_{k+1} + K_{k+1}\left[Z_{k+1} - h(\tilde{x}_{k+1})\right] \tag{14}$$

$$K_{k+1} = \Sigma_{k+1} H_{k+1}^T R_{k+1}^{-1} \tag{15}$$

$$\Sigma_{k+1} = \left[H_{k=1}^T R_{k=1}^{-1} H_{k=1} + M_{k+1}^{-1}\right]^{-1} \tag{16}$$

where K is called the gain matrix $(n \times m)$ dimensioned. Perform only one iteration in Equation (15).

3. The Implementation

The steps of the dynamic state estimation algorithm are described above. The covariance matrix R of the measurement error is assumed to be calculated online.

$$\text{where } \sigma_i = \left(a_{mi}|s_{ti}| + b_{mi}s_{fi}\right)/3$$

where a_m & b_m are the manufacture factors. s_t represents the real value of the measurement. s_f is the maximum value of the measurement. As mentioned before, for predicting state we used Holt's 2-exponential smoothing, the values of the smoothing parameters α and β, are fixed at 0.7 and 0.45, and for the filtering state we used Extended Kalman Filter. The elements of the covariance matrix Q of the system, is set at 10^{-6}. The load curve at each bus was composed of a linear trend and random fluctuation (jitter).

In this paper, the Dynamic State simulation is studied over a period of 20 time sample intervals. with increasing of constant value 5% of the load at all the buses at each period. Once the load is changed the load flow is ready to update all the real and reactive power and injection power on the lines, voltage magnitudes and angles on the system. In this paper, the actual values of the state vectors are obtained from the 14-bus IEEE standard data for the base case and load flow for rest of the time samples [11]. The test cases and their results are shown in tables and figures in the next sections.

3.1. Case Study

In this paper, we used standard IEEE 14-bus test system [11]. The measurement vector consists of 34 observations distributed as in **Table 1**.

The measurement value was simulated by adding random errors to the true values represented by normally distribution with zero mean and standard deviation.

3.2. Performance Indices

The performance of the algorithm in the simulation studies was obtained by comparing the forecasted and estimated values at time $k + 1$ with the actual values. The average performance indices for voltage magnitude and voltage angular forecasted and estimated are given as.

$$\varepsilon_{v\_fore} = \frac{1}{n}\sum_{i=1}^{n}\left|\frac{\left(v'_{i_{forecasted}} - v'_{i_{true}}\right)}{v'_{i_{true}}}\right| \times 100\% \tag{17}$$

$$\varepsilon_{\theta\_filt} = \frac{1}{n}\sum_{i=1}^{n}\left|\frac{\left(\theta'_{i_{filteded}} - \theta'_{i_{true}}\right)}{\theta'_{i_{true}}}\right| \times 100\% \tag{18}$$

$$J_k = \frac{\sum_{j=1}^{n}\left|\hat{Z}_k(i) - Z_k^{\mathrm{T}}(i)\right|}{\sum_{j=1}^{n}\left|Z_k(i) - Z_k^{\mathrm{T}}(i)\right|} \tag{19}$$

v'_{true} and θ'_{true} are the true value of voltage magnitude and angle, $v'_{forecased}$ and $v'_{filtered}$ are the forecasted and filtered voltage magnitude, $\theta'_{forecased}$ and $\theta'_{filtered}$ are the forecasted and filtered voltage angle, J_k is the performance index of the overall achievement, the ratio of estimated at time k and actual error of the measurement.

3.3. Results Discussion

In this paper, three test cases are performed, the normal operation, bad data and sudden load change.

Test 1: Normal operation case:

The normal operation case was illustrated by **Table 2, Table 3, Figures 1** and **2**. **Table 2** shows the perfor-

Table 1. The Measurements prepared by the load flow.

Type	Measurement vector
Active power flow	p(1-5), p(4-5), p(4-9), p(6-11), p(6-12), p(7-8), p(7-9), p(9-10) , p(13-14)
Reactive power flow	q(1-5), q(4-5), q(4-9), q(6-11), q(6-12), q(7-8), q(7-9), q(9-10), q(13-14)
Active and Reactive power injection	P_1, P_3, P_6, P_{10}, P_{12} Q_1, Q_3, Q_6, Q_{10}, Q_{12}
Voltages magnitude	V_1, V_3, V_8, V_{11}, V_{12}, V_{14}

Table 2. Performance indices under normal operation.

cases		Predicted		Filtered		J_k
		voltage	angle	voltage	angle	
Normal operation	Max	0.510	1.7546	0.499	1.0455	0.970
	Ave	0.1845	1.3450	0.178	0.6385	0.957

Table 3. The errors per bus under normal operation.

Bus number	predicted		Filtered	
	voltage	Angle	voltage	angle
1	0.2286	0.2198	0.000	0.0000
2	0.1977	0.1986	1.440	0.6792
3	0.2969	0.2954	1.666	0.9037
4	0.2480	0.2389	1.530	0.7662
5	0.2257	0.2168	1.458	0.6974
6	0.1560	0.1530	1.409	0.6483
7	0.2090	0.2080	1.457	0.6967
8	0.1621	0.1568	1.436	0.6754
9	0.2036	0.1995	1.444	0.6838
10	0.1953	0.1907	1.438	0.6780
11	0.1228	0.1207	1.395	0.6350
12	0.1420	0.1426	1.398	0.6375
13	0.0901	0.0870	1.389	0.6294
14	0.1055	0.0650	1.369	0.6093

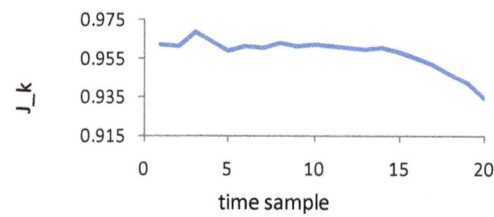

Figure 1. Performance index under normal operation.

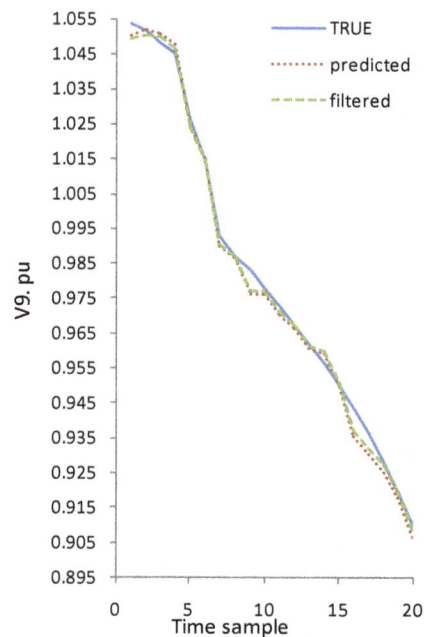

Figure 2. V9 under normal operation.

mance indices of the case. The maximum and average percentage error of voltage magnitudes and angles are made over 20 time sample. From this table, the algorithm has achieved very high performance.

Figure 1 describes the Performance index of the system dynamic nature under normal operation case corresponding to filtered indices calculated at each time sample.

Table 3 shows the percentage for estimated errors per bus for predicted and filtered state with regard to the voltage magnitudes and voltage angles.

In both states, the maximum error occurred in busbar 3 for voltage magnitudes and voltage angles. Furthermore the average errors are equal to the average error in **Table 2**. Bus 1 is the reference bus, so it is angle is outside of state variables. **Figure 3** represents the time behavior for true, forecasted and filtered values of voltage magnitude at bus 9. **Figures 3** and **4** show the graphs of the **Table 3**.

Test 2: Bad data case:

In this test, the simulation was carried out under bad data conditions with three different cases.

- Single bad data was considered. Active power flow pf(4 - 9) was suspected in error of 10% at the 6th time sample.
- Two measurements were suspected in error pf(9 - 10) of 20% at the 11th time sample. And also pf(13 - 14) of 50% decrement at the 11th time sample.
- Single bad data was considered. Reactive power flow qf(4 - 5) was suspected in error of 20% at the 18th time sample.

Suppose that no work is taken to eliminate these errors.

The result of these tests is shown in **Tables 4** and **5**, **Figures 5-7**.

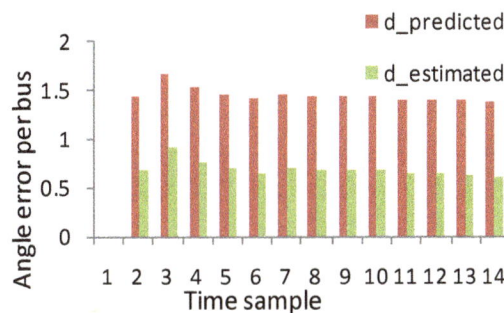

Figure 3. Average voltage error per bus *100%

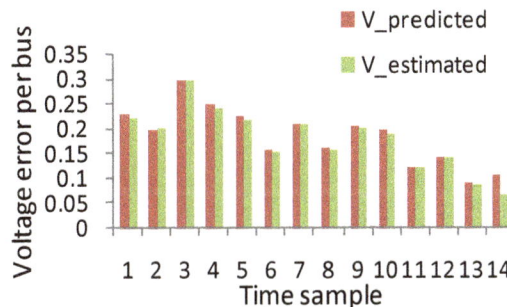

Figure 4. Average voltage angle error per bur*100%

Table 4. Performance indices under bad data condition.

cases		Predicted		Filtered		J_k
		voltage	angle	voltage	angle	
Bad date	Max	0.501	3.2594	0.484	2.539	4.3631
	Ave	0.210	2.501	0.206	1.786	1.6755

Table 5. The percentage error per bus under bad data conditions.

Bus number	Predicted		Filtered	
	Voltage	angle	voltage	angle
1	0.2264	0.0000	0.2283	0.0000
2	0.2172	1.5183	0.2179	0.7570
3	0.3045	1.6960	0.3039	0.9337
4	0.2500	1.5674	0.2563	0.8061
5	0.2730	1.4960	0.2578	0.7351
6	0.1571	3.1423	0.1533	2.3690
7	0.2581	3.1045	0.2637	2.3318
8	0.2271	3.0803	0.2144	2.3103
9	0.2418	2.8826	0.2536	2.1113
10	0.2171	2.9726	0.2266	2.2005
11	0.1176	3.1732	0.1227	2.3997
12	0.1352	3.3294	0.1352	2.5548
13	0.1606	3.6942	0.1470	2.9168
14	0.1462	3.5880	0.1025	2.5840

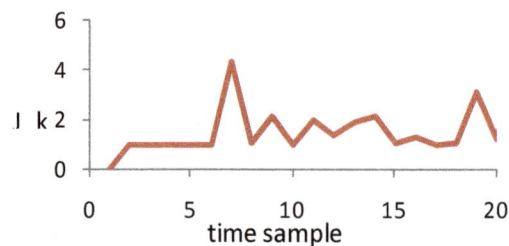

Figure 5. Performance index under bad data condition

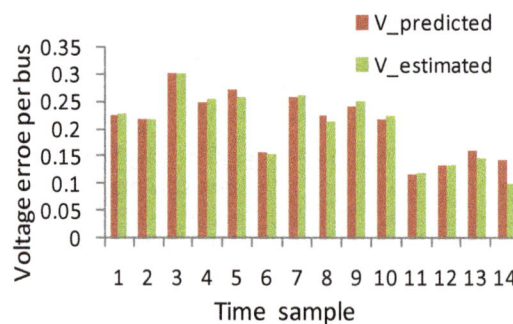

Figure 6. Average voltage error per bus *100%

Table 4 shows the performance indices of these tests. From this results the average and maximum error becomes larger comparing with normal operation test due to these bad data in measurements.

Figure 5 describes the Performance index of the system dynamic nature under bad data condition's corresponding to filtered indices calculated at each time sample.

Figures 6 and **7** show the percentage error per bus of the voltage magnitudes and voltage angle's respectively based on **Table 5**.

Test 3: Sudden load change:

In the case, the injected load at busbar 3, busbar 9 and busbar 13 are assumed to be sudden changed.

Busbar 3 and busbar 13, 50% of their values are cut at the 6th time sample and 15th time sample respectively. For busbar 9, we assumed that the load is increased to 40% at the 10th time sample.

The result of this case is shown in **Tables 6-8**, and also shown in **Figures 8-10**. **Table 6** shows the performance indices of the case of sudden load change. From this results the average and maximum error has become difference of normal operation and bad data test due to these changes in load.

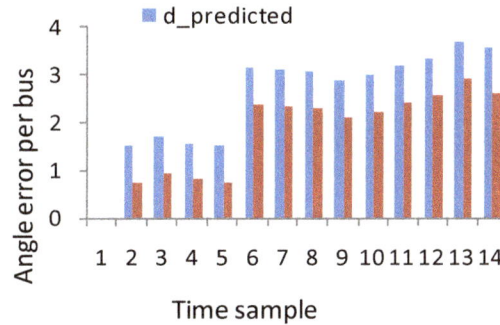

Figure 7. Average voltage angle error per bus *100%

Table 6. Performance indices under sudden load change.

cases		Predicted		Filtered		J_k
		voltage	angle	voltage	angle	
Sudden Load change	Max	0.5145	1.6942	0.4951	0.9857	0.9866
	Ave	0.195	1.35	0.1922	0.642	0.9748

Table 7. The percentage error per bus under sudden load change.

Bus number	Predicted		Filtered	
	voltage	angle	voltage	angle
1	0.2546	0.0000	0.2465	0.0000
2	0.2258	1.4344	0.2268	0.6735
3	0.2860	1.6345	0.2910	0.8728
4	0.2525	1.5383	0.2541	0.7771
5	0.2318	1.4830	0.2341	0.7226
6	0.1710	1.4125	0.1683	0.6522
7	0.2270	1.4660	0.2310	0.7057
8	0.1733	1.4453	0.1680	0.6847
9	0.2155	1.4520	0.2180	0.6912
10	0.2058	1.4460	0.2072	0.6852
11	0.1300	1.4002	0.1306	0.6400
12	0.1546	1.4025	0.1563	0.6422
13	0.0934	1.3900	0.0938	0.6296
14	0.1092	1.3712	0.0651	0.6113

Table 8. V13 under sudden load change.

Time sample	True	predicted	filtered
1	1.0501	1.0482	1.0478
2	1.0489	1.0485	1.0481
3	1.0474	1.0482	1.0479
4	1.0459	1.0467	1.0464
5	1.0246	1.0237	1.0235
6	1.0336	1.034	1.0338
7	1.0173	1.0165	1.0164
8	1.0056	1.0058	1.0058
9	0.9937	0.991	0.991
10	0.9827	0.9823	0.9824
11	0.9800	0.9791	0.9793
12	0.9769	0.9769	0.977
13	0.9738	0.9737	0.9739
14	0.9706	0.9723	0.9725
15	0.9798	0.9804	0.9805
16	0.9771	0.9745	0.9747
17	0.9744	0.9728	0.973
18	0.971	0.9708	0.971
19	0.9675	0.9681	0.9683
20	0.9641	0.9649	0.9652

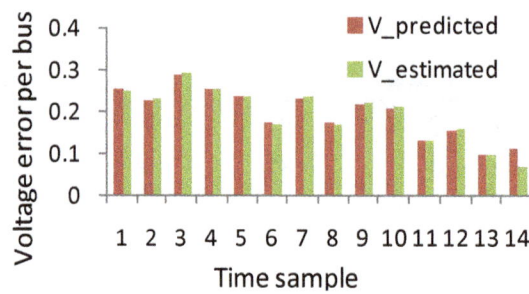

Figure 8. Average voltage error per bus *100%.

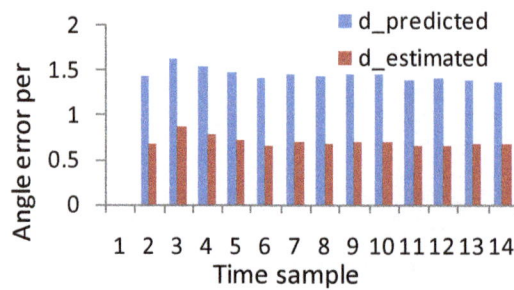

Figure 9. Average voltage angle error per bus *100%

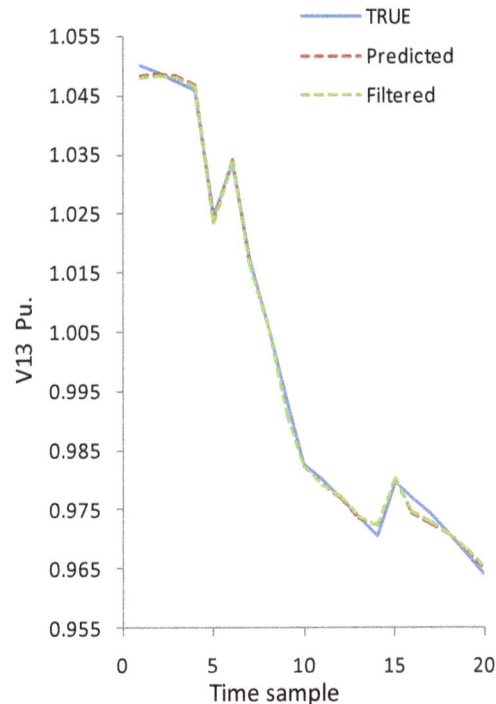

Figure 10. V13 under sudden load change condition.

Figure 10 represents the time behavior for true, forecasted and filtered values of voltage magnitude at busbar 13 based on **Table 8**. As they are shown in **Figure 10**, at time sample 6, 10 and 15 when sudden load change, the oscillation in the voltage is appeared very clear.

4. Conclusion

The dynamic state estimator technique has been made based on Holt's exponential smoothing and Extended Kalman Filter for forecasting and filtering state respectively. The system dynamic was simulated over 20 time samples, with increasing the load at all busbars 5% during any time sample. The algorithm of DSE has been simulated through these 20 time intervals used standard IEEE 14_bus test system under normal and abnormal operation. In this paper, the algorithm gave very good performance results through the normal operation and abnormal operation (bad data and sudden load change) conditions. The error per bus for voltage magnitude and voltage angle has been calculated over these 20 time intervals in both predicted and filtered states.

References

[1]   Shivakumar, N.R. and Jain, A.M. (2008) A Review of Power System Dynamic State Estimation Techniques. *Joint International Conference on Power System Technology and IEEE Power India Conference*, New Delhi, 15 October 2008, 1-6.

[2]   Jegatheesa, R. and Leean, K. (1991) Dynamic State Estimation. *TENCON'91.1991 IEEE Region 10 International Conference on EC3-Energy, Computer, Communication and Control Systems*, New Delhi, 28-30 Aug 1991, 492-496.

[3]   Debs, A.S. and Larson, R.E. (1970) A Dynamic Estimator for Tracking the State of a Power System. *IEEE Transactions on Power Apparatus and Systems*, **PAS-89**, 1670-1678.

[4]   Jain, A.M. and Shivakumar, N.R. (2008) Phasor Measurement in Dynamic State Estimation of a Power System. *TENCON 2008, 2008 IEEE Region 10 Conference*, Hyderabad, 19-21 November 2008, 1-6.

[5]   Huang, S.J. and Shih, K.S. (2002) Dynamic-State-Estimation Scheme including Nonlinear Measurement-Function Considerations. *Generation, Transmission and Distribution, IEE Proceedings*, **149**, 673-678.

[6]   Ali, A. and Antonio, G.E. (2004) Power System State Estimation Theory and Implementation. CRC Press, New York.

[7]   Mukhtar, A. (2013) Power System State Estimation. Artech House, London.

[8]   Leite da silva, A.M., Coutto Filho, M.B. and de Queiroz, J.F. (1976) State Forecasting in Electric Power Systems. *Generation, Transmission and Distribution, IEE Proceedings C*, **130**, 79-87.

[9]   Li, H. and Li, W.G. (2009) Estimation and Forecasting of Dynamic State Estimation in Power Systems. *International Conference on Sustainable Power Generation and Supply*, Nanjing, 6-7 April 2009, 1-6.

[10]   Prajakta, S.K. (2004) Time Series Forecasting Using Holt-Winters Exponential Smoothing. *Kanwal Rekhi School of Information Technology*, **4329008**, 1-13.

[11]   University of Washington Electrical Engineering (1993) Power Systems Test Case Archive. http://www.ee.washington.edu/research/pstca/

18

On Designing of the Main Elements of a Hybrid-Electric Vehicle Driving System

Petre-Marian Nicolae, Ileana-Diana Nicolae, Ionuț-Daniel Smărăndescu

Department of Electrical Engineering, Energetics and Aeronautics University of Craiova, Faculty of Electrical Engineering, Craiova, Romania
Email: pnicolae@elth.ucv.ro, smarandescu.ionut@yahoo.com

Abstract

The paper deals with the designing of an electric drive system used for hybrid electric vehicles. The driving system is realized with an induction motor and a voltage source inverter. Specifically, the application is for a series hybrid vehicle powered by electric storage batteries charged by solar batteries. In the first part of the paper the designing of the electric storage batteries and of the photoelectric system is presented. In the second part of the paper some aspects regarding the designing of the induction motor are presented. Then some aspects concerning the voltage source inverter designing are exposed.

Keywords

Hybrid Electric Vehicle; Drive System; Designing; Electric Storage Batteries; Photoelectric System; Induction Motor; Voltage Source Inverter

1. Introduction

At present, the main problems facing humanity are pollution, as that produced by industrial activities and that produced by motor vehicles, providing food for all mankind, as well as finding new solutions for the production of energy, because conventional sources (non-renewable) and crude oil are running low.

According to [1-3], oil is a limited source of energy, and at the current rate of exploitation of oil resources around the globe is expected to use this resource still about 46 years [4-5]. A possible solution to reduce the consumption of oil could be the use of alternative drive systems as hybrid electric vehicles that use electric storage batteries [6-7].

The ever increasing traffic volume has produced high levels of carbon dioxide emissions from the conventional gasoline and diesel-fueled vehicles, which will contribute to a substantial increase of pollution in the major cities all over the world, causing heavy consequences to the communities [8-11].

It has been recognized that hybrid electric vehicles are the only viable solution in order to reduce air pollution, in particular, in large urban areas. In a hybrid electric vehicle, the electric propulsion system is intended to provide advantages over conventional vehicles equipped with an internal combustion engine. With the help of the electric drive, the life of the internal combustion engine can be optimized, which often means low fuel consumption and low emissions.

In this context the induction motor offers weight and efficiency advantages over the more conventional DC motors, besides its traditional advantages of robustness, low cost and well established manufacturing techniques. These motors have comparable torque and efficiency, along with a rugged, durable design. They don't have a drag loss when the motor turns on and they don't lose their efficiency during high speed or low torque conditions. This makes them well suited for hybrid electric vehicles.

This work presents the design of a propulsion system for a series hybrid electric vehicle powered by an induction motor. For the operation of the hybrid electric vehicle, electric storage batteries charged by solar batteries are used.

The used block diagram of the hybrid electric vehicle is depicted in **Figure 1** [12].

The parameters of the chosen hybrid electric vehicle are:

- Rated power: $P_n = 60$ kW;
- Range: $L = 75$ km;
- Induction motor pole pairs: $p_1 = 1$;
- Vehicle speed: $v = 75$ km/h.

2. Sizing of the Electric Storage Batteries

For the operation of the hybrid electric vehicle are used electric storage batteries charged by solar batteries [13].

Hybrid electric vehicle running time to a speed of 50 km/h and at a distance of 160 km is:

$$t = \frac{L}{v} = 1\,h \tag{1}$$

The energy required to operate the hybrid electric vehicle at a power of 8 kW and with a running time of 1 h is:

$$E_n = P \cdot t = 216\,MJ \tag{2}$$

To have this energy it can be used electric accumulators. To do this, one chooses an accumulator that has a capacity C of 150 Ah:

$$C = 150\,Ah = 150 \cdot 1 = 150 \cdot 3600 = 54 \cdot 10^4\,As \tag{3}$$

The energy stored in the accumulator is:

$$EN_{AE} = U \cdot I \cdot t = 648 \cdot 10^4\,J \tag{4}$$

The number of electric accumulators is:

$$N_{AE} = \frac{E_n}{EN_{AE}} = 30 \tag{5}$$

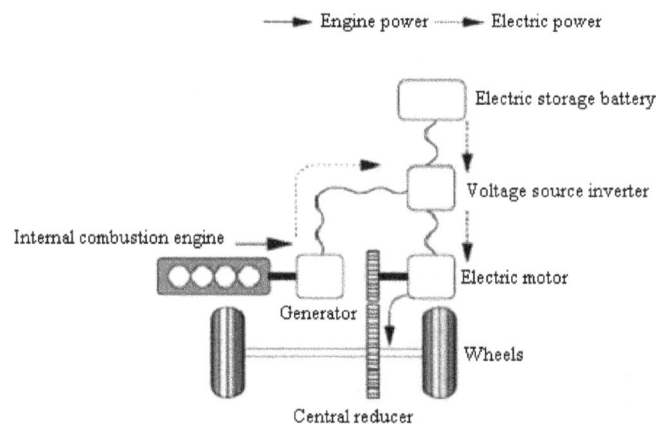

Figure 1. Structure and power flow diagram for a series hybrid electric vehicle.

The energy of a battery is:

$$E_{AE} = \frac{E_n}{N_{AE}} = 7.2\ MJ\ .$$ (6)

The Ag-Zn electric storage batteries disposal is depicted in **Figure 2** [14].

3. Photoelectric System Sizing

Charging of the electric storage batteries is made from a solar cells system. These batteries will not give the same energy throughout the day [15]. The surface of the solar battery for charging the electric storage batteries is $S = 20\ m^2$. For a system of fixed and mobile solar panels, the power can be calculated as:

$$P = U \cdot I \cdot no.hours \cdot \eta$$ (7)

For the fixed system the efficiency is $\eta = 0.4$ and for the mobile system $\eta = 1$.

The coordinates of the operation points from the operation characteristic of the batteries (**Figure 3**) are calculated as follows:

$$P_1 \begin{cases} P_1 = 20 \cdot 1 \cdot 10^3 \cdot 0.4 = 8000\ W \\ U_1 = 100\ V \\ I_1 = P_1 / U_1 = 80\ A \end{cases} \Rightarrow P_1(100, 80)$$ (8)

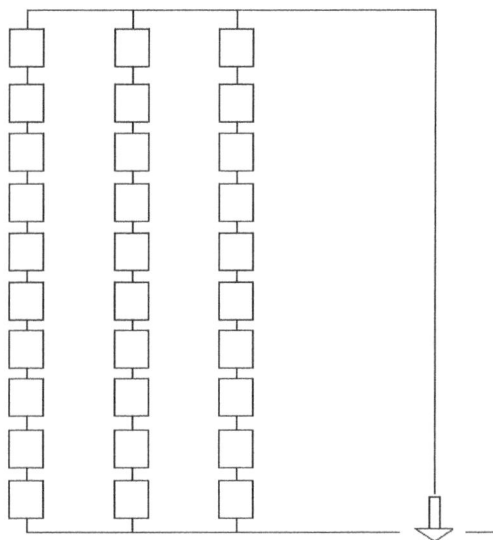

Figure 2. Electric storage batteries disposal.

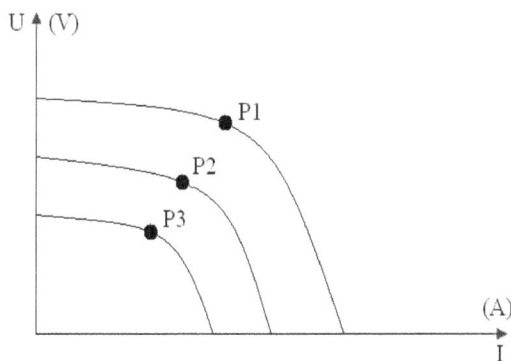

Figure 3. The operation characteristic of the batteries.

$$P_2 \begin{cases} P_2 = 20 \cdot 0.8 \cdot 10^3 \cdot 0.4 = 6400 \ W \\ U_2 = 90 \ V \\ I_2 = P_2 / U_2 = 71.11 \ A \end{cases} \Rightarrow P_2(90, 71.11) \qquad (9)$$

$$P_3 \begin{cases} P_3 = 20 \cdot 0.7 \cdot 10^3 \cdot 0.4 = 5600 \ W \\ U_3 = 80 \ V \\ I_3 = P_3 / U_3 = 70 \ A \end{cases} \Rightarrow P_2(80, 70) \qquad (10)$$

The energy given by the mobile system is:

$$E = E_1 + E_2 + E_3 = 80 \cdot 100 \cdot 6 \cdot 3600 + 71.11 \cdot 90 \cdot 2 \cdot 3600 + 70 \cdot 80 \cdot 1 \cdot 3600 = 239 \ MJ \qquad (11)$$

The energy of the fixed system is:

$$E = 239 \cdot 0.45 = 107.55 \ MJ \qquad (12)$$

From these calculations it is observed that the mobile system provides more energy, because this system follows the sun and provides the maximum energy in the range of given hours.

4. Aspects Concerning Induction Motor Designing

The induction motor was designed considering the following rated data:
- Rated power: P_n = 60 kW;
- Rated voltage: U_n = 230/400 V;
- Speed: n_N = 2985 rot/min;
- Number of pole pairs: p_1 = 1.

Further in the paper, the most representative parameters and results yielded from the induction motor designing algorithm are exposed.

The rated current is given by:

$$3 \cdot U_N \cdot I_N \cdot \cos \varphi \cdot \eta = P_N \qquad (13)$$

where $\cos\varphi$ = 0.85 and η = 0.91. In this case I_N = 112 A.

Next, one determines the induction motor parameters. These are: the resistances R_1 and R_2, the inductances L_1 and L_2, the coupling inductance M, and the idling current I_0 [16].

Determination of the internal resistance of the motor is calculated considering the power losses as:

$$0.1 \cdot P_N = 3 \cdot (R_1 + R_2) \cdot I_N \qquad (14)$$

It results that: $R_1 = R_2 = 0.08 \ \Omega$.

The idle current is given by:

$$I_0 = 0.3 \cdot I_N = 33.6 \ A \qquad (15)$$

The inductance is determined as:

$$L = \frac{U_N}{I_0 \cdot \omega_1} \qquad (16)$$

where ω_1 = 314 rad/s. In this case, L = 0.021 H.

The coupling inductance is:

$$M = 0.9 \cdot L \qquad (17)$$

In our case M = 0.0189 H.

Steady state equations of the induction motor are:

$$U_1 = R \cdot I_d - \omega_1 \cdot (L \cdot I_q + M \cdot I_{qr}) \qquad (18)$$

$$U_2 = R \cdot I_q + \omega_1 \cdot (L \cdot I_d + M \cdot I_{dr}) \qquad (19)$$

$$0 = R \cdot I_{dr} - (\omega_1 - \omega) \cdot (L \cdot I_{qr} + M \cdot I_q) \qquad (20)$$

$$0 = R \cdot I_{qr} + (\omega_1 - \omega) \cdot (L \cdot I_{dr} + M \cdot I_d) \tag{21}$$

Control of the algorithm takes into account the electromagnetic torque function in the form:

$$M_{elmg} = p_1 \cdot M \cdot (I_{dr} \cdot I_q - I_d \cdot I_{qr}) \tag{22}$$

The maximum of the function is analyzed by considering the method of Lagrange multipliers. One imposes a maximum stator current:

$$I_d^2 + I_q^2 = I_N^2 = 16,307.2 \ A^2 \tag{23}$$

For the stator flux Ψ_s one imposing a value closes to the magnetic saturation (Ψ_s=1.3 Wb) [17]:

$$\Psi_s^2 = (L \cdot I_d + M \cdot I_{dr})^2 + (L \cdot I_q + M \cdot I_{qr})^2 \tag{24}$$

For the induction motor starting, the voltages U_d and U_q are written as:

$$U_d = R_1 \cdot I_d - \omega_1 \cdot (L \cdot I_q + M \cdot I_{qr}) \tag{25}$$

$$U_q = R_1 \cdot I_q + \omega_1 \cdot (L \cdot I_d + M \cdot I_{dr}) \tag{26}$$

and the following values are obtained: ω_1= 8.21 rad/s; U_d = 0.064 V; U_q = 9.626 V.

Hybrid electric vehicle operation at rated speed is given by:

$$\begin{cases} 400 = R \cdot I_d - \omega \cdot (L \cdot I_q + M \cdot I_{qr}) \\ 0 = R \cdot I_q + \omega \cdot (L \cdot I_d + M \cdot I_{dr}) \\ 0 = R \cdot I_{dr} - (\omega - \omega_N) \cdot (L \cdot I_{qr} + M \cdot I_q) \\ 0 = R \cdot I_{qr} + (\omega - \omega_N) \cdot (L \cdot I_{dr} + M \cdot I_d) \\ M_N = M \cdot (I_q \cdot I_{dr} - I_d \cdot I_{qr}) \\ \omega_N = (2 \cdot \pi \cdot n_N / 60) \cdot 3 \end{cases} \tag{27}$$

and has the next results: I_d = −31.409 A; I_q = 298.02 A; I_{qr} = −328.49 A; I_{dr} = 32.253 A; M_N = 13.331 Nm; ω_N = 312.4 rad/s.

The obtained voltages are: U_d = 15.17 V; U_q = 25.36 V.

Regenerative braking of the hybrid electric vehicle is given by:

$$\begin{cases} 400 = R \cdot I_d - \omega \cdot (L \cdot I_q + M \cdot I_{qr}) \\ 0 = R \cdot I_q + \omega \cdot (L \cdot I_d + M \cdot I_{dr}) \\ 0 = R \cdot I_{dr} - (\omega - \omega_N) \cdot (L \cdot I_{qr} + M \cdot I_q) \\ 0 = R \cdot I_{qr} + (\omega - \omega_N) \cdot (L \cdot I_{dr} + M \cdot I_d) \\ M_F = M \cdot (I_q \cdot I_{dr} - I_d \cdot I_{qr}) \end{cases} \tag{28}$$

The obtained results are:

$$\begin{cases} I_d = 7.557 \ A \\ I_q = -306.76 \ A \\ I_{qr} = 273.54 \ A \\ I_{dr} = -4.262 \ A \\ M_F = -14.36 \ Nm \end{cases} \tag{29}$$

5. Designing of the Voltage Source Inverter

5.1. Choosing the Power Electronics for the Converter

The designing of the VSI was performed in a MATLAB module and is based on the following rated data:
- Rotor AC line voltage: $U_{aclinerot}$ = 400 V;

- Rotor rated current: $I_{nrot} = 112$ A;
- Fundamental frequency: $f = 25$ Hz;
- Effective value of the rectified output voltage: $U_{ef} = 653.2$ V;
- Rated current amplitude for the rotor: $I_{namrot} = \sqrt{2} \cdot I_{nrot} = 158.4$ A;
- Rated average current: $I_{med} = 0.9 \cdot I_{med} = 100.8$ A;
- Ripple current, considering a peak of 5%: $I_{Lripple} = 0.05 \cdot I_{namrot} = 7.91$ A.

5.2. Characteristic Parameters of the Voltage Source Inverter

The designing method is used for a wide range of engines and voltages. The sizing of the voltage source inverter was made for a traction motor with the power of 60 kW and a voltage of 400 V_{AC}. The maximum voltage that the voltage source inverter can supply for the motor is determined by the main supply voltage.

Characteristic and functional parameters, from which the sizing of the voltage source inverter was performed, are the following:

- Supplying voltage: $U_a = 110$ V ($-30\% \div +20\%$);
- Input filter parameters: $C_f = 1000$ μF; $L_f = 3.96$ mH;
- Operating frequency: $f_L = 2$ kHz;
- Induction motor rated current: $I_n = 112$ A.

The simplified diagram of the inverter is depicted in **Figure 4**.

5.3. Calculation of the Conduction Time

Full conduction time t_1 is determined from the condition that the average voltage at the motor terminals not to exceed its rated voltage, respectively $U_n = 400$ V.

For the switching period $T = 1 / f_L = 0.0005 \cdot 10^{-3}$ s and the supplying voltage $U_a = 110$ V, results:

$$t_1 = \frac{T \cdot U_n}{U_a} = 0.0037 \cdot 10^{-3} \, s \tag{30}$$

Changing of the voltage average value is achieved by changing the time length of the conduction, while maintaining a constant switching period (corresponding to an operating frequency of 2 kHz).

5.4. Calculation of the Average Current through the IGBT

The average current through the static contactor is a fundamental criterion for choosing it. The average current value is computed by mediation instantaneous values corresponding to a period:

$$I_{med} = \frac{1}{T} \int_0^T i_T dt = \frac{1}{2\pi} \int_0^\pi I_n \cdot \sin(\omega t) d\omega t \tag{31}$$

The current through the IGBT I_{igbt} is obtained with:

$$I_{igbt} = \frac{\sqrt{2} \cdot I_n}{\pi} \tag{32}$$

In this case, $I_{igbt} = 38.98$ A.

Since the induction motor is 60 kW/480 V, then the maximum current of the motor I_{max} is given by:

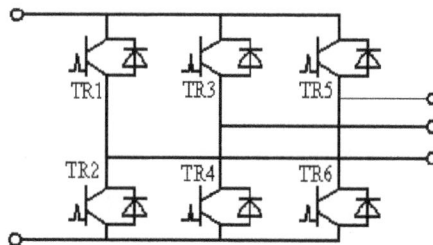

Figure 4. Voltage source inverter diagram.

$$I_{max} = 1.5 \cdot I_n . \tag{33}$$

In our case I_{max} = 129.90 A.

The average value corresponding to the maximum motor current is:

$$I_{med1} = \frac{\sqrt{2} \cdot I_{max}}{\pi} \tag{34}$$

In this case, I_{med1} = 58.47 A.

5.5. Network Filter Sizing

The network filter limits to a very large extent the switching surges and protects the equipment against network's short duration voltage variations. Short-term variations of the voltage applied to the equipment are largely caused by instability of the galvanic connection between the sensors and the contact line [18].

Natural frequency of resonance (f_0) for the filter is given by:

$$f_0 = \frac{1}{2 \cdot \pi \sqrt{L_f \cdot C}} \tag{35}$$

In this case, f_0 = 79.97 Hz.

In order to not fall into the spectrum fn of the harmonics produced by the inverter, f_0 must be lower than the lowest operating frequency of the inverter, meaning $f_0 < f_{min}$. Because $f_{min} = f$ = 2 kHz and f_0 = 79.97 Hz, the above condition is satisfied.

5.6. Choosing the Switching Transistors

Depending on the parameters I_{max} and the rated voltage imposed, from the catalog of SEMIKRON manufacturer of electronic components was chosen the following IGBT model: SKM900GA12E4 (SEMITRANS), which satisfies the conditions [19-20]:

$$I_{max} \leq I_{max\,cata\log} = 900A \tag{36}$$

$$U_a \leq U_{ncata\log} = 1200V \tag{37}$$

5.7. Choosing and Verification of the Cooling System

The dissipation of the heat that is generated in the semiconductor during conduction is made via an aluminum radiator that is cooled with forced air flow. Radiator sizing criteria is that the semiconductor junction temperature does not exceed the limit specified by the manufacturer, namely 175 °C under long-term.

The radiator is a subset of the equipment, with the aim of maintaining the temperature within limits for the proper functioning of the equipment, especially IGBT.

The thermal effects of the electric current peak by conduction state represent the worst case during operation.

Other important computed parameters are:

- Switching losses of the transistor:

$$P_{comT} = \left(E_{on} + E_{off} \right) \cdot \frac{I_{medT}}{I_{ccref}} \cdot f_{com} = 52.37\,W \tag{38}$$

- Conduction losses of the transistor:

$$P_{condT} = \frac{t_1}{T} \cdot U_{CEsat} \cdot I_{cmed} = 216.48\,W \tag{39}$$

- Average current through the antiparallel diode:

$$I_{med} = \frac{1}{T}\int_0^T i_T dt = \frac{1}{2\pi} \int_0^{2\pi/3} I_n \cdot \sin(\omega t) d\omega t \tag{40}$$

$$I_{med} = \frac{\sqrt{2} \cdot 3 \cdot I_n}{4\pi} = 60.42 \; A \tag{41}$$

- Antiparallel diode switching losses:

$$P_{comD} = \left(E_{on} + E_{off}\right) \cdot \frac{I_{medD}}{I_{ccref}} \cdot f_{com} = 8.73 \; W \tag{42}$$

One used data catalog for voltage drops at saturation. The total average power generated inside the case by a module is:

$$P = P_T + P_D = 257.92 \; W \tag{43}$$

Total losses on all three phases [21]:

$$P_{phase} = 3 \cdot P = 737.77 \; W \tag{44}$$

Was made the preliminary selection of the radiator type P16_300_16B, with the technical characteristics specified in the catalog tab, cooled by an air flow produced by the fan type SK-Heat Sink P16_300_16B, with the technical characteristics specified in the catalog of the manufacturer.

Junction over-temperature corresponding to the average power previously computed is checked within the chosen cooling conditions:

$$\Delta T = T_j - T_0 = P \cdot \sum R_{th} \tag{45}$$

The absolute temperature of the junction depends on the maximum ambient temperature specific to the temperate climate: $T_0 = 40°C$.

The thermal resistances involved in this calculation are specified in the power transistor catalog, respectively the radiator catalog:

- Thermal resistance junction—capsule for IGBT:

$$R_{thjcT} = 0.035 \; K/W;$$

- Thermal resistance junction—capsule for diode:

$$R_{thjcD} = 0.041 \; K/W;$$

- Thermal resistance capsule—radiator:

$$R_{thcr} = 0.038 \; K/W;$$

- Thermal resistance of the radiator, at forced cooling with an air flow of 295 m^3/h:

$$R_{thr} = 0.031°C \; /W.$$

The average temperature of the junction is computed both for IGBT, and for antiparallel diode located in the same capsule.

The general relation for calculating the junction temperature is:

$$T_j = T_0 + P \cdot (R_{thjc} + R_{thcr} + R_{thr}) \tag{46}$$

Junction temperature for IGBT is obtained with:

$$\begin{aligned} T_{jT} &= T_0 + (P_{comT} + P_{condT}) \cdot (R_{thjcT} + R_{thcr} + R_{thr}) \\ &= 341.11 \; K < 448.15 \; K \end{aligned} \tag{47}$$

In this case $T_{jT} = (341.11 - 273.15) = 67.96°C$.

For the antiparallel diode, the junction temperature is computed with:

$$\begin{aligned} T_{jD} &= T_0 + (P_{condD} + P_{comD}) \cdot (R_{thjcD} + R_{thcr} + R_{thr}) \\ &= 320.12 \; K < 448.15 \; K. \end{aligned} \tag{48}$$

In this case: $T_{jD} = 320.12 - 273.15 = 46.97°C$.

The results show that the cooling condition is fulfilled, namely:

$$T_j \le T_{jad} = 175 \, °C \; (or \; 448,15 \; K) \tag{49}$$

Results a factor of safety in heating of 67.96/175 = 0.388, which shows that the sizing of cooling conditions has been rigorously made, without large reserves and so without undue consumption of materials and within a minimum possible size.

For reliability an IGBT transistor was chosen with rated data higher than those considered to be adequate to the rated data of the induction motor.

Evaluating the results obtained by the sizing methods is observed that an optimal configuration was used [22].

6. Conclusions

The paper was intended for a study on hybrid electric vehicle driving systems and on electric storage batteries they use. The advantages of the electric drive system with the induction motor and voltage source inverter revealed that it satisfies the requirements of the present application. The drive systems based on the induction motor will eliminate the most disadvantages of DC drive systems.

The results obtained from the designing algorithm of the induction motor show that an optimal configuration was used.

In the paper was also presented a sizing algorithm for the power components of the voltage source inverter. Choosing of the power semiconductor devices was made from the data catalog of a producer of electronic components and the results obtained by computing were in good agreement with those from the catalog.

One can conclude that is possible to design propulsion system for a hybrid electric vehicle with the given rated data for the specified application and the required performance.

Given that the conventional means of transport (with internal combustion engine) are the main source of chemical and noise pollution on the planet, hybrid electric transportation systems are a more viable alternative for transporting people and goods.

References

[1] EIA (2012) Short Term Energy Outlook March 2012. U.S. Energy Information Administration.

[2] IEA (2010) World Energy Outlook 2010. International Energy Agency, Paris.

[3] Klass, L.D. (1998) Biomass for Renewable Energy, Fuels and Chemicals. Elsevier Inc., Philadelphia.

[4] BP (2011) BP Statistical Review of World Energy June 2011.

[5] IEA (2011) Key World Energy Statistcs 2011. International Energy Agency, Paris.

[6] Brown, S., Pykea, D. and Steenhof, P. (2010) Electric Vehicles: The Role and Importance of Standards in an Emerging Market. *Energy Policy*, **38**, 3797-3806. http://dx.doi.org/10.1016/j.enpol.2010.02.059

[7] Woods, L.R. and Lawrance, K.L. (2013) U.S. Department of Energy—Energy Efficiency and Renewable Energy. http://www1.eere.energy.gov

[8] Hori, Y. (2004) Future Vehicle Driven by Electricity and Control-Research on Four-Wheel-Motored UOT Electric March II. *IEEE Transactions on Industrial Electronics*, **51**, 954-962. http://dx.doi.org/10.1109/TIE.2004.834944

[9] Chan, C. (2002) The State of the Art of Electric and Hybrid Electric. *Proceedings of the IEEE*, **90**.

[10] Naito, S., Mutoh, N., Takagi, T. and Kouchi, Y. (1995) AC Drive Systems for Electric Vehicles. *Hitachi Review*, **77**.

[11] Yamamura, H., Masaki, R., Koizumi, O., Naoi, K. and Naito, S. (1992) Development of Powertrain System for Nissan FEV. *Proceedings of the 11th Electric Vehicle Symposium*, Florence.

[12] Katrasnik, T. (2007) Hybridization of Powertrain and Downsizing of IC Engine—A Way to Reduce Fuel Consumption and Pollutant Emissions—Part 1. *Energy Conversion and Management*, **48**, 1411-1423. http://dx.doi.org/10.1016/j.enconman.2006.12.004

[13] Clondescu, Gh. and Tomuta, O.D. (1977) Electric Accumulators, Maintenance and Repair (Acumulatoare Electrice, Intretinere si Reparare—In Romanian). Universitatea Tehnica de Constructii Bucuresti.

[14] Corrigan, D., Menjak, I. and Dhar, S. (2000) Nickel-Metal Hydride Batteries for ZEV-Range Hybrid Electric Vehicle. PNGV Future Truck Technical Report, University of California.

[15] Noreus, D. (2000) Substitution of Rechargeable NiCd Batteries, a Background Document to Evaluate the Possibilities of Finding Alternatives to NiCd Batteries. Arrhenius Laboratory, Stockholm University.

[16] Câmpeanu, A., Vlad, I. and Enache, S. (2011) Aided Design of Electrical Machines (Proiectarea Asistata a Masinilor Electrice—In Romanian). Universitaria, Craiova.

[17] Buja, G., Casadei, D. and Serra, G. (1998) Direct Stator Flux and Torque Control of an Induction Motor: Theorethical

Analysis and Experimental Results. *Conference of IEEE Industrial Electronics Society, IECON'98*, 50-64.

[18] Choi, B. (2007) Analysis of Input Filter Interactions in Switching Power Converters. *IEEE Transactions On Power Electronics*, **22**, 452-460. http://dx.doi.org/10.1109/TPEL.2006.889925

[19] http://www.semikron.com/

[20] http://www.semikron.com/products/data/cur/assets/SKM900GA12E4_22892130.pdf

[21] Pou, J., Osorno, D., Zaragoza, J., Ceballos, S. and Jaen, C. (*2011*) Power Losses Calculation Methodology to Evaluate Inverter Efficiency in Electrical Vehicles. *Proceedings of the 7th International Conference opn Workshop Compatibility and Power Electronics, CPE*, Article No. 5942269, 404-409.

[22] http://iota.ee.tuiasi.ro

19

A Novel Idea for Self-Balancing Car Parks for Plug in Electric Vehicle Charging

Andrew D. Clarke, Elham B. Makram

Department of Electrical and Computer Engineering, Clemson University, Clemson, USA
Email: adclark@g.clemson.edu, makram@clemson.edu

Abstract

Many existing studies seek to examine and mitigate possible impacts which plug in vehicle (PEV) charging will have on electric utilities. As PEVs increase in popularity, car parks will have to be built in order to allow for charging while away from home. Existing studies fail to consider the unbalance conditions in the distribution feeders to car parks during PEV charging. This paper presents an innovative idea to improve the unbalance conditions caused by a car park by reconfiguring PEV charger connections to a three-phase system. Although the developed algorithm is simple, results show its ability to balance the power among the phases. A car park is used as an example to show that balancing of the real power drawn by PEV chargers in a parking structure is a success.

Keywords

Unbalance, Electric Vehicles, Power System Control, Smart Grid

1. Introduction

Some forecasts predict that plug in electric vehicles (PEVs) will account for nearly one third of all new vehicle purchases by 2020 [1]. PEVs can be further classified as plug in hybrid electric vehicles (PHEVs), battery electric vehicles (BEVs), and extended range electric vehicles (EREVs). PHEVs and EREVs have both an electric drivetrain as well as an internal combustion engine while BEVs have only an electric drivetrain [2]. With the steadily increasing popularity of PEVs, charging infrastructure is needed to allow owners the security of being able to recharge when it becomes necessary, especially in the case of BEVs, where the vehicle has no propulsion source once the battery is depleted. PEV owners are therefore expected to desire charging stations installed both at home as well as large car parks located at commercial buildings.

In large commercial car parks, due to the number of PEV chargers that will be present, it is expected that power will be fed from a three-phase distribution system. AC PEV chargers are typically divided into three levels, shown in **Table 1** [3]. Currently available PEV chargers fall into the classification of the standardized levels 1 and 2 [4]. Due to the higher energy transfer capability of level 2 chargers compared to level 1, which corresponds to a shorter charge time, it is expected PEV owners will desire these chargers if available.

Many existing studies attempt to examine the effects of PEVs on a distribution system. These studies focus primarily on transformer overheating, system overloading, and ancillary services that can be provided through vehicle to grid (V2G) [5]-[9]. Existing studies fail to consider the problem of power unbalance in the distribution system however. In distribution systems, balancing of power between phases is attempted by distributing single phase loads equally between the three phases. In a car park, it is expected that one third of the chargers will be connected to each phase in order to attempt balancing. With varying loads due to individual PEV charging rates and intermittent loads from vehicles connecting and disconnecting, this becomes impossible using a static phase connection with single phase PEV chargers however. It is very likely that a PEV car park will therefore exacerbate the unbalance conditions that already present in most distribution systems by drawing uneven power from each phase.

These conditions may be harmful to the distribution system due to zero and negative sequence currents that flow in the system when unbalance is present. Unbalanced conditions lead to higher losses, higher temperatures in transformers and motors, diminished power transfer limits of devices, and the potential to inhibit correct protection device operation [10] [11]. The goal of this work is to expand on work in [12] and present an idea that can significantly decrease or eliminate unbalanced operation of a car park by balancing real power. In this work, the idea is to actively change the phase that each PEV charger is connected to in order to balance the power drawn from the distribution system by the car park.

This paper is organized in the following sections: Section 2 details the model used to represent each PEV; Section 3 has a description of the self-balancing car park; Section 4 contains the results of a simulation run with a sample algorithm to show the potential of using PEVs for balancing; and Section 5 has the final conclusion of the paper.

2. PEV Charger Model

PEV chargers typically consist of an AC-DC converter followed by a DC-DC converter. The AC-DC converter is responsible for the interaction of the charger with the electric grid and the DC-DC converter is responsible for ensuring the battery is charged according to a constant current and constant voltage profile [13] [14]. One control topology that can be used to control the PEV charger's AC-DC converter is current control. In this method of control, the current flowing into the charger is directly controlled [13]. It is shown in [15] that a single phase PEV charger can be developed that will accurately draw or supply a commanded real power based on this control methodology. A controlled current source is therefore used to model each PEV charger in place of the full detailed model. The required current can be calculated based on Equation (1) by using the measured voltage at the terminal of each PEV charger and the desired real and reactive powers. Based on the limits prescribed by standardized level 1 and 2 chargers, the real power is limited to a maximum of 7 kW per PEV [3].

$$I_{peak} = \frac{2(P + jQ)^*}{V_{peak}^*}$$

(1)

3. Self-Balancing Car Park

The goal of this paper is to present the idea of moving PEV chargers between phases automatically to balance a car park and show the viability of using such a scheme in a realistic example of a car park. This idea works by

Table 1. PEV charging levels [3].

Charging level	Nominal supply voltage	Maximum current	Continuous input power
1	120 V, 1 phase	12 A	1.44 kW
2	208 - 240 V, 1 phase	32 A	6.66 - 7.68 kW
3	208 - 600 V, 3 phase	400 A	>7.68 kW

moving PEVs drawing power from heavily loaded phases to lightly loaded phases and doing the opposite for PEVs supplying power through V2G. This has the effect of increasing the power drawn by the lightly loaded phase while simultaneously reducing the power drawn by the heavily loaded phase. Because PEV charging rates are not altered using this idea, PEV charging is allowed to continue in almost the same way as with a static phase connection. The impact on PEV owners is therefore expected to be unnoticeable, while the overall car park is expected to have less impact on the distribution system than an equivalent car park without phase balancing present. In order to be successful, each PEV must have the ability to switch between all three phases. This is accomplished by adding a switching element between the terminals of the PEV charger and the distribution grid. In this study, single phase breakers are used, however power electronic solid state switches or any other switch that allows disconnection from one phase and connection to another could also be used for this purpose. To illustrate the feasibility of this idea, a sample heuristic algorithm that requires minimal computations is developed and used to control the phase each PEV charger is connected to.

The algorithm works by first receiving data on how much power each PEV charger is currently drawing or supplying. It then checks to see if a predefined time has passed since the last placement of vehicles. For this study, the time was chosen as 10 seconds. This time represents the minimum time between switching a vehicle from one phase to another. By decreasing the time, the algorithm is able to operate faster to keep up with changing PEV loads. Wear on switching elements will be increased however. By increasing the time, wear on switching elements will be reduced. The algorithm will wait longer to adapt to changing PEV loads and will require a vehicle to wait longer upon plugging in before charging is allowed to commence.

After this time delay has passed since the last calculation, the algorithm checks if any vehicles have been connected or disconnected or if any vehicle charging rates have changed since the last calculation. If none of these have happened, the calculation time is stored and the algorithm exits and waits for the time delay again before resuming. No PEV chargers are moved between phases in this scenario. If at least one of these has happened however, the algorithm continues calculating a new assignment. The PEV charger powers are first sorted into a list from largest to smallest by absolute value. The initial power on each phase is considered to be 0 W. The list is then stepped thorough one entry at a time, starting with the first PEV charger in the list, which corresponds to the PEV charger with the maximum absolute value of power. The PEV charger is checked to see if it is drawing or supplying power. If it is drawing power, it will be assigned to the phase drawing the minimum power. If it is supplying power, it will be assigned to the phase drawing the maximum power. If it is doing neither, it will be assigned to disconnect from all three phases. If two phases or more phases are drawing the same power, the priority of the equal phases, is phase A, phase B, and finally phase C. The new power on each phase is recalculated using Equation (2) based on the assignment. The assignment continues until all PEV chargers have been assigned a phase or commanded to disconnect. After all assignments are complete, the switching commands are sent and the PEV chargers automatically switch to the assigned phase. A graphical representation of the algorithm can be found in **Figure 1**.

$$P_{\text{Phase}} = \sum_{x \in \text{Phase}} P_{\text{PEV}}(x) \tag{2}$$

4. Results

To test the sample algorithm, it is applied to a simulated car park containing 300 level 2 PEV chargers and different scenarios are tested to show its operation and the feasibility of this idea. For the first half of all scenarios, balancing using PEVs does not occur and represents a traditional car park with static phase connections. One third of all chargers are connected to each phase. Balancing is then allowed using PEVs and the new powers drawn from each phase are shown. The first scenario represents the most extreme case possible. It is extremely unlikely to occur, however it is possible. In this case, a total of 100 PEVs are each connected to a charger and all PEVs are considered to have parked at a charger connected to phase A while drawing the maximum power of 7 kW. **Figure 2** shows that before PEV balancing occurs, the power drawn by the car park is extremely unbalanced and will likely have negative effects on the distribution system supplying power to it. After PEV balancing occurs however, the power drawn by the car park very close to balanced. This is because vehicles are switched from phase A to phases B and C. It should be noted that the power on phase A remains slightly higher than phases B and C because after all phases have an equal number of PEV chargers assigned to them, there is still one remaining to be assigned and based on the priority described earlier, it is placed on phase A. Also, the

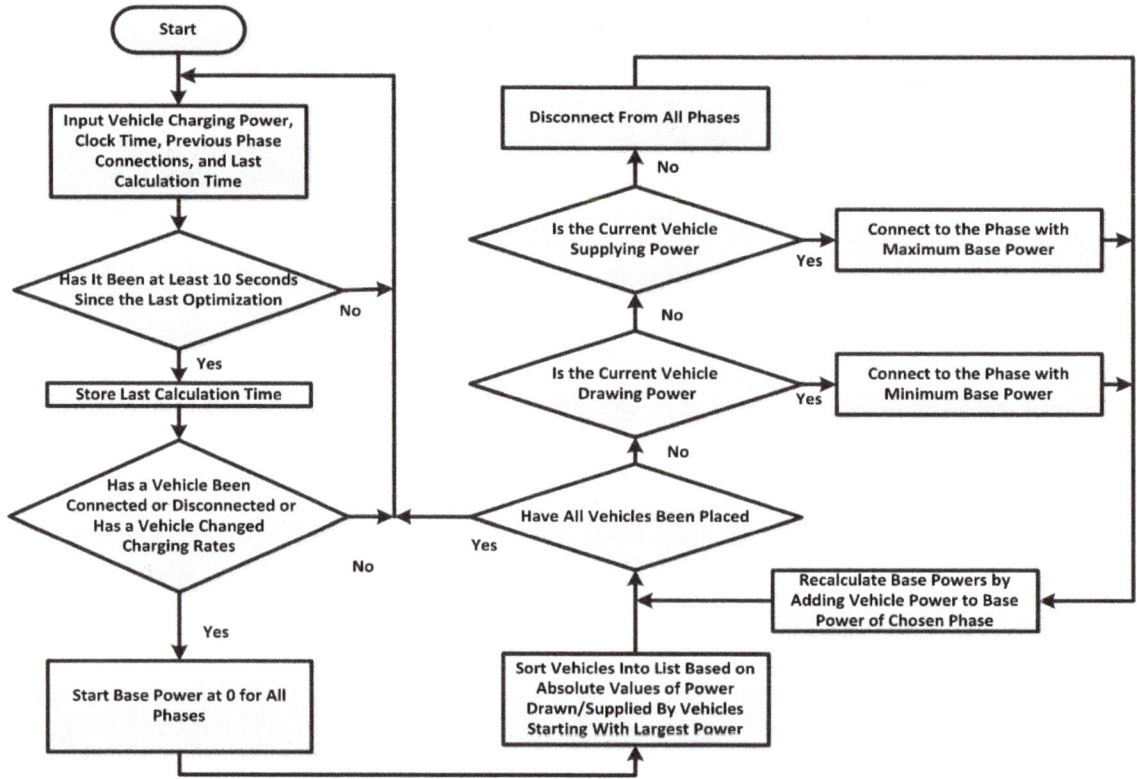

Figure 1. Self-balancing car park algorithm.

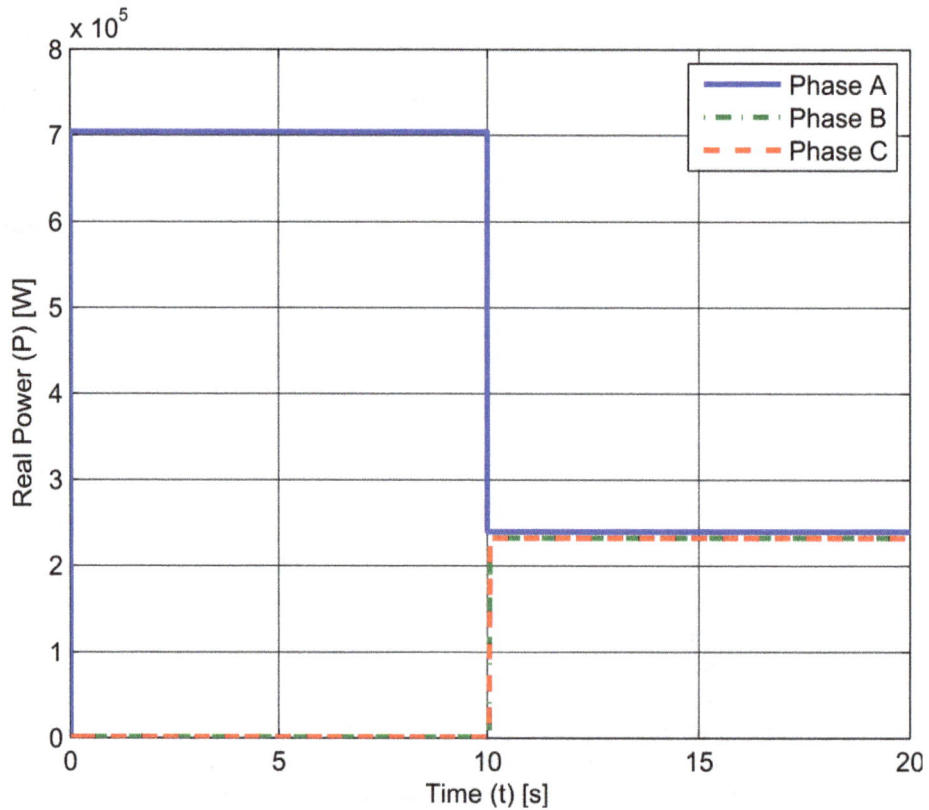

Figure 2. Results from first scenario.

power for phases B and C is the same so the lines overlap in this figure.

Another scenario is conducted in which all 300 PEV charger powers are chosen using a pseudorandom function in MATLAB. In this case, the initial period before balancing begins shows the powers of the three phases are closer to balanced however unbalance still exists. After the initial 10 seconds of unbalance conditions, PEVs are used for balancing and the power between phases becomes closely matched. **Figure 3** shows the results. It should be noted that there is a transient seen during the switching, as would be expected based on the instantaneous changes of the simple switching devices and current sources used in this paper, however because it is not of interest for showing the feasibility of this idea, the full magnitude is not shown.

In order to be practical in a real world car park, the balancing must be able to keep up with reconfiguration of the PEV load in the car park. As PEVs are plugged in or unplugged from chargers, the power drawn from the chargers will change. When this happens, the PEV chargers that are drawing power must be reconfigured to re-balance the power drawn by the car park. A third scenario is conducted where PEV charger powers are changed half way through the simulation. **Figure 4** shows the results of a changing load. The first 20 seconds of **Figure 4** is the same as shown in **Figure 3**. At 20.05 seconds, all PEV charger powers are again assigned using a pseudo-random function in MATLAB. The phase each PEV charger is connected to is not changed until 30 seconds however due to the 10 second delay between assignments. During this short period, new PEVs that are connected to the system will not begin to charge because the charger was previously disconnected from all phases. At 30 seconds, the balancing algorithm runs again and the assignments are updated. The powers of the three phases again approach balanced conditions.

5. Conclusion

Even with the implementation in this paper, which is accomplished through a simple heuristic algorithm that requires minimal computational power, the presented innovative idea is shown to be a viable way to mitigate

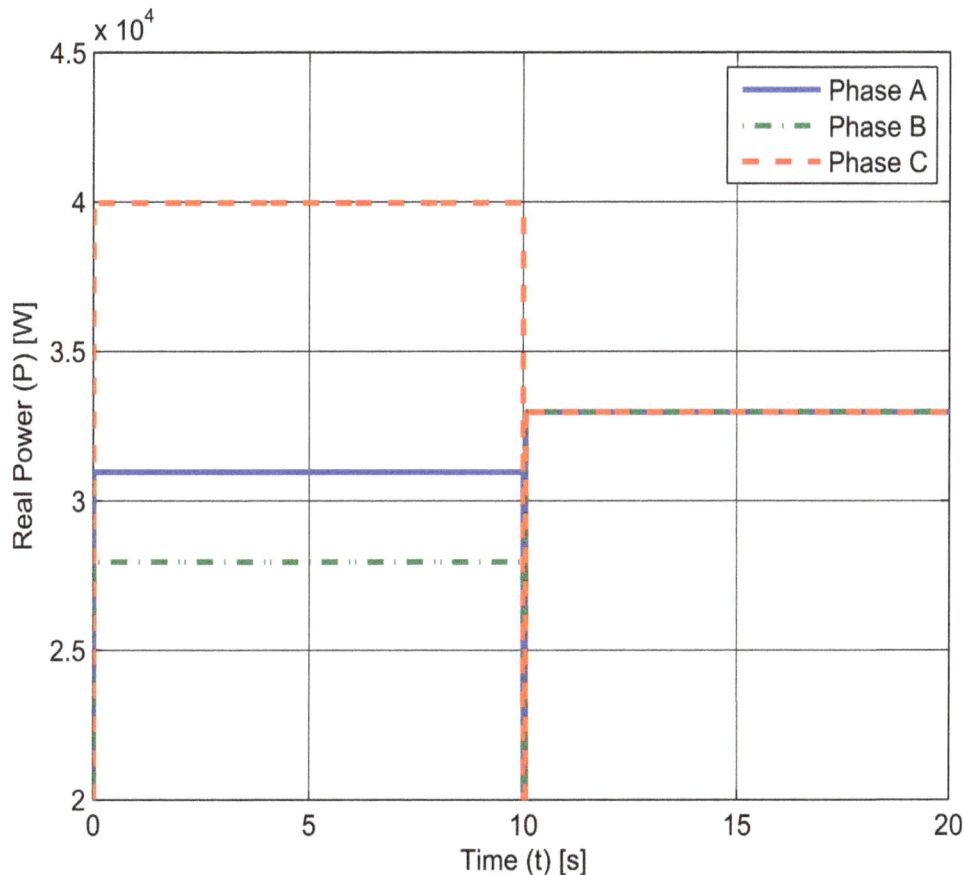

Figure 3. Results from second scenario.

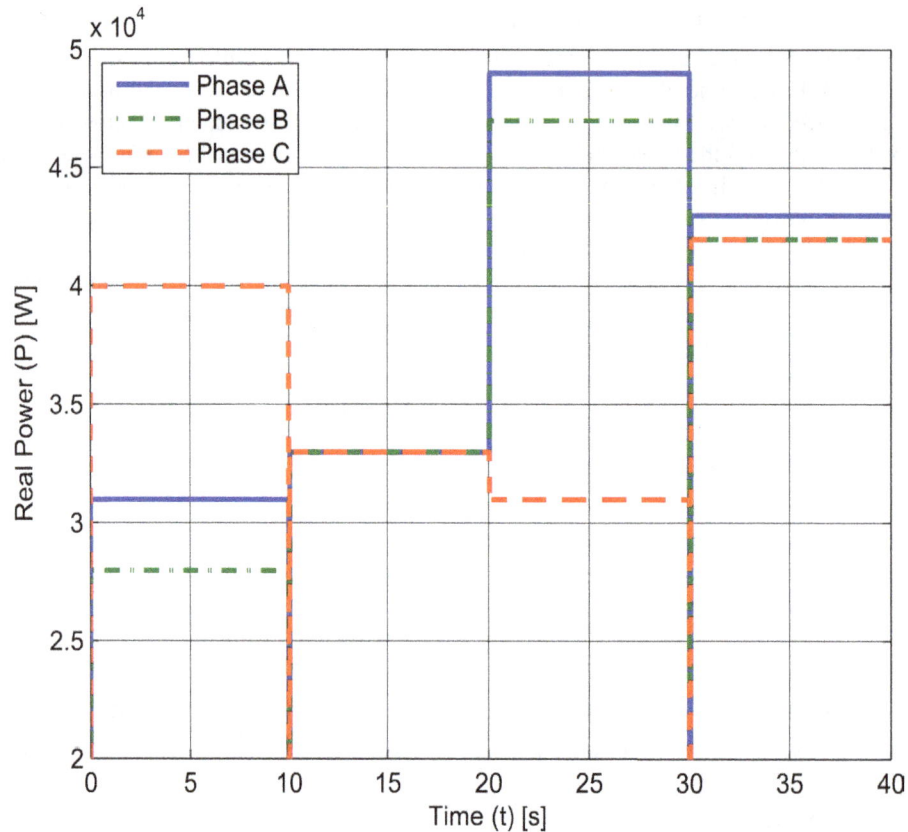

Figure 4. Results from third scenario.

unbalance in a PEV car park. It can be seen that by applying this method of moving PEV chargers between phases based on the current PEV car park configuration, balancing can be accomplished. Also, as the availability of PEVs in a car park changes, it is shown that balancing remains feasible by reassigning PEV charger phases after the change in PEV connections. By balancing the power drawn from each phase, the impact of PEV car parks on electric distribution systems will be decreased. PEV charging is also minimally affected due to the ability of this idea to balance power without changing the charging rate of each PEV charger. Because of these factors, consumers are expected to offer little resistance to the implementation of this idea in PEV car parks, making it valuable to both car park owners and electric utilities.

Acknowledgements

The authors would like to thank the members of the Clemson University Electric Power Research Association (CUEPRA) for their financial support and valuable discussion.

References

[1]   Voelcker, J. (2009) How Green Is My Plug-In? *IEEE Spectrum*, **46**, 42-58.

[2]   Bunga, S.K. (2013) Impact of Plug in Electric Vehicle Battery Charging on a Distribution System. ProQuest Dissertations and Theses, Ann Arbor, 84. http://search.proquest.com/docview/1356783945

[3]   Kisacikoglu, M.C., Ozpineci, B. and Tolbert, L.M. (2010) Examination of a PHEV Bidirectional Charger System for V2G Reactive Power Compensation. 25*th Annual IEEE Applied Power Electronics Conference and Exposition* (*APEC*), Palm Springs, 21-25 February 2010, 458-465.

[4]   Bosch (2014) Charging Stations. Bosch Electric Vehicle Solutions. http://www.pluginnow.com/charging_stations

[5]   Hilshey, A.D., Hines, P.D.H., Rezaei, P. and Dowds, J.R. (2013) Estimating the Impact of Electric Vehicle Smart Charging on Distribution Transformer Aging. *IEEE Transactions on Smart Grid*, **4**, 905-913. http://dx.doi.org/10.1109/TSG.2012.2217385

[6] Sortomme, E., Hindi, M.M., MacPherson, S.D.J. and Venkata, S.S. (2011) Coordinated Charging of Plug-In Hybrid Electric Vehicles to Minimize Distribution System Losses. *IEEE Transactions on Smart Grid*, **2**, 198-205. http://dx.doi.org/10.1109/TSG.2010.2090913

[7] Pieltain Fernández, L., Román, T.G.S., Cossent, R., Domingo, C.M. and Frías, P. (2011) Assessment of the Impact of Plug-In Electric Vehicles on Distribution Networks. *IEEE Transactions on Power Systems*, **26**, 206-213. http://dx.doi.org/10.1109/TPWRS.2010.2049133

[8] Han, S., Han, S. and Sezaki, K. (2010) Development of an Optimal Vehicle-to-Grid Aggregator for Frequency Regulation. *IEEE Transactions on Smart Grid*, **1**, 65-72. http://dx.doi.org/10.1109/TSG.2010.2045163

[9] Sortomme, E. and El-Sharkawi, M.A. (2012) Optimal Scheduling of Vehicle-to-Grid Energy and Ancillary Services. *IEEE Transactions on Smart Grid*, **3**, 351-359. http://dx.doi.org/10.1109/TSG.2011.2164099

[10] Chindris, M., Cziker, A., Miron, A., Balan, H., Iacob, A. and Sudria, A. (2007) Propagation of Unbalance in Electric Power Systems. *9th International Conference on Electrical Power Quality and Utilisation*, Barcelona, 9-11 October 2007, 1-5.

[11] Makram, E., Zambrano, V., Harley, R. and Balda, J. (1989) Three Phase Modeling for Transient Stability of Large Scale Unbalanced Distribution Systems. *IEEE Transactions on Power Systems*, **4**, 487-493. http://dx.doi.org/10.1109/59.193820

[12] Clarke, A. and Makram, E. (2014) An Innovative Approach in Balancing Real Power Using Plug in Hybrid Electric Vehicles. *Journal of Power and Energy Engineering*, **2**, 1-8. http://dx.doi.org/10.4236/jpee.2014.210001

[13] Ko, S.-H., Lee, S.R., Dehbonei, H. and Nayar, C.V. (2006) Application of Voltage- and Current-Controlled Voltage Source Inverters for Distributed Generation Systems. *IEEE Transactions on Energy Conversion*, **21**, 782-792. http://dx.doi.org/10.1109/TEC.2006.877371

[14] Kisacikoglu, M.C., Ozpineci, B. and Tolbert, L.M. (2011) Reactive Power Operation Analysis of a Single-Phase EV/PHEV Bidirectional Battery Charger. *IEEE 8th International Conference on Power Electronics and ECCE Asia (ICPE & ECCE)*, Jeju, 30 May-3 June 2011, 585-592. http://dx.doi.org/10.1109/ICPE.2011.5944614

[15] Clarke, A., Bihani, H., Makram, E. and Corzine, K. (2013) Fault Analysis on an Unbalanced Distribution System in the Presence of Plug-In Hybrid Electric Vehicles. *Clemson Power System Conference*.

20

Power Stabilization System with Counter-Rotating Type Pump-Turbine Unit for Renewable Energy

Toru Miyaji[1*], Risa Kasahara[1], Toshiaki Kanemoto[2], Jin-Hyuk Kim[3], Young-Seok Choi[3], Toshihiko Umekage[2]

[1]Graduate School of Engineering, Kyushu Institute of Technology, Fukuoka, Japan
[2]Faculty of Engineering, Kyushu Institute of Technology, Fukuoka, Japan
[3]Green Energy System Technology Center, Korea Institute of Industrial Technology, Cheonan, Korea
Email: [*]n344154t@tobata.isc.kyutech.ac.jp

Abstract

Traditional type pumped storage system contributes to adjust the electric power unbalance between day and night, in general. The pump-turbine unit is prepared for the power stabilization system, in this serial research, to provide the constant power with good quality for the grid system, even at the suddenly fluctuating/turbulent output from renewable energies. In the unit, the angular momentum changes through the front impeller/runner must be the same as that through the rear impeller/runner, that is, the axial flow at the outlet should be the same to the axial flow at the inlet. Such flow conditions are advantageous to work at not only the pumping mode but also the turbine mode. This work discusses experimentally the performance of the unit, and verifies that this type unit is very effective to both operating modes.

Keywords

Power Stabilization; Pump-Turbine; Counter-Rotation; Impeller; Generator Motor

1. Introduction

To contribute preventing the global warming, the efficient use of clean energy resources, such as a solar power, a wind power and so on, is indispensable, while the electric power is provided steadily for the grid system.

In this serial research, a power stabilization system with a counter-rotating type pump-turbine unit is developed for providing instantaneously the constant output to the grid system. In the pump-turbine unit, front and rear impellers/runners rotate conversely, inner and outer armatures also counter-rotate at the same torque. In the pumping mode, the rotating speeds of both the impellers are adjusted automatically by the smart control. Hence, the unstable characteristic and cavitation can be suppressed effectively [1]. Moreover, the unit in the turbine mode has fruitful advantages that not only the induced voltage is sufficiently high without supplementary

---

[*]Corresponding author.

equipment such as a gearbox, but also the rotational moment hardly acts on the mounting bed because the rotational torque is counter-balanced in the armatures/runners [2]. In this paper, the hydrodynamic performance of the pump-turbine unit with the counter-rotating type impellers designed exclusively for pump mode is investigated experimentally in the power stabilization system.

2. Power Stabilization System with Pumped Storage

Figure 1 shows the diagram of the power stabilization system with the counter-rotating type pump-turbine unit. The system is mainly composed of the electric accumulator with the minimal capacity, the power control device, and counter-rotating type pump-turbine unit.

As the wind power station is the most distinctive for the unstable and fluctuating output among the renewable energy resources, it was applied as an example for the power stabilization system of this work. **Figure 2** shows the output from field tests of the intelligent wind power unit in house works [3], and its output was averaged every one minute as shown in **Figure 3**, where the average output is arbitrarily 16,000 times as high as one in **Figure 2**. It is assumed conveniently, here, that the wind power station should provide the constant power P_G of 1 MW for the grid system. The output from the wind power unit connects directly to the electric accumulator, and the power control device detects that the output is higher or lower than P_G. The power control device demands the electric accumulator not only to provide the constant power P_G for grid system but also to operate the pump-turbine unit at the pumping mode by the surplus power, while the wind velocity is faster than that giving P_G, as shown in **Figures 1** and **3**. That is, the surplus output is stored, at once, as the potential energy by the pumping mode. On the contrary, as shown in the same figures, the power control device demands the pump-turbine unit to operate at the turbine mode converting from the stored potential energy to the hydroelectric output, so as to take P_G with accompanying the shortage output from the wind power unit. Then, the pumping

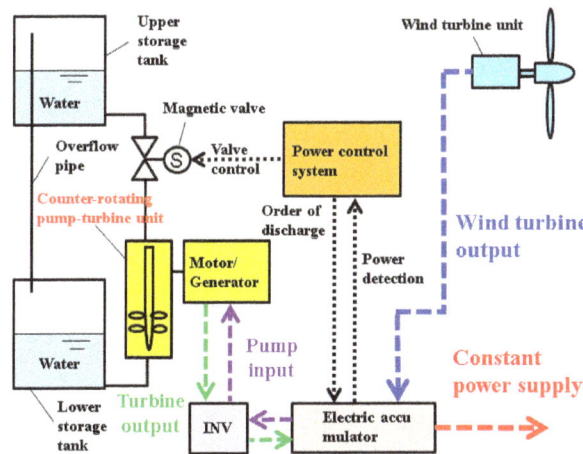

Figure 1. Power stabilization system.

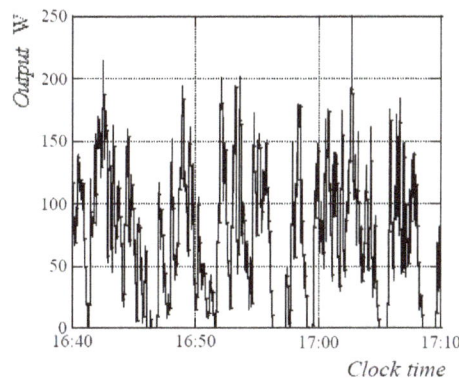

Figure 2. Output from the wind power unit in house.

mode or the turbine mode is instantaneously operated in the alternative every one minute in **Figure 3**, judging the output from the wind power unit. The final target of this serial work is to make average time short as possible.

Figure 4 shows the water volume of the upper and the lower storage tanks in the power stabilization system while operated just above. Moreover, **Figure 5** shows the integrated wattmeter in the accumulator, where the pumping/turbine head is 15 m, the input is 625 kW at the pumping mode, and the hydroelectric output is adjusted to guarantee P_G in response to the output from the wind power unit. The electrical accumulator re- quires the capacity up to 24 kWh as shown in **Figure 5(a)**, while the power stabilization system operates in **Figure 4**. The capacity 24 kWh is 1 - 8th of the capacity of the electrical accumulator installed traditionally in the wind power station without the proposed stabilization system, as confirmed in **Figure 5**. To put the above stabilization system into the practical use, the suitable pump-turbine unit suitable for above operations should be prepared.

3. Counter-Rotating Type Pump-Turbine Unit

3.1. Model Unit

The unit designed exclusive for the pumping mode, which is very important for the pumped storage, was

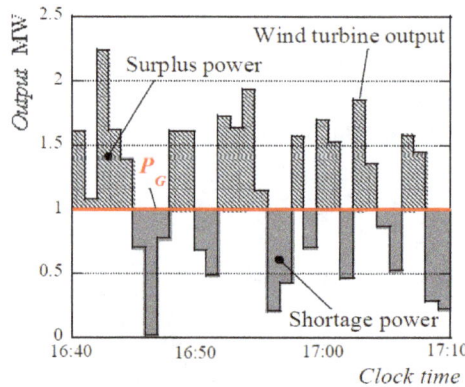

Figure 3. Resultant output averaged every 1 minute.

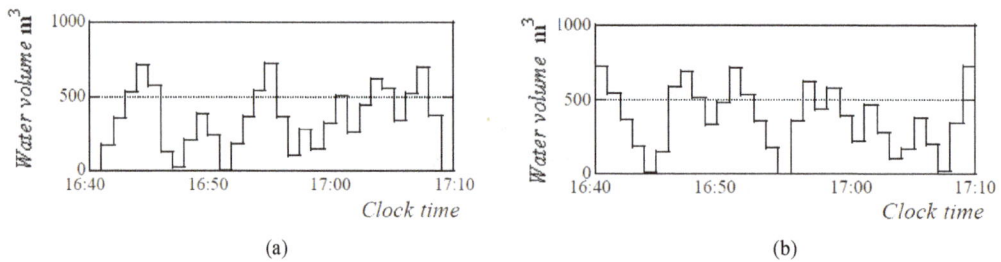

(a)   (b)

Figure 4. Water volume in the storage tanks. (a) Upper storage tank; (b) Lower storage tank.

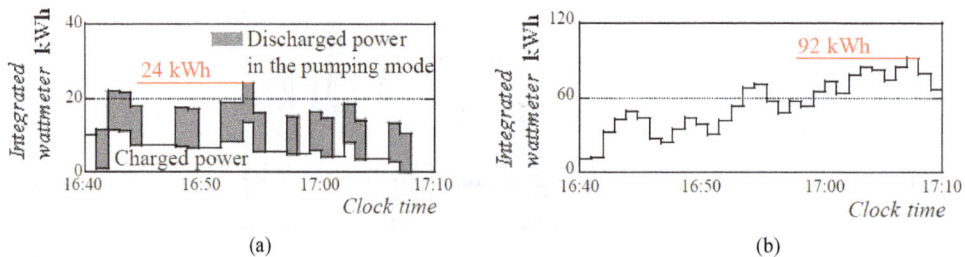

(a)   (b)

Figure 5. Capacity in the electric accumulator. (a) With power stabilization system; (b) Without power stabilization system.

prepared for the preliminary experiments. **Figure 6** shows the counter-rotating type axial-flow pump unit. The major specifications at the normal operation are the theoretical head H_{ET} = 4.4 m, the discharge Q = 1.78 m³/min, the individual impeller speed $n_F = n_R$ = 1500 min⁻¹ (subscripts F and R denote the front and the rear impellers). The specific speed of the individual impeller is N_S = 1100 m, m³/min, min⁻¹, and the specific speed as the pumping unit with the counter-rotating type impellers is N_{ST} = 1320 m, m³/min, min⁻¹. The impeller diameters D are 150 mm, and the boss ratio is 0.4.

3.2. Counter-Rotating Type Impellers

The blade profiles of this unit are shown in **Figure 7** and its dimensions are given in **Table 1**, where RQ and Z are the distances in the circumferential/tangential and the axial directions divided by D, and b_d is the inlet and the outlet angles of the blades measured from the axial direction (subscript 1 to 4 denote the inlet and the outlet of the front and rear impellers, respectively). The numbers of the front and rear blades are 5 and 4 in the pumping mode (the numbers of the front and the rear blades are 4 and 5 in the turbine mode). Impeller/Runner B used in the experiments was designed by the three-dimensional inverse method to improve the exclusively pump performance [4].

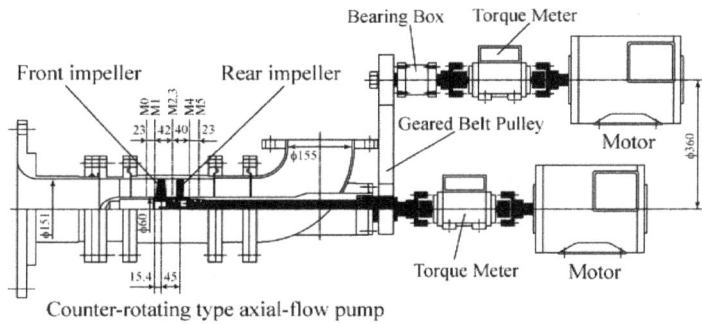

Figure 6. Trail model of the counter-rotating type pump turbine unit.

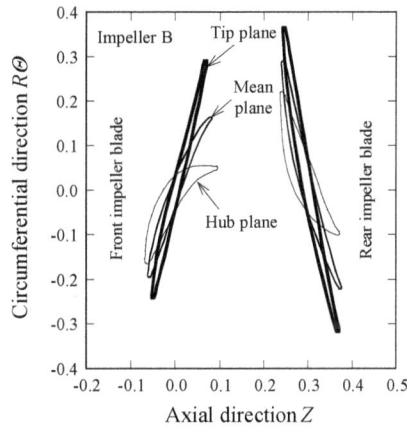

Figure 7. Blade profiles of the counter-rotating type impeller.

Table 1. Blade angles and solidities.

Impeller		Blade angle (deg.)				Solidity	
		β_{d1}	β_{d2}	β_{d3}	β_{d4}	Front	Rear
	Hub	−76.9	−14.9	71.9	42.9	0.99	0.89
B	Mean	−79.2	−58.0	77.2	70.1	0.82	0.82
	Tip	−80.4	−79.0	79.9	75.9	0.81	0.78

4. Performance

Figure 8 shows the unit discharge $Q_{11}[= Q/(D^2 H^{1/2})]$, unit torque $M_{11}[= M/(D^3 H)]$, unit input or the output $P_{11}[= P/(D^2 H^{3/2})]$, the hydraulic efficiencies in the pumping mode $h_h[=rgQH/P]$, and the turbine mode $h_h[= P/(rgQH)]$, the unit rotational speeds of the front and the rear impellers/runners $N_{11\,F,R}(= n_{F,R}D/H^{1/2})$, and the unit relative rotational speed $N_{11}(= nD/H^{1/2})$, while the head is kept constant $H = 2$ m. The water is pumped up while the discharge Q_{11} is positive at the positive N_{11} and Q_{11}, which is called the pumping area (I). It is necessary to make the rotational speed N_{11} faster than 150 m, min^{-1}, for pumping up. The slower rotational speed cannot pump up, because the pump head is in proportion to the square rotational speed. Therefore, the braking area (II) exists where Q_{11} is negative even at the positive N_{11}. The negative rotational speed is in the turbine mode (III). That is, the rotational speed must be increased rapidly when the operation changes from the turbine to the pumping

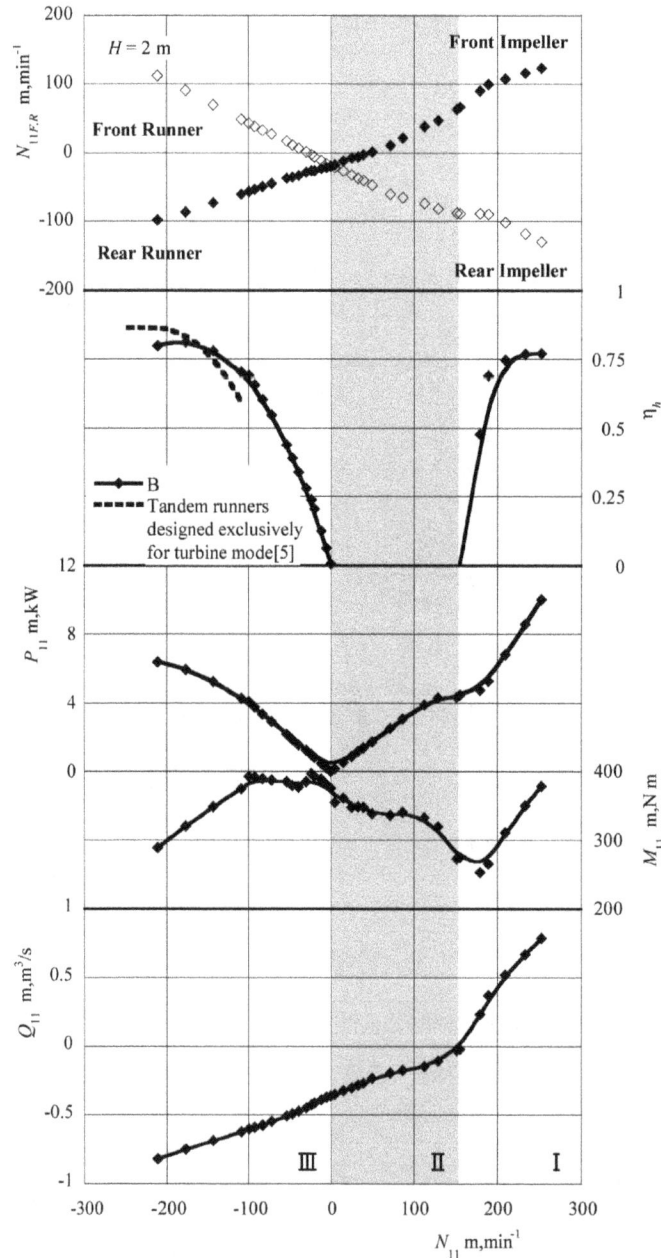

Figure 8. Overall performances of the counter-rotating type pump-turbine unit.

modes. In $Q_{11} = -0.45 \sim -0.35$ m, min^{-1} at the turbine mode, the front runner N_{11F} rotates in the same direction as the rear runner N_{11R} but faster than N_{11R} to take the output. The hydraulic efficiency is maximum at the moderate rotational speed, and the value nearly corresponds to the efficiency (the dot line) of the counter-rotating type hydroelectric unit designed exclusively for turbine mode [5].

These results suggest that this type unit has greatly effective performance for both pumping and the turbine modes.

5. Conclusion

The power stabilization system with the counter-rotating type pump-turbine unit was proposed to provide the constant power with high quality for the grid system. The hydrodynamic performance of the counter-rotating type pump-turbine unit with the impellers designed exclusively for pumping mode was also investigated experimentally, and proved beneficial operation at not only the pumping but also the turbine modes.

References

[1] Kanemoto, T., Kimura, S., Ohba, S. and Satoh, M. (2000) Smart Control of Axial Flow Pump Performances by Means of Counter-Rotating Type (1st Report, Counter-Rotating Type and Performances). *JSME, Series B*, **66**, 2927-2933. (in Japanese) http://dx.doi.org/10.1299/kikaib.66.651_2927

[2] Kanemoto, T., Tanaka, D., Kashiwabara, T., Uno, M. and Nemoto, M. (2001) Tidal Current Power Generation System Suitable for Boarding on a Floating Buoy. *International Journal of Offshore and Polar Engineering*, **11**, 77-79.

[3] Nunoya, T., Takata, C., Makita, K. and Kanemoto, T. (2011) Field Test of the Intteligent Wind Power Unit. 2011 COWEOE Conference Proceeding (International Conference on Offshore Wind Energy and Ocean Energy.

[4] Oba, S. and Kanemoto, T. (2004) Performance Comparison by Design Change of Counter-Rotating Impeller. *22th IAHR Symposium on Hydraulic Machinery and Systems*, Stockholm.

[5] Suzuki, T., Takano, G., Nakamura, Y. and Kanemoto, T. (2011) Counter-Rotating Type Hydroelectric Unit (Effects of Blade Numbers on On-Cam Operations). *ASME-JSME-KSME* 2011 *Joint Fluids Engineering Conference*, Hamamatsu.

Permissions

All chapters in this book were first published by Scientific Research Publishing; hereby published with permission under the Creative Commons Attribution License or equivalent. Every chapter published in this book has been scrutinized by our experts. Their significance has been extensively debated. The topics covered herein carry significant findings which will fuel the growth of the discipline. They may even be implemented as practical applications or may be referred to as a beginning point for another development.

The contributors of this book come from diverse backgrounds, making this book a truly international effort. This book will bring forth new frontiers with its revolutionizing research information and detailed analysis of the nascent developments around the world.

We would like to thank all the contributing authors for lending their expertise to make the book truly unique. They have played a crucial role in the development of this book. Without their invaluable contributions this book wouldn't have been possible. They have made vital efforts to compile up to date information on the varied aspects of this subject to make this book a valuable addition to the collection of many professionals and students.

This book was conceptualized with the vision of imparting up-to-date information and advanced data in this field. To ensure the same, a matchless editorial board was set up. Every individual on the board went through rigorous rounds of assessment to prove their worth. After which they invested a large part of their time researching and compiling the most relevant data for our readers.

The editorial board has been involved in producing this book since its inception. They have spent rigorous hours researching and exploring the diverse topics which have resulted in the successful publishing of this book. They have passed on their knowledge of decades through this book. To expedite this challenging task, the publisher supported the team at every step. A small team of assistant editors was also appointed to further simplify the editing procedure and attain best results for the readers.

Apart from the editorial board, the designing team has also invested a significant amount of their time in understanding the subject and creating the most relevant covers. They scrutinized every image to scout for the most suitable representation of the subject and create an appropriate cover for the book.

The publishing team has been an ardent support to the editorial, designing and production team. Their endless efforts to recruit the best for this project, has resulted in the accomplishment of this book. They are a veteran in the field of academics and their pool of knowledge is as vast as their experience in printing. Their expertise and guidance has proved useful at every step. Their uncompromising quality standards have made this book an exceptional effort. Their encouragement from time to time has been an inspiration for everyone.

The publisher and the editorial board hope that this book will prove to be a valuable piece of knowledge for researchers, students, practitioners and scholars across the globe.

List of Contributors

Karel Fleurbaey, Noshin Omar, Mohamed El Baghdadi, Jean-Marc Timmermans and Joeri Van Mierlo
Department of Electric Engineering and Energy Technology, Vrije Universiteit Brussel, Brussels, Belgium

Roger Ellman
The-Origin Foundation, Inc., Santa Rosa, CA, USA

Liankai Chen, Wenqing Lai, Guoyi Jiang, Yan Zhou and Haibo Liu
East Inner Mongolia Electric Power Company Limited, Inner Mongolia, Hohhot, China

Jun Wang, Yong Chen, Zhaoyu Qin, Lei Ke, Lei Wang and Yang Shen
Wuhan NARI Limited Liability Company of State Grid Electric Power Research Institute, Wuhan, China

Nfally Dieme and Moustapha Sane
Laboratory of Semiconductors and Solar Energy, Department of Physics, Faculty of Science and Technology, Cheikh Anta Diop University, Dakar, Senegal

Hongxu Yin and Fenfei Lv
The Power Company of Dezhou, Shandong, Dezhou, China

Rui Xiao
College of Mechanical & Electrical Engineering, Jiaxing University, Jiaxing, China

Andrew D. Clarke and Elham B. Makram
Department of Electrical and Computer Engineering, Clemson University, Clemson, USA

Andrew D. Clarke and Elham B. Makram
Department of Electrical and Computer Engineering, Clemson University, Clemson, USA

Andrew D. Clarke, Himanshu A. Bihani, Elham B. Makram and Keith A. Corzine
Department of Electrical and Computer Engineering, Clemson University, Clemson, USA

Zhebin Sun, Kang Li and Zhile Yang
School of Electronics, Electrical Engineering and Computer Science, Queen's University Belfast, Belfast BT9 5AH, UK

Qun Niu
School of Mechatronic Engineering and Automation, Shanghai Key Laboratory of Power Station Automation Technology, Shanghai University, Shanghai 200072, China

Aoife Foley
School of Mechanical and Aerospace Engineering, Queen's University Belfast, Belfast BT9 5AH, UK

Abdullah M. Al-Shaalan
Electrical Engineering Department College of Engineering King Saud University, Riyadh, Saudi Arabia

Hanmei Hu, Bo Hong and Qinfeng Li
College of Electrical Engineering and New Energy, China Three Gorges University, Yichang, China

Ting Chen
College of Electrical Engineering, Wuhan University, Wuhan, China

Mayoro Dieye and Gregoire Sissoko
Laboratory of Semiconductors and Solar Energy, Department of Physics, Faculty of Science and Technology, Cheikh Anta Diop University, Dakar, Senegal

Senghane Mbodji and Biram Dieng
Department of Physics, Alioune DIOP University of Bambey, Bambey, Senegal

Martial Zoungrana and Issa Zerbo
Laboratoire d'Energies Thermiques et Renouvelables (L.E.T.RE), Departement de Physique, U.F.R-S.E.A, Universitede Ouagadougou, Ouagadougou, Burkina Faso

Miguel Edgar Morales Udaeta, Carolina Attas Chaud, André Luiz Veiga Gimenes, Luiz Claudio Ribeiro Galvao
Energy Group of the Electrical Power and Automation Engineering, Department of the Polytechnic, University of São Paulo—GEPEA/EPUSP, São Paulo, Brazil

K. R. Allaev
Power Engineering Department, Tashkent State Technical University, Tashkent, Republic of Uzbekistan

A. M. Mirzabaev
Mir Solar LLC, Tashkent, Republic of Uzbekistan

T. F. Makhmudov
Power Engineering Department, Tashkent State Technical University, Tashkent, Republic of Uzbekistan

T. A. Makhkamov
Research and Development Department, Tecon Groups, Moscow, Russian Federation

Syed Ali Raza Shah and Bashir Ahmed Leghari
Department of Mechanical Engineering, Balochistan University of Engineering and Technology, Khuzdar, Pakistan

Zahoor Ahmed, Wazir Muhammad Laghari and Attaullah Khidrani
Department of Electrical Engineering, Balochistan University of Engineering and Technology, Khuzdar, Pakistan

Cristina Guzman, Kodjo Agbossou and Alben Cardenas
Hydrogen Research Institute and The "Département de Génie Électrique et Génie Informatique", Université du Québec à Trois-Rivières, Trois-Rivières, Canada

Sideig A. Dowi and Amar Ibrahim Hamza
School of Electrical and Electronics Engineering, North China Electric Power University, Beijing, China

Petre-Marian Nicolae, Ileana-Diana Nicolae and Ionuţ-Daniel Smărăndescu
Department of Electrical Engineering, Energetics and Aeronautics University of Craiova, Faculty of Electrical Engineering, Craiova, Romania

Andrew D. Clarke and Elham B. Makram
Department of Electrical and Computer Engineering, Clemson University, Clemson, USA

Toru Miyaji and Risa Kasahara
Graduate School of Engineering, Kyushu Institute of Technology, Fukuoka, Japan

Toshiaki Kanemoto and Toshihiko Umekage
Faculty of Engineering, Kyushu Institute of Technology, Fukuoka, Japan

Jin-Hyuk Kim and Young-Seok Choi
Green Energy System Technology Center, Korea Institute of Industrial Technology, Cheonan, Korea

www.ingramcontent.com/pod-product-compliance
Lightning Source LLC
Chambersburg PA
CBHW050444200326
41458CB00014B/5054